The Pesticide Chemist and Modern Toxicology

S. Kris Bandal, EDITOR
3M Company

Gino J. Marco, EDITOR
CIBA–GEIGY Corporation

Leon Golberg, EDITOR
*Chemical Industry Institute
of Toxicology*

Marguerite L. Leng, EDITOR
The Dow Chemical Company

Based on a symposium
sponsored by the ACS Division
of Pesticide Chemistry at
a Special Conference at
Downington, PA, June 26, 1980.

ACS SYMPOSIUM SERIES **160**

AMERICAN CHEMICAL SOCIETY
WASHINGTON, D. C.   1981

Library of Congress CIP Data

The pesticide chemist and modern toxicology.
 (ACS symposium series; 160 ISSN 0097–6156)

 Includes bibliographies and index.

 1. Pesticides—Congresses.
 I. Bandal S. Kris. II. American Chemical Society.
Division of Pesticide Chemistry. III. Series.

TP248.P47P48   615.9'02   81–10790
ISBN 0–8412–0636–8 AACR2 ACSMC8 160 1–582
                                                 1981

ACS Symposium Series

M. Joan Comstock, *Series Editor*

(7948

FOREWORD

The ACS SYMPOSIUM SERIES was founded in 1974 to provide a medium for publishing symposia quickly in book form. The format of the Series parallels that of the continuing ADVANCES IN CHEMISTRY SERIES except that in order to save time the papers are not typeset but are reproduced as they are submitted by the authors in camera-ready form. Papers are reviewed under the supervision of the Editors with the assistance of the Series Advisory Board and are selected to maintain the integrity of the symposia; however, verbatim reproductions of previously published papers are not accepted. Both reviews and reports of research are acceptable since symposia may embrace both types of presentation.

CONTENTS

PREFACE

The Division of Pesticide Chemistry of the American Chemical Society has held three conferences in lieu of ACS national spring meetings in 1972, 1975, and 1980. The principal goal of these conferences has been the presentation of in-depth, high-quality programs on subjects of current and universal importance in a friendly, small-group atmosphere to promote maximum exchange of ideas.

The 1972 workshop was held in Fargo, ND and discussed various experimental techniques involved in metabolism, residue, and analytical chemistry. The second workshop, held in 1975 in the beautiful town of Vail, CO, dealt with "Bound and Conjugated Pesticide Residues," a matter of great importance to most chemists concerned with pesticide metabolism, analyses, and residues. The proceedings of this second conference were published in 1976 as ACS Symposium Series Volume 29, edited by D. D. Kaufman, G. G. Still, G. D. Paulson, and S. K. Bandal.

Our third conference on topics of great importance to pesticide chemists on a timely basis was held in Downingtown, PA in June, 1980. We felt that the current concern about the safety evaluation of pesticide chemicals, and the toxicological significance of nanogram amounts of pesticides that can be detected using sophisticated analyitcal techniques, has given a new and broader dimension to the sciences of pesticide chemistry and toxicology. Our perception of the toxicological problems due to chemicals has changed radically.

The objective of the Downington Special Conference was to provide a means for the disciplines of toxicology and pesticide chemistry to interact in a direct and personal way. The number of participants was limited to less than 300 to afford an opportunity for personal discussions on how these two disciplines influence each other, to better understand similarities and differences, and to learn from one another about data gathering and interpretation. We put special emphasis on recent developments in toxicology, especially as it is related to carcinogenicity. The metabolism and analytical studies needed to support safety evaluation of pesticides were discussed with ample focus on the recently promulgated proposals for good laboratory practice. During the latter part of the conference, a symposium was held on the regulatory aspects of pesticide safety evaluation, not only for those in the United States, but also in Europe, Canada, and Asia. A number of informal workshops also were organized on topics proposed by registrants, ranging from the United States Environmental Protection Agency guidelines for hazard evaluation to the importance of accurate and timely communication of technical information.

We, the conference organization committee and the editors of this publication, believe that the Downingtown Conference was successful in achieving the above-mentioned goals by permitting a structured, formal, technical program while encouraging spontaneous interactions among pesticide chemists, biochemists, analytical chemists, regulatory scientists, and toxicologists. We sincerely thank the conference speakers and the participants for their contributions in achieving these goals.

S. Kris Bandal
Agricultural Products/3M
230-B 3M Center
St. Paul, MN 55144

Gino J. Marco
CIGA-GEIGY Corporation
P.O. Box 11422
Greensboro, NC 27409

Leon Golberg
Chemical Industry Institute of Toxicology
P.O. Box 12137
Research Triangle Park, NC 27709

Marguerite L. Leng
The Dow Chemical Company
1803 Building
Midland, MI 48640

March, 1981

TOXICOLOGICAL ASPECTS

Toxicological Aspects: An Introduction

LEON GOLBERG[1]

Chemical Industry Institute of Toxicology, P.O. Box 12137,
Research Triangle Park, NC 27709

The objective of this Conference is to delineate the interaction of Pesticide Chemistry with Toxicology. The first point
to be stressed is the fact that the broad, basic principles of
Toxicology are applicable to all chemicals - no matter what their
structure or intended application may be - and even to physical
agents acting on man and the biota. Toxicology is an Esperanto
in the universe of biological effects exercised by chemicals.

In 1969 I had the privilege to serve as a member of the
Secretary's Commission on Pesticides and their Relationship to
Environmental Health, the so-called Mrak Commission (1). The
fundamentals that were spelled out at that time concerning effects
of pesticides on man are in many respects still valid today.
We may ask: how far have we progressed in the intervening decade?
Some of the answers will be forthcoming in the course of this
Conference.

What has changed most radically is our perception of the
toxicological problems in the field of chemicals, taken as a
whole. Ten years ago we still tended to segregate the various
categories of chemicals into separate compartments, based on
their perceived end-uses. Today a far more catholic view pre-
vails. Increasing consciousness of the huge universe of chemicals
is coupled to an awareness of our state of ignorance of the properties of a great many of them. A truly enormous task lies ahead,
to bring the toxicology of even the more important chemicals
to the level of the present state of the art. To advance our
understanding of mechanisms of toxic action is an ever greater
challenge.

Compounding the problem is the realization that the background of "natural" chemicals in the environment, in food and
within our bodies includes a remarkably high proportion of toxic,
mutagenic and hence potentially carcinogenic agents. The work
of Sugimura and his colleagues (2,3,4) has served to throw some
light on this subject. Stich and coworkers (5) have demonstrated
that the intestinal contents and feces of man, animals and birds
contain mutagens, even in their volatile components. Such is
the present faith in positive results of mutagenesis tests as

[1] Current address: 2109 Nancy Nanam Drive, Raleigh, NC 27607.

predictors of carcinogenic potential that compounds found to
be negative in the NCI Carcinogenesis Bioassay program are to
be retested in those instances where such a conflict arises.
These compounds include 3-methyl-4-nitroquinoline-N-oxide, azoxy-
benzene, diphenylnitrosamine, 1-naphthylamine and methyl orange
(6).

The increasing control exercised by the federal bureaucracy
over chemicals of all kinds, and the current moves towards international testing standards, have spearheaded the trend towards
rigidly-fixed protocols for toxicological studies. While lip
service is paid to the need for frequent updating, the provision
for such necessary changes is quite inadequate. Above all, the
flexibility that is essential to ensure meaningful risk assessment is being totally eliminated in the drive towards uniformity
and standardization of protocols. The baby is in danger of being
thrown out with the bathwater, if the investigator is not encouraged to tailor the studies to the problems posed by the specific
test material. Such purposeful flexibility is particularly desirable if full advantage is to be taken of newly-developed techniques, for instance in immunology, genotoxicity and neurobehav-
ioral studies.

With increasing concern about the numbers of chemicals wait-
ing to be tested, national and international pressure has developed to produce a "quick fix" that will solve these problems.
As the International Agency for Research on Cancer has expressed
the issue: "In principle, test systems should be cheap, and
the results obtained should be relevant to man". Our audience
today ought to be aware of the fact that the rat is still the
biggest bargain available to meet this challenge, for example
with regard to the amount of information that can be derived
from a single test such as the subchronic study conducted over
a period of 1, 3 or 6 months (7,8). By the end of this Conference we plan to have covered both the current strategy with respect to screening tests and the broad perspective of tests intended to achieve an assessment of risk under defined conditions
of exposure.

Literature Cited

1. Report of the Secretary's Commission on Pesticides and their
    Relationship to Environmental Health (1969). U. S. Department
    of Health, Education, and Welfare, Washington, D.C.
2. Sugimura, T., Nagao, M., Kawachi, T., Honda, M., Yahagi,
    T., Seino, Y., Sato, S., Matsukura, N., Matsushima, T.,
    Shirai, A., Sawamura, M. and Matsumoto, H. (1977). Mutagen-
    carcinogens in food with special reference to highly mutagenic
    pyrolytic products in broiled foods. In Origins of Human
    Cancer, Book C, Human Risk Assessment. Ed. by H. H. Hiatt,
    J. D. Watson and J. A. Winsten, pp. 1561-1577, Cold Spring
    Harbor Laboratory.

3. Sugimura, T., Nagao, M., Matsushima, T., Yahagi, T., Seino, Y., Shirai, A., Sawamura, M., Natori, S., Yoshihira, K., Fukuoka, M. and Kuroyanagi, M. (1977). Mutagenicity of flavone derivatives. Proc. Jap. Acad. 53:194-197.
4. Sugimura, T. and Nagao, M. (1979). Mutagenic factors in cooked foods. CRC Crit. Rev. Toxicol. 6:189-209.
5. Stich, H. F., Stich, W. and Acton, A. B. (1980). Mutagenicity of fecal extracts from carnivorous and herbivorous animals. Mutat. Res. 78:105-112.
6. Pesticide & Toxic Chemical News (June 11, 1980). Mutagenicity tests lead to bioassays for chemicals considered non-carcinogenic, pp. 5-7.
7. Golberg, L. (1975). Safety evaluation concepts. J. Ass. Official Analyt. Chemists 58:635-644.
8. Scientific Committee, Food Safety Council. (1978). Subchronic Toxicity Studies. In "Proposed System for Food Safety Assessment". Fd Cosmet. Toxicol. Suppl. 2, 16:83-96.

RECEIVED February 11, 1981.

The Revolution in Toxicology: Real or Imaginary

LEON GOLBERG[1]

Chemical Industry Institute of Toxicology, P.O. Box 12137,
Research Triangle Park, NC 27709

Toxicology has traditionally been concerned with the effects of chemical or physical agents in bringing about alterations of structure, function or response of living organisms. The higher organisms used by the toxicologist are never devoid of spontaneous disease, especially as they age, so that it is against this background that toxic changes attributable to a test compound have to be gauged. As Salsburg (1) has pointed out:

> "When groups of animals are exposed to any biologically active substance over a long period of time, there will be a shift in patterns of lesions that will be dose related".

The traditional task of the toxicologist has been to identify the nature of that shift in lesions, to characterize the dose-response relationships for each major change, and to elucidate the mechanism of toxic action - in other words, to determine the basis of that shift.

An appropriate point of departure for considering toxic effects is the topic of homeostasis, the ensemble of defensive mechanisms that Nature has built into every organism. Homeostasis comprises the responses to changes, both external and internal, physiological adjustments (2, 3) that help to maintain what Claude Bernard termed "the stability of the internal medium", in other words the balance between the needs of the cell and the needs of the organism (4). Thus homeostasis can be considered in terms of three components, one concerned with the normal internal composition and function of the cell, another with the intercellular integration of function within a multicellular organism and the third being the gamut of compensating mechanisms that come into play when the organism is stressed by any of a multitude of physical or chemical agents such as hypoxia, extremes of temperature or the action of toxicants.

The concept of homeostasis is important to the toxicologist

---

[1] Current address: 2109 Nancy Nanam Drive, Raleigh, NC 27607.

because it prescribes the limits within which the body can adjust to toxic effects with no apparent deviation of normal function, other than perhaps temporary perturbations. From this concept is derived the so-called "No observed effect level" of exposure to a toxicant. In some instances the organism can meet the challenge and stress of toxic exposure by adaptations that involve the development of tolerance, provided that time is afforded for the organism to change in this way. When exposure is excessive in degree or too abrupt or both, the physiological defense mechanisms prove inadequate and pathological disturbances ensue. Even at this point, however, when damage has been done to one or more target organs, repair mechanisms are available that come into play at many levels from DNA on up. Provided that the onslaught by the toxicant abates for a sufficient length of time to permit repair of structure and restoration of function to take place, the condition of the organism may return to apparent normality. Evidence on this score will be provided by long-term follow-up, or by further challenges with observation of the responses (2, 5, 6, 7).

Over and above acute and subchronic effects, there may be changes of more subtle character, occurring early in the course of exposure as so-called "silent" lesions but making themselves manifest much later in the lifetime of the organism as frank pathological changes, including neoplasia (8). The consequences of genetic toxicity affecting germinal cells may only become apparent in subsequent generations. Finally, the aging process itself may reflect the accumulation of toxic insults, and failure to achieve perfection in the restoration of damage, over the course of a lifetime.

The provision made to protect cells against oxygen toxicity illustrates some of the principles mentioned above (9). The biological reduction of oxygen by the monovalent pathway proceeds through superoxide radicals (O_2^-), hydrogen peroxide and hydroxyl radicals ($OH^{\cdot}$), possibly to singlet oxygen ($^1O_2$). Hydroxyl radicals are so dangerous to the cell that very efficient mechanisms exist to limit their formation by scavenging the superoxide radicals, by means of superoxide dismutases, and destroying H_2O_2 by catalases and peroxidases (10, 11). Another and partly related toxic phenomenon is lipid peroxidation which is capable of causing damage to cell membranes. The toxic effects of many compounds are mediated, at least in part, through lipid peroxidation. Again, the cell possesses defenses in the form of antioxidants, super-oxide dismutases, carotenoids and the enzymes glucose-6-phosphate dehydrogenase, glutathione peroxidase and glutathione reductase acting together (12, 13). Beyond its role in the action of these last two enzymes, glutathione and kindred non-protein sulfhydryl compounds afford biological protection against electrophiles, epoxides and other highly-reactive potential toxicants through the action of glutathione S-transferases (14, 15, 16).

The Nature and Dimensions of Toxicity

Three aspects of toxic action need to be defined as accurately as possible. In the first place the toxic potentialities of
the test compound should be explored with a view to pinpointing
one or more target organs that are revealed in the course of
acute, subchronic, long-term, reproductive and other studies.
This information affords a bird's eye view of the overall land-
scape. Once the intrinsic capacity to cause injury to a specific
target organ or system has been characterized, some measure of the
potency of the substance is essential, preferably in the form of
dose-response data in appropriate test systems. Thus the potential for neurotoxicity, myelotoxicity, mutagenicity or carcinogenicity is spelled out in terms of a specific bracket within the
range of 10^7 of possible potency. Naturally this definition
applies only to a given set of experimental circumstances: particular species, strain, sex and age of animals derived from a
particular stock at a particular source, housed under particular
defined conditions, and given a diet of specified composition.
Air and water, in common with many other details, require close
attention. Any one of these and numerous other minutiae of the
testing protocol can influence the outcome of the test, and hence
merits close attention. Given a defined potency and a dose level
at which no adverse effect is observed (in comparison with controls), one is in a position to draw a comparison with the actual
or anticipated levels of exposure of people or other species to
the test material. Here we have a possible range of at least 10^8;
so that the product of potency and exposure (which are, of course,
independent of each other) is 10^{15}. For purposes of risk assessment the all-important question then is: where, within this vast
range, does a given chemical or pesticidal ingredient lie when it
is used in its intended applications? Anyone tempted to adopt the
popular expressions "toxic" or "non-toxic" should bear in mind the
fact that, like sinners, none of us is perfect: it is the nature
and extent of our sins that matter.

Individual Susceptibility to Toxic Effects

Pesticides encounter susceptible or resistant target species.
The range of susceptibility to toxic action is often very broad in
man and laboratory animals. Host susceptibility is predominantly
determined by genetic background but may be profoundly influenced
also by diet, human lifestyle (including consumption of alcohol,
tobacco, drugs) age, sex, state of health and numerous environmental factors. Pregnancy and infancy are examples of conditions
in which special susceptibility may exist.
Genetic control of susceptibility to toxicants operates
through a variety of mechanisms. One of these is metabolic. In
animals, the murine Ah complex represents a "cluster" of genes
exercising temporal control on tissue-specific regulatory genes

controlling monooxygenase activities mediated by cytochrome P-450
(17). In man and animals the phenotypes determining acetylator
(18, 19, 20, 21) and methylator (22) status have a powerful influence on drug metabolism and toxicity. Human cancer susceptibility is based on "ecogenetics" of the individual's background
and environmental exposures (23).

Toxic Interactions

Beside the influence of inadvertent exposures to environmental toxicants at home or in the workplace, the deliberate use
of therapeutic agents, "street" drugs, solvent "huffing" and other
sources of a multiplicity of toxic agents may impinge on the
effects of pesticidal exposure, in the field or elsewhere. While
the Washington Post (June 20, 1980) may have gone too far in
describing Agent Orange as "just one garnish in a toxic cocktail",
attention does need to be directed to the possibilities of additive, synergistic or antagonistic interactions between several
chemicals acting simultaneously or sequentially.

This issue was addressed by the Mrak Commission (24) under
three headings: inhibition of esterases, alteration of microsomal
enzyme activity, and target-level interactions. Also taken into
account were the influences exercised by tissue storage of persistent compounds, and by exogenous physical factors such as diet,
temperature and radiation. Much more is now known about each of
these topics, particularly the induction of hepatic mixed function
oxidase activity (25, 26) or its inhibition by exposure to heavy
metals such as cadmium (27). There is a prevalent tendency to
emphasize the possibility of additive and synergistic toxic (especially carcinogenic) effects of simultaneous exposures, but not
to mention the well-documented fact that antagonistic interactions
between the biological effects of the components may render a
mixture less toxic or even non-carcinogenic (2, 5).

One of the important spheres of interaction lies in possible
modification of the immune status and responses of test organisms,
including man. The field of immunotoxicology, like that of
behavioral toxicology, is still in its infancy. A penetrating
analysis of the problems inherent in premature efforts to pre-
scribe mandatory tests in this area (28) concludes as follows:

"There is no way of knowing what tests are more sensitive,
representative of effects, and would provide consistent
conclusions if a number of test chemicals were examined.
Toxicology is becoming more and more a regulatory discipline
and the trend of looking for new tests that would evaluate
untoward health effects seems strong. In this perspective
we should realize that adding more tests in toxicity testing
schedules, particularly with respect to immunotoxicity
evaluation, may not offer much advantage.
"The need of basic science investigations in toxicologic

research needs no further emphasis. This might be the time
to divert our attention into looking more for the mechanisms
rather than merely the effects of chemicals on the immune
system. Only then can we make more objective judgments on
the risks and benefits of environmental chemicals, particularly when the chemical exposures are low but prolonged, and
the system in question is the one that generally expresses
its deficiencies or modifications only when challenged by
an unwanted invader. Only after we have a better understanding of these mechanisms can we appropriately understand
the species differences, mechanisms of immune tolerance, if
any, and even the toxicologic effect that might be mediated
via immune modifications."

Selection of Test Material

The foundations of effective toxicological assessment may be
undermined if insufficient attention is devoted to a variety of
chemical aspects of the problem. The decision whether to study a
technical product or a purified material (and, if so, what degree
of purity) is, of course, fundamental and often very difficult.
Beyond that, one has to realize that the specification of a compound is usually drawn up for technical purposes rather than as a
basis for toxicological investigation (29). Consequently, a
number of critical safety issues may be overlooked. Time and
again, much toxicological effort has been set at nought by failure
to pay attention to what appeared to be unimportant "trace" impurities or added stabilizers in commercial products. There is a long
history of mistakes, and current concerns about the presence of
dioxin in 2,4,5-T, pentachlorophenol, hexachlorophene, and a
variety of other chlorinated compounds illustrate the principle.
On the other hand, the search for traces of trace impurities in
the pursuit of an explanation for the alleged carcinogenicity of
saccharin has gone to increasing extremes (30).
     One has also to take into consideration the changes which the
compound may undergo before it finally enters the body of the
individual of interest. For example: interaction with food
components; degradation during the course of formulation or storage
or in the environment, including photochemical oxidation; and
biotransformation in a variety of organisms, from bacteria to
plants and animals, including the intestinal flora of man.
     Particularly with a technical product comprising numerous
components and impurities, the objection is often raised that
investigations of the sort recommended involve an extraordinary
effort which is not justified. Obviously, a reasonable balance
has to be maintained between effort involved and significance of
the results. A decision on how much effort should be necessary
will in part be based on the levels of exposure to be used in
toxicity tests. If these are high, then it may well happen that
trace impurities can assume considerable importance in determining
the biological outcome.

Hierarchy of Hazard Evaluation

Predictive toxicity involves much more than studies in animals. A logical hierarchical approach to the evaluation of hazard presented by a test material is illustrated in Table I. The sequence is not intended to imply a series of watertight compartments: the art of modern toxicology lies in the skillful deployment of the most appropriate procedures, severally or in combination, to answer specific questions.

Table I.   Hierarchy of Hazard Evaluation

1.	Structure-activity correlation
2.	Physical and chemical properties
3.	In vitro and other short-term tests
4.	Screening procedures
5.	Animal studies
6.	Human studies
7.	Risk assessment

Analysis of quantitative structure-activity relationships (QSAR) has become increasingly important as a means of predicting likely biological activity on the basis of the vast store of existing information on SAR. The traditional approach has been Hansch analysis, incorporating independent variables and physicochemical parameters, and involving regression analysis of partition coefficients, electronic effects of substituents (Hammett sigma parameter), steric parameters (Taft steric constants, Verloop parameters) and indicator variables (31). More recently, pattern recognition techniques have come to the fore, in which a computer generates, on the basis of the structure of the compound, molecular structure descriptors to be used for mathematical analysis of QSAR. Remarkable predictive accuracy has been achieved, for instance with various classes of carcinogens (32, 33). Two problems exist: the reliability of the data base, and the need to incorporate metabolic information. As to the first, the weaknesses of the standard carcinogenesis bioassay are not as well-recognized as they should be (2, 5), but some effort is at last under way to try to overcome them (34). Introduction of metabolic information has to be very selective, concentrating on metabolic activation rather than the multiplicity of detoxication products.
    In a document entitled "Proposed System for Food Safety Assessment", the Scientific Committee of the Food Safety Council (35) has attempted to delineate the steps by which decisions on safety or toxicity are arrived at. What is interesting about this approach is the departure from the traditional sequence of tests

by inserting at an early stage in the investigations two elements
of particular importance: tests of genetic toxicity, and metabolic and pharmacokinetic studies. The sequence of the main
groups of investigations may vary according to the nature of the
test material and the purposes for which it is intended. The
emphasis on these two groups of scientific procedures is all the
more welcome because of the reluctance on the part of both Industry and Regulatory authorities to accept the key role and fundamental importance of metabolism and pharmacokinetics, in relation
to the contribution that such data can make to the design of
protocols, as well as to the understanding of effects and overall
interpretation of toxicological data. The decision-tree approach
(35) makes limited provision for the study of transplacental,
prenatal and postnatal events and omits detailed consideration of
behavioral and immunological aspects of toxic action. (It is not
intended to cover environmental considerations.) The outlook is
pragmatic. Inevitably there is no expressed interest in elucidating the mechanism of toxic action of the test compound. Nevertheless, the information intended to be gathered in the course of
the study of metabolism and pharmacokinetics and genetic toxicology, when skillfully combined, may well throw considerable
light on the basic biological properties of the compound.

One difficulty in achieving such understanding is the fact
that, if the decision-tree approach is adhered to rather rigidly,
toxicological properties and target organs will not have been
revealed at the time that the investigations on metabolism and
pharmacokinetics are being carried out. While whole-body autoradiography is a useful guide in directing such investigations,
there is no substitute for knowledge of the site(s) and dose-
response relationships of toxic action.

In view of the detailed attention that will be paid to metabolism later in the Conference, metabolic activation will be the
main focus of discussion here, since it gives rise to electrophilic alkylating or arylating intermediates capable of inducing
damage to critical cellular macromolecules. A diagrammatic view
is provided in Fig. 1 of the obstacle course faced by such an
electrophile in reaching a nucleophile at the target site. Of
particular importance in determining toxicity is the delicate
balance between, on the one hand, electrophile production and, on
the other, electrophile destruction, or other biotransformations
that serve the purpose of deactivation (36). A host of species-
specific and organ-specific factors exercise their influence on
this balance.

The availability of rapid tests of mutagenic potential has
facilitated the detection of activated metabolites. An elegant
demonstration of the use of the Ames test for this purpose is
provided by the work of Casida and his colleagues (37, 38, 39, 40)
who tracked down the formation of a mutagenic activation product,
2-chloroacrolein, from S-chloroallylthiocarbamate herbicides
(diallate, triallate and sulfallate). Metabolic activation by

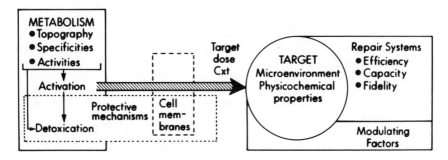

Figure 1.   Diagram of the chain of events attendant upon metabolic activation

intestinal bacteria and by plant extracts has also been shown to occur (41).

Human Studies

Studies carried out in human volunteers, or with tissues of human origin have great potential value (42). The contribution of epidemiological studies is discussed later in this Conference by Dr. M. W. Palshaw. In certain instances, such as anticholinesterases, only human experience can serve to define no-effect and minimum-effect levels for man. In other situations where exposure is low, indices of effect may be hard to come by. The alkylation of histidine and cysteine in the globin moiety of hemoglobin has been suggested as an index of effects of alkylating agents (43) but has not found general application.

In contrast to these problems, evidence of exposure to pesticides is often much more readily available by analysis of excreta, body fluids and expired air (44). The power of modern analytical procedures, a topic to be addressed later in this Conference, is exemplified by the characterization of 115 organic compounds in samples of breath from 54 subjects (44). Exhaled ethane and n-pentane in mice, rats and monkeys (45) has proved to be a useful index of lipid peroxidation, these gases being derived from $\omega3$- and $\omega6$-fatty acid hydroperoxides (12, 13). Non-invasive measures of drug metabolizing capacity have been developed, using ^{14}C-phenacetin or ^{14}C-aminopyrine; hepatic dysfunction can be assessed in an analogous manner (46, 47, 48, 49). On the horizon is the exciting promise of the application of nuclear magnetic resonance to monitor metabolite concentrations, non-invasively, in human subjects (50).

Finally, human tissues are finding increasing use for metabolic and other studies. A human liver bank has been established in Sweden for storage of liver microsomal suspensions (51). Human lymphocytes, monocytes and fibroblasts have found extensive application, particularly in mutagenic studies (52, 53, 54, 55).

Summing up, while the principal emphasis in this Conference will necessarily be placed on animal studies, the toxicologist should seize every opportunity to secure human data, for which animal results are at best an imperfect substitute.

A Revolution in Toxicology?

Advances in toxicological methodology in recent years, notably better analytical methods and the procedures made available by molecular biology and genetic toxicology are making possible an understanding of the mechanisms of toxic action. This in itself is an autocatalytic process: as we gain better and better understanding of such mechanisms it becomes easier to deal with the next problem in the same category. In the process of understanding mechanism, one has to take into account the influence of exposure to the compound on homeostatic processes of the body, and of the defensive and adaptive limits which the body can attain in response to toxic exposure. Thus there evolves a comprehension of the dose–response relationship for that particular compound under the conditions of testing. Evaluation of safety involves the conceptual integration and interpretation of the information gained from physical measurements, knowledge of chemical structure and properties, and the study of biological effects in relation to doses used. Thus an intellectual activity enters into evaluation of safety which transcends the mere assembly of data. Interpretation of risk assessment involves further a thorough knowledge and understanding of the nature, uses and exposure levels (existing or anticipated) of a chemical or mixture of chemicals in a product.

The new concepts, techniques and approaches that are creating a ferment in Toxicology, taken together with the impetus fuelled by accelerating advances in the basic sciences, bid fair to revolutionize the practice of risk assessment. Whether this very real promise will be translated into concrete achievements in terms of greater safety depends on the freedom and encouragement afforded to the toxicologist to participate in and advance the revolution. In the short term, the prospects do not appear favorable; but History teaches us that powerful forces working for change do ultimately find expression, despite bureaucratic defense of the status quo.

References Cited

1. Salsburg, D. (1980). The effects of lifetime feeding studies on patterns of senile lesions in mice and rats. Drug Chem. Toxicol. 3:1-33.
2. Golberg, L. (1979). Toxicology: Has a new era dawned? Pharmacol. Rev. 30:351-368.
3. Robertson, B. (1980). Basic morphology of the pulmonary defence system. Europ. J. Respir. Dis., Suppl. 107, 61:21-40.
4. Yabrov, A. (1980). Adequate function of the cell: Interactions between the needs of the cell and the needs of the organism. Med. Hypotheses 6:337-374.
5. Golberg, L. (1979). The Dangers of New Discoveries and the Discovery of New Dangers. In Human Health and Environmental Toxicants: Royal Society of Medicine International Congress and Symposium Series No. 17. Academic Press, Inc. (London) Ltd., and the Royal Society of Medicine, pp. 19-43.
6. Golberg, L. (1980). Rapid Tests in Animals and Lower Organisms as Predictors of Long-Term Toxic Effects. In Current Concepts in Cutaneous Toxicity. Ed. V. Drill, Academic Press, N.Y., pp. 171-212.

7. Golberg, L. (1979). Implications for human health. Environ.
    Health Perspect. 32:273-277.
8. Hard, G. C., King, H., Borland, R., Stewart, B. W. and
    Dobrostanski, B. (1977). Length of in vivo exposure to
    a carcinogenic dose of dimethylnitrosamine necessary for
    subsequent expression of morphological transformation by
    rat kidney cells in vitro. Oncology 34:16-19.
9. Ciba Foundation Symposium 65. (1979). Oxygen Free Radicals
    and Tissue Damage. Excerpta Medica, New York.
10. Fridovich, I. (1979). Superoxide dismutases: defence
    against endogenous superoxide radical. In Oxygen Free
    Radicals and Tissue Damage. Excerpta Medica, New York,
    pp. 77-94.
11. Flohé, L. (1979). Glutathione peroxidase: Fact and fiction.
    In Oxygen Free Radicals and Tissue Damage. Excerpta Medica,
    New York, pp. 95-122.
12. Tappel, A. L. (1980). Measurement of and protection from
    in vivo lipid peroxidation. In Free Radicals in Biology,
    Vol. IV. Ed. W. A. Pryor, Academic Press, New York,
    pp. 1-47.
13. Bus, J. S. and Gibson, J. E. (1979). Lipid peroxidation
    and its role in toxicology. In Reviews in Biochemical
    Toxicology 1. Eds. E. Hodgson, J. R. Bend and R. M. Philpot,
    Elsevier/North-Holland, New York, pp. 125-149.
14. Reed, D. J. and Beatty, P. W. (1980). Biosynthesis and
    regulation of glutathione: Toxicological implications. In
    Reviews in Biochemical Toxicology 2. Eds. E. Hodgson, J.
    R. Bend and R. M. Philpot, Elsevier/North-Holland, New
    York, p. 213.
15. Chasseaud, L. F. (1979). The role of glutathione and
    glutathione S-transferases in the metabolism of chemical
    carcinogens and other electrophilic agents. In Advances
    in Cancer Research, Vol. 29. Eds. G. Klein and S. Weinhouse,
    Academic Press, New York, pp. 175-274.
16. Berrigan, M. J., Gurtoo, H. L., Sharma, S. D., Struck, R.
    F. and Marinello, A. J. (1980). Protection by N-acetylcy-
    steine of cyclophosphamide metabolism - related in vivo
    depression of mixed function oxygenase activity and in
    vitro denaturation of cytochrome P-450. Biochem. Biophys.
    Res. Commun. 93:797-803.
17. Kahl, G. F., Friederici, D. E., Bigelow, S. W., Okey, A.
    B. and Nebert, D. W. (1980). Ontogenetic expression of
    regulatory and structural gene products associated with
    the Ah locus. Dev. Pharmacol. Ther. 1:137-162.
18. Timbrell, J. A. (1979). The role of metabolism in the
    hepatotoxicity of isoniazid and iproniazid. Drug Metab.
    Rev. 10:125-147.
19. Reece, P. A., Cozamanis, I. and Zacest, R. (1980).
    Kinetics of hydralazine and its main metabolites in slow
    and fast acetylators. Clin. Pharmacol. Ther. 28:769-778.

20. Shepherd, A.M.M., Ludden, T. M., McNay, J. L. and Lin, M.-S. (1980). Hydralazine kinetics after single and repeated oral doses. Clin. Pharmacol. Ther. 28:804-811.

21. Tannen, R. H. and Weber, W. W. (1980). Inheritance of acetylator phenotype in mice. J. Pharmacol. Exp. Ther. 213:480-484.

22. Weinshilboum, R. M. (1980). 'Methylator status' and assessment of variation in drug metabolism. Trends in Pharmacol. Sci. 1:378-380.

23. Harris, C. C., Mulvihill, J. J., Thorgeirsson, S. S. and Minna, J. D. (1980). Individual differences in cancer susceptibility. Ann. Intern. Med. 92:809-825.

24. Report of the Secretary's Commission on Pesticides and their Relationship to Environmental Health (1969). "Interactions". U. S. Department of Health, Education, and Welfare, Washington, D.C., pp. 509-564.

25. Fabacher, D. L., Kulkarni, A. P. and Hodgson, E. (1980). Pesticides as inducers of hepatic drug-metabolizing enzymes--I. Mixed function oxidase activity. Gen. Pharmac. 11:429-435.

26. Kulkarni, A. P., Fabacher, D. L. and Hodgson, E. (1980). Pesticides as inducers of hepatic drug-metabolizing enzymes--II. Glutathione S-transferases. Gen. Pharmac. 11:437-441.

27. Chadwick, R. W., Faeder, E. J., King, L. C., Copeland, M. F., Williams, K. and Chuang, L. T. (1978). Effect of acute and chronic Cd exposure on lindane metabolism. Ecotoxicol. Environ. Safety 2:301-316.

28. Sharma, R. P. and Zeeman, M. G. (1980). Immunologic alterations by environmental chemicals: Relevance of studying mechanisms versus effects. J. Immunopharmacol. 2:285-307.

29. Scientific Committee, Food Safety Council. (1978). The importance of specifications for substances in their safety evaluation in foods. In "Proposed System for Food Safety Assessment". Fd Cosmet. Toxicol., Suppl. 2, 16:17-24.

30. National Academy of Sciences Committee Saccharin Report No. 1. (1978). "Saccharin: Technical Assessment of Risks and Benefits". National Research Council/National Academy of Sciences, Washington, D. C. pp. 3-44 to 3-61.

31. Stuper, A. J., Brügger, W. E. and Jurs, P. C. [Eds.] (1979). Computer Assisted Studies of Chemical Structure and Biological Function. John Wiley & Sons, New York, pp. 2-14.

32. Jurs, P. C., Chou, J. T. and Yuan, M. (1979). Computer-assisted structure-activity studies of chemical carcinogens. A heterogeneous data set. J. Medicinal Chem. 22:476-483.

33. Jurs, P. C., Chou, J. T. and Yuan, M. (1979). Studies of chemical structure-biological activity relations using pattern recognition. In Computer-Assisted Drug Design. Eds. E. C. Olson and R. E. Christoggersen, American Chemical Society, Washington, D.C., pp. 103-129.

34. National Institute of Environmental Health Sciences. Request for Research Grant Applications: RFA (NIH–NIEHS–EP–81–1). NIH Guide for Grants and Contracts, Vol. 9, No. 12, October 10, 1980.

35. The Scientific Committee, Food Safety Council. (1978). "Proposed System for Food Assessment". Fd Cosmet. Toxicol., Suppl. 2, 16:1–136.

36. Wright, A. S. (1980). The role of metabolism in chemical mutagenesis and chemical carcinogenesis. Mutat. Res. 75:215–241.

37. Schuphan, I., Rosen, J. D. and Casida, J. E. (1979). Novel activation mechanism for the promutagenic herbicide diallate. Science 205:1013–1015.

38. Schuphan, I. and Casida, J. E. (1979). S-chloroallyl thiocarbamate herbicides: Chemical and biological formation and rearrangement of diallate and triallate sulfoxides. J. Agric. Food Chem. 27:1060–1067.

39. Rosen, J. D., Schuphan, I., Segall, Y. and Casida, J. E. (1980). Mechanism for the mutagenic activation of the herbicide sulfallate. J. Agric. Food Chem. 28:880–881.

40. Rosen, J. D., Segall, Y. and Casida, J. E. (1980). Mutagenic potency of haloacroleins and related compounds. Mutat. Res. 78:113–119.

41. Wildeman, A. G., Rasquinha, I. A. and Nazar, R. N. (1980). Effect of plant metabolic activation on the mutagenicity of pesticides. Amer. Ass. Cancer Res., Abstract No. 357, p. 89.

42. Golberg, L. (1975). Safety evaluation concepts. J. Ass. Official Analyt. Chem. 58:635–644.

43. Calleman, C. J., Ehrenberg, L., Jansson, B., Osterman-Golkar, S., Segerbäck, D., Svensson, K. and Wachtmeister, C. A. (1978). Monitoring and risk assessment by means of alkyl groups in hemoglobin in persons occupationally exposed to ethylene oxide. J. Environ. Path. Toxicol. 2:427–442.

44. Krotoszynski, B. K., Bruneau, G. M. and O'Neill, H. J. (1979). Measurement of chemical inhalation exposure in urban population in the presence of endogenous effluents. J. Analyt. Toxicol. 3:225–234.

45. Dumelin, E. E., Dillard, C. J. and Tappel, A. L. (1978). Breath ethane and pentane as measures of vitamin E protection of Macaca radiata against 90 days of exposure to ozone. Environ. Res. 15:38–43.

46. Desmond, P. V., Branch, R. A., Calder, I. and Schenker, S. (1980). Comparison of [^{14}C]phenacetin and amino[^{14}C]-pyrine breath tests after acute and chronic liver injury in the rat. Proc. Soc. Exp. Biol. Med. 164:173–177.

47. Roots, I, Nigam, S., Gramatzki, S., Heinemeyer, G. and Hildebrandt, A. G. (1980). Hybrid information provided by the ^{14}C-aminopyrine breath test studies with ^{14}C-mono-methylaminoantipyrine in the guinea pig. Naunyn-Schmiedeberg's Arch. Pharmacol. 313:175–178.

48. Henry, D. A., Sharpe, G., Chaplain, S., Cartwright, S.,
     Kitchingman, G., Bell, G. D. and Langman, M.J.S. (1979).
     The [^{14}C]-aminopyrine breath test, a comparison of different
     forms of analysis. Br. J. clin. Pharmac. 8:539-545.
49. Sotaniemi, E. A., Pelkonen, R. O. and Puukka, M. (1980).
     Measurement of hepatic drug-metabolizing enzyme activity
     in man. Eur. J. Clin. Pharmacol. 17:267-274.
50. Griffiths, J. R. and Iles, R. A. (1980). Nuclear magnetic
     resonance - a 'magnetic eye' on metabolism. Clin. Sci.
     59:225-230.
51. Von Bahr, C., Groth, C.-G., Jansson, H., Lundgren, G., Lind,
     M. and Glaumann, H. (1980). Drug metabolism in human liver
     in vitro: Establishment of a human liver bank. Clin.
     Pharmacol. Ther. 27:711-725.
52. Albertini, R. J. (1980). Drug-resistant lymphocytes in
     man as indicators of somatic cell mutation. Teratogenesis,
     Carcinogenesis, and Mutagenesis 1:25-48.
53. Lake, R. S., Kropko, M. L., McLachlan, S., Pezzutti, M.
     R., Shoemaker, R. H. and Igel, H. J. (1980). Chemical
     carcinogen induction of DNA-repair synthesis in human
     peripheral blood monocytes. Mutat. Res. 74:357-377.
54. Yang, L. L., Maher, V. M. and McCormick, J. J. (1980). Error-
     free excision of the cytotoxic, mutagenic N^2-deoxyguanosine
     DNA adduct formed in human fibroblasts by (±)-7β,8α-dihydroxy-
     9α,10α-epoxy-7,8,9,10-tetrahydrobenzo[a]pyrene. Proc. Natl.
     Acad. Sci. USA 77:5933-5937.
55. Vigfusson, N. V. and Vyse, E. R. (1980). The effect of the
     pesticides, dexon, captan and roundup, on sister-chroma-
     tid exchanges in human lymphocytes in vitro. Mutat. Res.
     79:53-57.

RECEIVED February 11, 1981.

Widening Concepts of Toxicology

BERNARD A. SCHWETZ

Director of the Toxicology Research Laboratory, Health & Environmental Sciences,
Dow Chemical U.S.A., Midland, MI 48640

During the past decades, the science of Toxicology has
undergone a continuous evolutionary process of increasing sophistication. Early efforts often involved evaluating hazards of
gross proportions - such as survival itself of individuals exposed
to a toxin. Today's efforts have been extended to the most
subtle effects measurable by modern technology - the impact of
which is sometimes more theoretical than real. This evolution has
been brought about by a number of events and stimuli - increased
understanding of basic biological processes, better equipment,
social and political pressures, better application of knowledge
to the solution of problems, and the development of better test
methods to predict hazard. Our concerns have expanded in many
dimensions - time, space, species, nature of the toxins, exposure
levels, as well as the parameters about which we're concerned.
Each of these will be discussed separately.

Time

In earlier generations when the primary concern was survival,
time considerations were limited to the immediate event. Delayed
effects or the consequence of repeated exposure was of little
concern. In time, we became more concerned about the effects of
repeated exposure to levels which were not an immediate threat to
survival but may eventually become a threat to life - still of
the individual exposed. As we became more aware of long-range
effects of overexposure, we learned that exposure to toxins can
affect us later and even in future generations. Thus, the focus
of concern has extended from an immediate effect on the exposed
person, to an effect on the exposed person at a later time, to a
possible effect on some member of a future generation completely
removed from the toxin in question.

Toxicological test procedures have evolved with these changing concerns. When survival was the primary concern, tests were
developed to assess the effect of acute overexposure - the LD50
and LC50 (lethal dose or concentration to 50% of exposed animals),

0097–6156/81/0160–0021$05.00/0
© 1981 American Chemical Society

and the effect of dermal exposure. The concern over repeated
exposure to sublethal quantities led to subchronic tests (longer
than acute but less than life-time) designed to identify target
organs, sensitive species, and effect/no adverse effect dose
levels. The effect of chronic exposure, including oncogenicity,
has been sought during the past 40 years. Other endpoints of
repeated exposure have been effects on reproduction and embryonal/
fetal development. The concern for future generations led to the
development of tests for mutagenic potential. Thus, the tests
used by Toxicologists have evolved parallel with our time-frame
of concern.

Space

Concern over the localization and distribution of toxins has
changed dramatically over the decades. Time and space concerns
changed simultaneously. At a time when the greatest concern was
for acute hazard the only concern for the distribution was the
concentration in the immediate locale. Control was easier to
manage when the area of concern was so small. As we learned that
smaller and smaller amounts of toxins caused detectable changes,
interest surged in the distribution of chemicals away from the
immediate source. We are now concerned about any measurable
amount of all chemicals, even in the most remote recesses of the
universe. The stratosphere and ionosphere, polar ice caps,
depths of the oceans and the earth are all being sampled regularly
for analysis for the presence of contaminants. Once the chemicals
reach these remote parts of the environment, control is essentially lost. The litany of chemicals which fall in this category is
very familiar to most chemists and biologists – DDT and its
metabolites, PCB's, PBB, HCB, etc.

Species

Human beings used to be the only species of real concern.
Now every species identified is important to somebody and is
protected to varying degrees. For generations our concern beyond
people was limited to beasts of burden and plants or animals that
were food sources. Now we include all kinds of pets and wild
animals in all areas of the world. For a variety of reasons,
some physical, some chemical, species such as California seals,
the brown pelican, certain penguins, the Kirtland warbler, the
alligator, and the snail darter have all been in the news in the
past few years as endangered species.

Toxins of Concern

As with the aspects already discussed, there have also been
changes in the toxins over which we are concerned. Probably one
of the earliest classes of agents of toxicological concern was

those used for homocidal purposes. The toxicity of acute overdose
with arsenic and cyanide was appreciated long before toxicology
was a science. The effects of medicaments and tonics such as
quinine and peyote were taken advantage of and overused for many
generations. The consequences of occupational exposure to toxins
has been realized during the past 100 years or so, starting with
such examples as scrotal cancer among chimney sweeps from soot
and damage to the central nervous system from mercury (mad
hatter's disease).

Toxins occurring naturally in our environment, such as
selenium and many plant alkaloids, were well known even before
their chemical identity was confirmed.

The greatest visibility today is given to materials related
to modern technology, primarily contaminants and by-products as
well as products themselves. "Environmental" factors of current
health concern include not only such things as heavy metals,
PCB's, DDT metabolites, and TCDD, but also smoking and our habits
of eating and drinking. Thus, the nature of the toxins over
which we've been concerned during the years has changed with our
social awareness and scientific understanding.

Concentrations of Toxins

The amount or concentration of any given toxin in the environment that was considered to be important has been diminishing
rapidly. The analytical chemist has obviously been at the forefront of this evolution. The qualitative determinations of the
past have given way to quantitative analyses down to the level of
counting molecules. The analytical chemists and toxicologists
seem almost to have been competing during the past few decades
for lower levels of sensitivity. At a time when toxicologists
were limited in sophistication to merely counting the number of
live and dead animals after acute exposure to an agent, chemists
were making qualitative analyses or were measuring chemicals at
the percent level. As toxicologists progressed beyond the whole-
animal level of observation and began to look for grossly visible
organ changes, the analytical chemists were detecting parts per
million. The use of light microscopy and biochemical measures
was the next level of discrimination by toxicologists; while this
technology evolved, analytical chemists penetrated to parts per
billion. Today's toxicologists are using electron microscopy to
assess structural changes and are detecting chemical changes at
the molecular level of cell organization. Chemists are now
detecting chemicals at the parts per trillion level and below.
At what level of detection is it no longer important to know if a
chemical is present? In many cases, the detection limit is below
one which the toxicologist would predict to have adverse effects
in tissue, soil, water, air, etc. As the level of exposure to a
toxin decreases, the effect of exposure diminishes until it
cannot be discerned from the normal background noise - biological
variability.

The body has a wide array of very effective defense mechanisms which protect it from toxins. This initial response of the body is a process of physiological adaptation. This includes such phenomena as enzyme induction, hormonal changes, alterations in blood flow, blood cell distribution, energy utilization, immunologic responses, as well as the rate of cell division and cell destruction. This process of adaptation is not considered to be evidence of toxicity; it's the body's normal response to an insult. When the amount of insult exceeds the body's ability to maintain itself, toxicity exists. The transition from physiological adaptation to a toxic response can be considered a threshold.

Unfortunately, detection of chemicals at any level in the environment or in animal tissues is considered by many people today to constitute a problem which must be alleviated, whether that level is above or below the threshold for an adverse effect. Is it really in the best interest of good science to continue to push the level of detection lower and lower if we've already gone below a level which is of realistic concern? Before we push the limit of detection below the level of biological meaningfulness, we should ask ourselves this question - what are the numbers going to mean when we get them?

We know that for many chemicals (vitamins, trace minerals, hormones, amino acids, electrolytes, etc.) there are optimal levels in the body which are required for normal function. Too little is injurious to health, as is too much. Any diabetic is only too aware of the delicate balance of insulin required to maintain health. Too little sodium in the body interferes with the transmission of nerve impulses; hypertensive patients know the consequence of too much sodium. Many other examples are well described in the medical literature.

Parameters of Concern

The endpoints of toxicity which have caused the greatest concern have changed along with the factors already discussed. Today's concerns center around the quality of life and zero risk. Both of these are very difficult to define and measure. The current efforts of the news media and the pressures involved in obtaining financial support for research programs has provided an unprecedented visibility for toxicological findings. What society is concerned about is not necessarily the same as what the toxicologists and the medical profession are concerned about. The discrepancy is relatable, primarily, to the nature of the information being given to the general public.

In addition to these societal influences, significant changes have occurred in toxicology over the past twenty years; the state-of-the-art will continue to evolve. Some of these are summarized as follows. Acute toxicology data will continue to be as important in the future as they have been in the past; these

data continue to be very important in the hazard assessment of
chemicals. Subchronic studies have played a very important role
in toxicology in the past and will probably continue as such.   In
contrast, chronic studies for the purpose of assessing chronic
toxicity and oncogenicity, cannot continue to be run as a screening test; the time, space, and person-power required to do so is
a luxury that we cannot afford. We cannot accelerate our efforts
at a rate sufficient to meet our needs for this type of testing.
The decisions that have previously been made from life-time study
data will have to be made from data that are easier to collect.
The primary limitations on expanding our capacity to do chronic
studies are space and the availability of qualified pathologists.
In the future, chronic studies will be done to define the slope
of the dose-response curve for toxic effects rather than as a
screen for the potential to cause adverse effects. This applies
a lot of pressure for toxicologists to work smarter, not just
harder.

Efforts in the area of reproductive toxicity are likely to
increase in the future. Screening for teratogens (agents which
cause birth defects) has seemed to plateau. Chemicals with the
thalidomide-type of hazard appear to be very rare. More subtle
effects on reproduction and development should be sought with
greater discrimination. Chemicals which affect the development
of sperm and ova must be identified better than in the past.
Better animal models need to be developed to accurately detect
subtle changes in reproductive performance. Chemicals such as
diethylstilbesterol and 1,2-dibromo-3-chloropropane have height-
ened the concern of toxicologists and the public in this important
area of research.

Studies of the metabolism and pharmacokinetics of chemicals
seemed to reach a peak of activity in the late 1970's. Such
studies will continue to be very significant in our efforts to
improve our evaluation of the data we collect in animal studies.

Studies in the areas of metabolism/pharmacokinetics and
molecular interactions between toxins and target molecules is
clearly an area where we can work smarter rather than harder to
make progress in understanding mechanisms of toxicity.

In the past few years, mutagenesis has been used as an
indicator of carcinogenic potential; there has been less emphasis
placed on mutagenesis as an endpoint in itself. To the present,
a gametic chemical mutagen has not been clearly identified in
humans. When the first chemical mutagen is identified in humans,
this area of research will probably assume a new role among toxicologists, via pressure from the public.

Mutagenesis tests as an indicator of carcinogenic potential
(such as the Ames' test) have clearly come into their own in the
past few years. It is expected that they will continue to play
an important role in setting priorities for conducting more
definitive studies in the areas of mutagenesis and carcinogenesis
on a more selective basis and thereby permit us to use our
resources more wisely.

Studies to identify the mechanisms of carcinogenesis will increase in importance in the next years. Such information will facilitate interpretation of data collected in other studies and will enhance our ability to predict hazard for man. Some of the data sets which currently appear to conflict with each other will undoubtedly be put into proper perspective through these studies in the future as we learn more about species differences, sex differences, and the role of dose level and route of exposure.

Behavioral toxicology is an area that is clearly coming into its own and will have more visibility and impact in the future. Toxicologists have been observing demeanor for years, but the more sophisticated methodologies of assessing behavior are just now being evaluated by toxicologists to determine their reliability and predictability. With the increasing concern being expressed by the public, behavioral toxicology will likely continue to grow in visibility and acceptance.

The toxicologic aspects of immunology are clearly coming up as a new area of toxicology. Certain chemicals unquestionably affect immunologic mechanisms or organs involved in the immunologic response. The integrity of our immune system is so critical to many body functions that the interrelationship between immunology and toxicology is without doubt going to be a critical area in the future. Certain chemicals clearly affect the immune system - stimulation and/or inhibition. The implications of such effects are not totally clear at present but are sure to become more important in the future.

In summary, toxicology is clearly a dynamic science. Despite its relative newness as a science, much change has occurred and is likely to continue to occur in the future. New directions are always being identified. The role of the chemist, and particularly the analytical chemist, in the evolution of toxicology is very important. Good science depends heavily on close coordination between these two areas of expertise. Especially in the area of environmental toxicants, one area cannot evolve without parallel developments in the other.

RECEIVED February 24, 1981.

Organ Specificity in Toxic Action: Biochemical Aspects

JOHN S. DUTCHER[1] and MICHAEL R. BOYD

Molecular Toxicology Section, Clinical Pharmacology Branch, National Cancer Institute, Bethesda, MD 20205

The pharmacologic and toxic properties of many organic compounds result from reversible interactions with biological systems. But some chemicals, including certain insecticides, allergens, cytotoxins, carcinogens and mutagens, produce their toxic manifestations by irreversible, covalent interactions with tissue constituents. Because of the importance of these toxicities to animal and human health, it is important to elucidate the mechanisms of toxication and detoxication of these agents in order to help predict and minimize risk and set exposure guidelines. One useful approach to this end is to investigate the factors which determine target organ toxicity, factors which either make certain tissues more sensitive to the toxic effects of chemicals and/or protect other tissues from their deleterious effects.

4-Ipomeanol [1-(3-furyl)-4-hydroxypentanone, IPO, Figure 1] is a naturally occuring, highly organ-specific toxin. It has proven to be a useful model compound for the study of certain biochemical factors which can influence organ-selective toxicity, and can be used to illustrate some of the important biochemical aspects of organ-selective toxicity.

An Example of Organ Specific Toxicity: 4-Ipomeanol

The injestion of mold-damaged sweet potatoes has been implicated for many years in outbreaks of poisoning in cattle (1,2,3). Affected animals suffer severe and often fatal respiratory distress. Pathological findings are usually restricted to the lungs; these include edema, congestion and hemorrhage(4). The major causative agent responsible for this pulmonary-specific toxicity is IPO (Figure 1)(5,6), one of a number of toxic 'stress metabolites' produced in sweet potatoes (Ipomoea batatas) infected with the common mold, Fusarium

[1] Current address: Lovelace Inhalation Toxicology Research Institute, Box 5890 Albuquerque, NM, 87115.

solani(7). Simple methods to prepare IPO and its [3]H- and
[14]C-labeled analógues are available (8,9,10,11) and have
facilitated investigations of its mechanism of toxicity.

Mechanism. Numerous studies on the mechanism of IPO
toxicity have supported the view that tissue damage by the
compound is due to a highly reactive, alkylating metabolite(s)
(Figure 2)(12). In vitro experiments demonstrated that this
metabolic activation is catalyzed by a cytochrome P-450 enzyme
system which is located in the endoplasmic reticulum of target
cells(10). This metabolite(s) forms covalent bonds with
cellular macromolecules, and it causes cell death (necrosis).
The amount of cellular necrosis (measured by microscopic
examination of the respective tissues 24 hours after exposure
to the toxin) and the extent of protein alkylation (assayed by
employing [14]C- or ^{3}H-IPO and measuring the amount of label
bound covalently to tissue proteins 2 hours after exposure to
the toxin) have proven to be useful measures of toxicity(13).
It is important to emphasize that alkylation of protein by IPO
metabolite(s) is used as an indirect measure of the amount of
reactive metabolite(s) present at a target site; protein
alkylation is not necessarily the primary event leading to
cell necrosis by IPO.

Organ-Specific Toxicity. IPO produces striking organ-
specific toxicity in a number of laboratory animals(12,13,14)
as well as in cattle. In the rat, cellular necrosis is seen
only in the lung after a single, intraperitoneal dose of the
toxin (Figure 3). Likewise, organ-specific alkylation by the
reactive metabolite(s) of IPO is predominantly in the lung,
with only a small amount in the liver and kidney. Other organs
have only background levels of alkylation. Both toxicity and
alkylation are dose dependent and the lung is the only site of
toxicity at any dose and after any of several different routes
of administration (intraperitoneal, intravenous or oral).
Guinea pigs and rabbits show a pattern of toxicity and
alkylation similar to the rat, the lung being the primary
target organ for toxicity and the major site of alkylation,
irrespective of dose. However, the hamster and the mouse show
somewhat different patterns of toxicity (Figure 4). Pulmonary
bronchiolar necrosis occurs in both of these species, but IPO
also produces renal tubular necrosis in the adult male mouse
and occasionally causes centrilobular hepatic necrosis in the
hamster. Organ specific alkylation corresponds to the sites
of toxicity; high levels of lung and kidney alkylation are seen
in the adult male mouse whereas the hamster shows high levels of
hepatic and pulmonary alkylation, especially when high doses
of IPO are administered (not shown in Figure 4, see ref 13).
Administration of IPO to avain species results in yet another
pattern of organ-specific toxicity(14). In birds, IPO produces

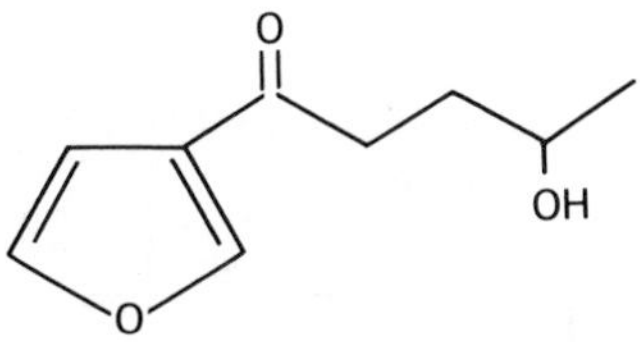

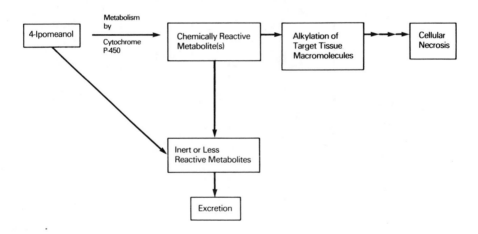

Figure 1. *Structure of 4-ipomeanol, a toxin isolated from sweet potatoes infected with the common mold* F. solani

Figure 2. *Role of metabolic toxication and detoxication in 4-ipomeanol tissue necrosis*

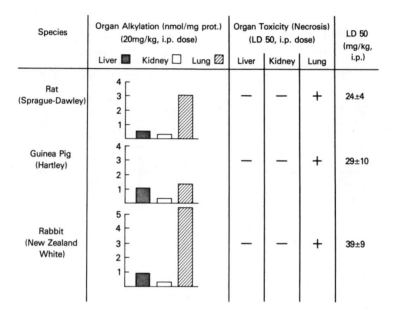

Figure 3.   Species differences in 4-ipomeanol organ alkylation, organ toxicity, and lethality, I

only hepatic necrosis. Hepatic alkylation by IPO is predominant, with relatively little binding to lung or kidney. Finally, although the patterns of target organ toxicity and alkylation show marked species differences, it should be emphasized that these patterns are remarkably consistent among different strains of a given test species(12,13,14).

These marked species differences in IPO toxicity, coupled with its striking organ-specific toxicity, have made it a useful model compound for studying the factors which influence organ-specific toxicity of metabolically activated toxins.

Factors Which May Influence Patterns of Organ-Specific Toxicity

Organ-specific toxicity may be the result of a complex set of interrelated events. Many factors, singly or in combination, can affect the sensitivity of a specific tissue to a toxin. These not only may be related to the nature of the toxin [i.e., site and degree of activation, stability of the reactive metabolite(s)], but also to the target tissue involved (i.e., selective exposure, protective and repair mechanisms present). At present, it is difficult or impossible to predict a priori which of these factors are of greatest importance for a specific toxin. Therefore, it is necessary to study each compound individually to determine its pattern of organ-specific toxicity and what factors underlie this specificity.

Formation of reactive metabolites in the target tissue. Since the liver contains large concentrations of enzymes responsible for xenobiotic metabolism, many compounds that are metabolized to alkylating agents or free radicals are hepatotoxic [e.g., bromobenzene (15), carbon tetrachloride (16), aflatoxin (17)]. But other organs besides the liver contain drug metabolizing activity (18), albeit to a lesser extent than the liver, and several examples of extrahepatic metabolic activation and toxicity are known (see refs. 19 and 20 for reviews). Besides in situ activation, another possible mechanism for extrahepatic toxicity is hepatic activation followed by transport of the reactive metabolite(s) to the extrahepatic target tissue by way of the circulation.

Experiments have been conducted to determine whether pulmonary or hepatic activation is responsible for the pulmonary toxicity of IPO. In the rat, both liver and lung microsomes have the ability to activate IPO. When animals are pretreated with 3-methylcholanthrene (3MC), an inducer of certain xenobiotic metabolizing enzymes, the in vitro rate of activation of IPO is increased in liver but not lung microsomes(10). As discussed previously, IPO produces high levels of pulmonary alkylation and bronchiolar necrosis in rats, but little hepatic alkylation and no hepatic necrosis.

When rats are pretreated with 3MC, the liver becomes the major
organ for toxicity (centrilobular necrosis) and alkylation by
IPO (Figure 5)(21). The fact that induction of the liver to
produce more reactive metabolite does not cause increased
alkylation and toxicity in the lung supports the concept that
the pulmonary toxicity of IPO is due to in situ metabolic
activation.

Similar experiments demonstrate that the renal alkylation
and toxicity of IPO seen in the adult male mouse is due to
formation of the toxic metabolites in the kidney(22). IPO
activation is markedly enhanced in liver microsome preparations
from C57BL/6J mice pretreated with 3MC, but not significantly
increased in microsome preparations from the lungs or kidneys.
3MC-pretreatment causes alterations in the in vivo target-organ
alkylation and toxicity of IPO; namely, alkylation is markedly
elevated in the livers, while actually decreased in lungs and
kidneys in comparison to nonpretreated controls. IPO
frequently causes hepatic necrosis in pretreated mice, but
never in controls, and renal and pulmonary toxicity are less
than in controls. In contrast, DBA/2J mice are not inducible
with 3MC and pretreatment with this agent has no significant
effect on microsomal activation or in vivo target organ
alkylation and toxicity by IPO in this strain.

Thus, the findings that 3MC greatly increases the
formation of reactive IPO metabolite(s) in the liver without
increasing in vivo alkylation of the lung (rats and mice) and
kidney (mice) supports the premise that the alkylating
metabolite(s) of IPO are formed in situ, and not in the liver
followed by transport to the lung and/or kidney.

Metabolism and distribution studies in rats have shown
that 3MC pretreatment decreases the plasma and lung
concentrations of IPO(23). Therefore, the decreases in the
levels of alkylation in rat lungs and mouse lungs and kidneys
when pretreated with 3MC (Figure 5) is probably due to the
decreased exposure of these organs to unmetabolized IPO.

Stability and Transport of Reactive Metabolites. In
tissues incapable of activating IPO, no tissue damage and
little alkylation is seen. This, coupled with the evidence
for in situ activation discussed above, indicates that the
alkylating metabolite(s) is too reactive and/or unstable to
escape the site of activation and circulate to other tissues.

The pyrrolizidine alkaloids (PA's, Figure 6), a group of
naturally occuring toxins, produce extrahepatic target organ
damage by a mechanism different than IPO. These compounds are
activated in the liver to chemically reactive pyrroles which
are transported to other tissues via the circulation(24). The
major pathological effect of PA ingestion is hepatotoxicity at
low doses; higher doses produce toxicity in many other organs
as well (i.e. lung, kidney, brain, muscle). Studies on the

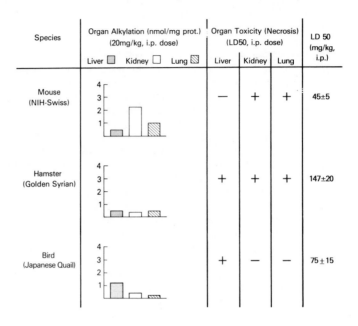

Figure 4.   Species differences in 4-ipomeanol organ alkylation, organ toxicity, and lethality, II

Species (Strain)	Effect on Alkylation			Effect on Toxicty		
	Lung	Liver	Kidney	Lung	Liver	Kidney
Rat (Sprague-Dawley)	↓	↑		↓	↑	
Mouse (C57BL/6)	↓	↑	↓	↓	↑	↓
Mouse (DBA/2)	NE*	NE	NE	NE	NE	NE

*No effect

Figure 5.   Effect of 3-methylcholanthrene pretreatment on the in vivo alkylation and toxicity of 4-ipomeanol

lung toxicity indicate that vascular walls (arteries,
capillaries and veins) are the primary sites of the toxic
lesions, as would be expected if pulmonary exposure to the
toxin is through the circulatory system. Furthermore,
intravenous administration of low doses of
dehydropyrrolizidines produces pulmonary damage similar to
that caused by much larger doses of the parent PA's. A few in
vitro studies indicate that lung preparations have little
ability to convert PA's to their toxic pyrrolic metabolites.
Also, phenobarbital pretreatment, an inducer of hepatic
activation of PA's, potentiates both the liver and the lung
toxicity. These and other studies provide evidence that the
toxicity of the PA's is due to pyrrolic metabolites formed
primarily in the liver. The active metabolites are stable
enough to be transported in the bloodstream to the organs
where toxicity is observed. The profile for target-organ
toxicity produced by this type of mechanism is much less
specific than that seen with in situ activation; the toxicity
is most pronounced in the organ where activation takes place,
but lesions also occur in many other organs which may not
posess the ability to activate the toxin.

An example of a mechanism which produces organ-specific
toxicity in a tissue distant from the site of primary
activation is the induction of bladder tumors by 2-naphthyl-
amine (Figure 7)(17). The ultimate carcinogen appears to be
the chemically reactive N-hydroxy-2-naphthylamine, but the
bladder does not contain enzymes capable of forming this
metabolite. Instead, it is formed in the liver, stabilized by
glucuronidation and transported to the kidneys via the bloodstream. The carcinogenic hydroxylamine is regenerated in
acidic urine by hydrolysis, resulting in selective exposure of
the bladder to the ultimate carcinogen.

Organ Structure and Cell Specific Activation. Although
the contribution of extrahepatic metabolism to the fate of a
particular xenobiotic may be quite small in comparison to
hepatic metabolism, the biologic consequences of extrahepatic
metabolic transformations leading to irreversible or cumulative
cellular changes may be substantial. This is important in
view of the generally greater cellular heterogeneity of many
extrahepatic tissues compared to the liver, and the likelihood
that extrahepatic xenobiotic-metabolizing activities are not
randomly distributed throughout all cells in these organs.
Specific cell types possessing enzyme activities required for
the metabolic activation of xenobiotic substances might
be extraordinarily susceptible to toxicities by those
agents, and this selective cellular toxicity may contribute
to the sensitivity of an organ to the effects of alkylation.
For instance, autoradiographic studies with IPO(25) show that
the toxin bound in liver is widely distributed throughout the

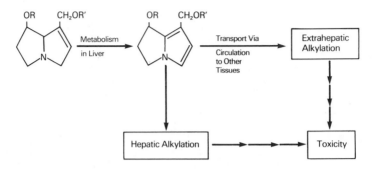

Figure 6. Role of metabolism and systemic transport in the toxicity of pyrrolizidine alkaloids

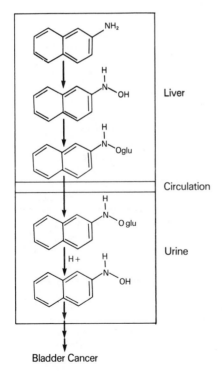

Figure 7. Role of metabolism and systemic transport in the formation of bladder tumors by 2-naphthylamine

organ, but that the material bound in the mouse kidney is located only in renal cortex, and only in the cortical tubules that become necrotic. The radioactivity bound in the lung is even more highly localized, being found in the nonciliated pulmonary bronchiolar cells, the major site of IPO toxicity in the lung.

    Location of activating enzymes. The liver contains the highest concentration of xenobiotic metabolizing enzymes of any organ, and is the organ responsible for the majority of foreign compound metabolism. It is not surprising, therefore, that many agents which produce their toxicity by metabolic activation are hepatotoxins. The in situ activation of IPO in liver, kidney and lung suggests that its striking pattern of organ-selective toxicity could be due to the relative ability of the target tissues to activate the toxin. Evidence supporting this premise has been obtained by comparing the ability of lung, liver and kidney microsomal preparations from various species with organ selectivity of IPO toxicity in vivo(12,13,14). In microsome preparations from all species except the hamster and the bird, alkylation in the lung preparations is equal to or greater than that with the corresponding liver preparations. Likewise, in all species except the hamster and the bird, the primary organ for toxicity and alkylation is the lung. In hamster liver microsomes, IPO covalent binding activity is unusually high, and likewise is exceptionally high in avian liver microsomes. Moreover, IPO covalent binding activity is very low or absent in bird lung microsomes. In vivo alkylation in the hamster is high in both the liver and the lung, and it is the only species in which IPO occasionally causes liver damage in addition to pulmonary injury. In vivo in the bird, alkylation is predominantly in the liver, and this is the only site of necrosis. Similarly, mouse kidney microsomes (from adult male animals) are highly active in mediating in vitro alkylation by IPO, but kidney microsomes from all other species tested are almost completely inactive in catalyzing IPO alkylation. This is of particular interest because the adult male mouse was the only species tested in which IPO consistently caused renal cortical necrosis in addition to pulmonary bronchiolar necrosis.
    These correlations between in vitro alkylation and in vivo alkylation and toxicity suggest that differences in patterns of tissue-specific toxicity of IPO are due, at least in part, to differences in the ability of target tissues to activate the toxin (i.e. activities are either present or absent in potential target tissues). There may be multiple reasons for these differences. For example, although a cytochrome P-450 appears to be required for the metabolic activation of IPO, there is not a good correlation between total microsomal content of this enzyme and the capacity of the microsomes to metabolize IPO(12,13,14). For instance, rat kidney microsome preparations

are incapable of activating IPO to alkylating metabolite(s), but
they do contain significant concentrations of cytochrome(s)
P-450. Tissues capable of metabolizing xenobiotics contain
multiple types of cytochrome(s) P-450(26,27) and these different
types may show very different substrate specificities for
metabolism. Therefore, it seems likely that only certain forms
of cytochrome P-450 are capable of supporting the metabolic
activation of IPO, and that this could be an important
determinant of tissue-specific metabolism of the compound.

Factors other than cytochrome P-450 also could be
responsible for differences in extrahepatic vrs hepatic
microsomal activities involved in IPO metabolism. For
example, one experiment that suggested this showed that an
antibody prepared against purified cytochrome b5 almost
completely inhibited the metabolism of IPO by rat pulmonary
microsomes, but it had little effect on IPO metabolism by rat
hepatic microsomes(28).

Factors Which Modify Target-Tissue Metabolic Activity:
Age, Sex, Inducers and Inhibitors of Metabolism. The ability
of an organ to activate a specific toxin is one explanation
of organ-selective toxicity. Factors such as age, sex,
circadian rhythms, nutritional status, and exposure to chemicals
are known to affect xenobiotic metabolizing enzymes, and
therefore might affect organ-specific toxicity of metabolically
activated toxins. Several of these factors have striking
effects on the organ-specific toxicity produced by IPO.

In C57BL/6J mice, there are striking, age-related
differences in target organ necrosis produced by IPO(29). In
adult male mice (>6 weeks of age), IPO produces both pulmonary
bronchiolar and renal tubular necrosis. But young mice (1.5
weeks old) show only pulmonary necrosis. In vivo studies
demonstrate that kidney alkylation is age dependent. In young
mice, lung binding is high and kindey binding is almost nonexistent. As age increases, lung binding decreases but kidney
binding increases markedly. This age-related effect could be
explained by either age-related differences in target organ
metabolism or age related differences in tissue distribution of
the parent compound. To investigate the latter possibility,
concentrations of unmetabolized drug were measured at varying
times in young (2.5 weeks of age) and old (11 weeks of age)
mice(30). Kidney, lung, liver and blood concentrations were
always higher in the young mice. If the age related kidney
alkylation and toxicity were due to tissue distribution, older
mice would be expected to have lower levels of kidney alkylation
and toxicity, which they do not. Therefore, changes in
distribution of the drug do not account for the age related
development of renal toxicity.

Age-related changes in kidney activation of IPO can be
demonstrated by incubating slices of kidney from mice of varying

ages with   IPO (31). The ability of kidney slices to activate
the drug is assayed by measuring the rate of kidney protein
alkylation. Alkylation of kidney slices from 10 week old mice
is strikingly higher than that from 2 week old mice.

Thus, in vivo and in vitro studies demonstrate that the
age-related differences in kidney alkylation and toxicity by IPO
in the mouse are due to changes in renal metabolism rather than
changes in the tissue distribution of the parent compound.
Histologic evaluation of lung, liver and kidney from the rat and
hamster show no age-related differences in target organ
alkylation or toxicity.

As opposed to the adult male mouse, the adult female mouse
is highly resistant to renal alkylation and toxicity by IPO
(32). Alkylation by reactive IPO metabolite(s) occurs
preferentially in the lungs of female mice over a wide range
of doses. Even near-lethal doses of IPO do not cause renal
necrosis in female mice. Little ability to produce alkylating
metabolites is seen in renal slice preparations from adult
female mice.

Many chemicals can alter the rate and/or the pathways of
xenobiotic metabolism, both in hepatic and extrahepatic tissues.
Exposure to these chemicals through environmental pollution,
agriculture, natural sources, modern medicine and personal
social habits is common. For toxins that are activated by
metabolism, exposure to metabolic inducers or inhibitors can
affect organ-specificity as well as overall sensitivity to a
toxin.

The studies with 3MC and IPO discussed earlier illustrate
this point. Pretreatment with 3MC shifts the target organ for
IPO alkylation and toxicity in rats from the lung to the liver
(21). In vitro studies demonstrate that 3MC pretreatment
increases the alkylation of liver microsomes from rats, but does
not affect alkylation of lung microsomes(10). This suggests
that the in vivo hepatic toxicity of IPO is due to increased
hepatic formation of the toxic metabolite. 3MC pretreatment
reduces the amount of circulating IPO, which probably accounts
for the decreased pulmonary alkylation and toxicity (23).

Many chemicals are known to inhibit the metabolism of
xenobiotics. Pyrazole, piperonyl butoxide, SKF-525A or
cobaltous chloride pretreatments (inhibitors of cytochrome
P-450 catalyzed metabolism) decrease the in vitro alkylation
both in rat liver and lung microsomes by IPO(10).
Correspondingly, pretreatment of rats with these compounds
also decreases the tissue alkylation and toxicity of IPO in
vivo(21).

These studies illustrate the effects of exogenous chemicals
on the organ-specificity and sensitivity of an organism to IPO
toxicity. It is striking that agents which primarily affect the
hepatic metabolism of IPO (3MC and phenobarbital) have such a
dramatic effect on extrahepatic toxicity, even when this
toxicity is due to in situ activation.

Detoxication Pathways. Since a toxic response to a
metabolically activated chemical depends upon the balance
between formation and detoxication pathways for the toxic
metabolite(s), differences in deactivation pathways between
tissues could also contribute to patterns of organ-selective
toxicity. Although tissue-specific activation of IPO roughly
correlates with the major differences in target organ
toxicity(13), potential differences in detoxication pathways
could explain the more subtle species differences in sensitivity
and organ-specificity of IPO toxicity.

As discussed earlier, pretreatment of rats with 3MC
increases alkylation of liver microsomes by IPO while not
affecting lung microsome alkylation, and results in increased
liver alkylation and toxicity in vivo. Pretreatment with
phenobarbital, a non-specific inducer of xenobiotic metabolism,
also increases alkylation of liver microsomes by IPO while not
affecting lung microsome alkylation(21). However, phenobarbital
pretreatment does not alter the target organ for alkylation and
toxicity of IPO in vivo and alkylation is decreased in both lung
and liver. In vivo metabolism studies offer a posible
explanation for this difference(33). Phenobarbital increases
glucuronide formation (as measured by the amount of IPO-
glucuronide excreted in urine) which decreases the exposure of
liver and lung to the parent compound. 3MC pretreatment, on the
other hand, does not increase glucuronide formation. Therefore,
glucuronidation appears to be a detoxication pathway for IPO,
and phenobarbital induces this pathway more than IPO toxication
pathways, whereas 3MC does not. This demonstrates that
differences in the rate of glucuronide formation, at least in
the liver, have the potential to alter the amount of toxic
metabolite produced.

As illustrated by Figure 2, detoxication potentially can
occur after, as well as before activation of the parent
compound. Glutathione (GSH) conjugation has been shown to
detoxify reactive metabolites of certain hepatotoxic substances
[e.g. acetaminophen(34), bromobenzene(35)] by reacting with
their electrophilic metabolites to form less toxic, readily
excretable glutathione conjugates. Recent studies implicate
IPO-GSH conjugate formation as a detoxication pathway for
reactive IPO metabolites. Alkylation of rat microsomes by IPO
is dramatically decreased by the addition of GSH to the
incubation mixtures(6,10). This reduction in alkylation is
not due to decreased production of the reactive metabolite(s) as
determined by substrate disappearance. The formation of at
least two IPO-GSH conjugates corresponding to the decrease in
alkylation can be demonstrated using an analytical method
based on high-pressure anion-exchange chromatography (6). The
role of GSH conjugation as a protective mechanism for IPO-

induced toxicity is indicated by results from in vivo
experiments(37,38,39). In the rat, where IPO is a selective
pulmonary toxin, depletion of pulmonary GSH by IPO is dose
dependent. GSH levels in the kidney and liver are not affected.
Likewise, in the mouse (where IPO is a nephrotoxin as well as
a pulmonary toxin), increasing doses of IPO lead to decreases
in GSH levels both in the lung and in the kidney, but have
little effect on the liver. Finally, in the bird, where IPO
is predominantly hepatotoxic, IPO selectively depletes GSH in
the liver. Pretreatment of rats, mice, or birds with a dose
of diethylmaleate (DEM) which depletes tissue GSH, increases
target organ alkylation and necrosis by IPO. In the rat,
pretreatment with piperonyl butoxide, an inhibitor of the
metabolic activation of IPO in the lung, reduces both the
pulmonary toxicity and depletion of lung GSH. Finally,
administration of alternate nucleophiles (cysteine or
cysteamine) decreases rat pulmonary toxicity and alkylation.
    Both the in vitro and in vivo studies of IPO-GSH
conjugation support the view that GSH plays a protective role
in lung, kidney and liver toxicity produced by IPO by serving
as an alternate nucleophile for the reactive metabolite(s)
formed in situ.

Conclusions

    Organ-specific toxicity by chemicals appears to result from
complex interactions between many biochemical, physiological and
chemical factors. The biochemical factors which contribute to
toxicity, and which are responsible for differences in
susceptibility among target organs, species, and strains are
only beginning to be understood. Differences in age, sex,
toxication and detoxication pathways, repair mechanisms, and
responses to metabolic inducers and inhibitors are potential
determinants of organ-specific toxicity. At present, it is
difficult or impossible to predict a priori which of these play
critical roles in determining the sensitivity and tissue-
specific toxicity for a specific chemical and a given organism.
A better understanding of the biochemical factors which
influence organ-specific toxicity is needed to improve our
ability to rationally extrapolate toxicity data from animals
to humans and to assist in predicting and assessing the health
hazards to humans from chemical exposure.

Literature Cited

1. Hansen, A. A. Potato poisoning. North Am. Vet., 1928, 9,
   31-34.
2. Abo, S.; Nomura, S. Sweet potato disease. J. Vet. Hyg.
   Assoc., 1942, 10, 17-25.

3. Monlux, W.; Fitte, J.; Kendrick, G.; Dubuisson, H. Progressive pulmonary adenomatosis in cattle. Southwest Vet., 1953, 6, 267-269.
4. Peckham, J. C.; Mitchell, F. E.; Jones, O. H.; Doupnik, B. Atypical interstitial pneumonia in cattle fed moldy sweet potatoes. J. Am. Vet. Med. Assoc., 1972, 160, 169-172.
5. Boyd, M. R.; Burka, L. T., Harris, T. M.; Wilson, B. J. Lung toxic furanoterpenoids produced by sweet potatoes (Ipomoea batatas) following microbial infection. Biochim. Biophys. Acta, 1974, 337, 184-195.
6. Boyd, M. R.; Wilson, B. J. Isolation and characterization of 4-ipomeanol, a lung toxic furanoterpenoid produced by sweet potatoes (Ipomoea batatas). J. Agric. Food Chem., 1972, 20, 428-430.
7. Wilson, B. J.; Yang, D. T. C.; Boyd, M. R. Toxicity of mold-damaged sweet potatoes (Ipomoea batatas). Nature, 1970, 227, 521-522.
8. Boyd, M. R.; Wilson, B. J.; Harris, T. M. Confirmation by chemical synthesis of the structure of 4-ipomeanol, a lung-toxic metabolite of the sweet potato, Ipomoea batatas. Nature New Biol., 1972, 236, 158-159.
9. Boyd, M. R.; Burka, L. T.; Wilson, B. J. Distribution, excretion and binding of radioactivity in the rat after intraperitoneal administration of lung-toxic furan, 4-ipomeanol-C^{14}. Toxicol. Appl. Pharmacol., 1975, 32, 147-157.
10. Boyd, M. R.; Burka, L. T.; Wilson, B. J.; Sasame, H. A. In vitro studies on the metabolic activation of the pulmonary toxin, 4-ipomeanol, by rat lung and liver microsomes. J. Pharmacol. Exp. Therap., 1978, 207, 677-686.
11. Boyd, M. R.; Dutcher, J. S. Convenient methods for the preparation of [5-^{14}C]-4-ipomeanol and [^{3}H(G)]-4-ipomeanol of high specific radioactivity. Labelled Compds. & Radiopharmaceuticals (in press).
12. Boyd, M. R.; Dutcher, J. S.; Buckpitt, A. R.; Jones, R. B.; Statham, C. N. Role of metabolic activation in extrahepatic target organ alkylation and cytotoxicity by 4-ipomeanol, a furan derivative from moldy sweet potatoes; possible implications for carcinogenesis. "Naturally-Occurring Carcinogens-Mutagens and Modulators of Carcinogenesis", E. C. Miller et. al., Eds.; Japan Sci. Soc. Press: Tokyo; Univ. Park Press: Baltimore; 1979; p. 35-56.
13. Dutcher, J. S.; Boyd, M. R. Species and strain differences in target organ alkylation and toxicity by 4-ipomeanol; predictive value of covalent binding in studies of target organ toxicities by reactive metabolites. Biochem. Pharmacol., 1979, 28, 3367-3372.
14. Buckpitt, A. R.; Boyd, M. R. Xenobiotic metabolism in birds, species lacking pulmonary Clara cells. The Pharmacologist, 1978, 20, 181.

15. Gillette, J. R.; Mitchell, J. R.; Brodie, B. B. Biochemical mechanisms of drug toxicity. Ann. Rev. Pharmacol., 1974, 14, 271-288.
16. Recknagel, R. O. Carbon tetrachloride hepatotoxicity. Pharmacol. Rev., 1967, 19, 145-208.
17. Miller, J. A.; Miller, E. C. Metabolic activation of chemicals to reactive electrophiles: an overview. "Proceedings of the Seventh International Congress of Pharmacology", vol. 9, Y. Cohen, Ed.; Pergammon Press: Oxford & New York, 1979; p. 3-12 and references cited therein.
18. Testa, B.; Jenner, P. "Drug Metabolism: Chemical and Biochemical Aspects"; Marcel Dekker, Inc.: New York, 1976; p. 419-424.
19. Boyd, M. R. Biochemical mechanisms in chemical-induced lung injury: roles of metabolic activation. CRC Crit. Rev. Toxicol., 1980, 7, 103-176.
20. Boyd, M. R. Effects of inducers and inhibitors on drug metabolizing enzymes and on drug toxicity in extrahepatic tissues. "Environmental Chemicals, Enzyme Function, and Human Disease", (Ciba Foundn. Symp. No. 76); Excerpta Medica: Amsterdam, 1980; p. 43-66.
21. Boyd M. R.; Burka, L. T. In vivo studies on the relationship between target organ alkylation and the pulmonary toxicity of a chemically reactive metabolite of 4-ipomeanol. J. Pharmacol. Exp. Therap., 1978, 207, 687-697.
22. Boyd, M. R.; Dutcher, J. S. Renal toxicity due to reactive metabolites formed in situ in the kidney: investigations with 4-ipomeanol in the mouse. J. Pharmacol. Exp. Therap., 1980, in press.
23. Statham, C. N.; Boyd, M. R. unpublished results.
24. Mattocks, A. R. Mechanisms of pyrrolizidine alkaloid toxicity. "Proc. 5th Int. Congr. Pharmacology", vol. 2; S. Karger: Basel, 1973; p. 114-123.
25. Boyd, M. R. Evidence for the Clara cell as a site of cytochrome P450-dependent mixed-function oxidase activity in lung. Nature, 1977, 269, 713-715.
26. Johnson, E. F. Multiple forms of cytochrome P-450: Criteria and significance. "Reviews in Biochemical Toxicology", Vol. 1; Hodgson, E.; Bend, J. R.; Philpot, R. M.; Eds.; Elsevier/North-Holland: New York, Amsterdam, Oxford, 1979; p. 1-26.
27. Guengerich, F. P. Preparation and properties of highly purified cytochrome P-450 and NADPH-cytochrome P-450 reductase from pulmonary microsomes of untreated rabbits. Mol. Pharmacol., 1977, 13, 911-933.
28. Sasame, H. A.; Boyd, M. R. Possible role of cytochrome b_5 as a rate-limiting factor in metabolic activation of 4-ipomeanol by lung microsomes. Fed. Proc., 1978, 37, 464.

29. Jones, R.; Allegra, C.; Boyd, M. Predictive value of
    covalent binding in target organ toxicity: Age-related
    differences in target organ alkylation and toxicity by
    4-ipomeanol. Toxicol. Appl. Pharmacol., 1979, 48, A129.
30. Jones, R. B.; Boyd, M. R. unpublished results.
31. Longo, N.; Boyd, M. R. unpublished results.
32. Longo, N. S.; Boyd, M. R. Sex differences in renal
    alkylation and toxicity by 4-ipomeanol in the mouse.
    "Abstracts of papers": Society of Toxicology, 19th annual
    meeting, Washington, D.C.; Academic Press, Inc.: N.Y.,
    1980; A130.
33. Statham, C. N.; Dutcher, J. S.; Boyd, M. R. Ipomeanol-
    4-glucuronide, a major urinary metabolite of 4-ipomeanol
    in rats. "Abstracts of papers": Society of Toxicology,
    19th annual meeting, Washington, D.C.; Academic Press, Inc.:
    N.Y., 1980; A45.
34. Mitchell, J. R.; Thorgeirsson, S. S.; Potter, W. Z.; Jollow,
    D. J., Keiser, H. Acetaminophen-induced hepatic injury:
    Protective role of glutathione in man and rational for
    therapy. Clin. Pharmacol. Ther., 1974, 16, 676-684.
35. Jollow, D. J.; Mitchell, J. R.; Zampaglione, N.; Gillette,
    J. R. Bromobenzene induced liver necrosis. Protective role
    of glutathione and evidence for 3,4-bromobenzene oxide as
    the hepatotoxic intermediate. Pharmacology, 1974, 11,
    151-169.
36. Buckpitt, A.; Dutcher, J.; Boyd, M. Determination of
    electrophillic metabolites produced during microsomal
    metabolism of 4-ipomeanol by high-pressure anion exchange
    chromatography of the glutathione adducts. Fed. Proc.,
    1979, 38, 692.
37. Boyd, M.; Statham, C.; Stiko, A.; Mitchell, J.; Jones, R.
    Possible protective role of glutathione in pulmonary
    toxicity by 4-ipomeanol. Toxicol. Appl. Pharmacol., 1979,
    48, A66.
38. Buckpitt, A. R.; Statham, C. N.; Boyd, M. R. Protective role
    of glutathione in the alkylation and hepatotoxicity by
    4-ipomeanol in the bird. "Abstracts of papers"; Society of
    Toxicology; 19th annual meeting, Washington, D.C.; Academic
    Press, Inc.: N.Y., 1980; A130.
39. Dutcher, J. S.; Boyd, M. R. unpublished results.

RECEIVED February 2, 1981.

Epigenetic Mechanisms of Action of Carcinogenic Organochlorine Pesticides

GARY M. WILLIAMS

Naylor Dana Institute for Disease Prevention, American Health Foundation, Valhalla, NY 10595

Many of the most widely used chlorinated cyclic hydrocarbon compounds have been found to be carcinogenic in experimental laboratory rodents (Table I).

Table I.   Carcinogenicity of Chlorinated Cyclic Hydrocarbon Pesticides

Compound	Principal Target Organ		
	Mouse	Rat	References
Aldrin	liver		1
Chlordane	liver, uterus		2
Chlorobenzilate	liver	NS[a]	3
DDT	liver, lung	liver	3,4,5
Dieldrin	liver	NS	6, 7
Heptachlor	liver	thyroid	8
Hexachlorobenzene	liver		9
Hexachlorocyclohexane (BHC), lindane	liver	liver	10,11
Kepone	liver	liver, thyroid	12
Mirex	liver	liver	3,13
PCB	liver	liver	14,15

[a] no significant increase in neoplasms

Cyclic hydrocarbons with chlorine substituents that block ring oxidation are resistant to biodegradation and thus accumulate in the environment and persist for long periods in animals once they are absorbed. The persistence of organochlorine pesticides

0097–6156/81/0160–0045$05.00/0
© 1981 American Chemical Society

together with their animal carcinogenicity has given
rise to concern that exposed humans would be at risk
of cancer development from these chemicals (16,17,
18). Indeed, extrapolation of dose-response effects
from rodents to humans predicts substantial cancer
causation (16). However, epidemiologic studies of
highly exposed groups have failed to reveal any significant increase in cancer occurrence (19,20) and no
increase in cancer incidence has been associated with
pesticide usage (21). Such a discrepancy suggests
that the mechanism of action of chlorinated cyclic
hydrocarbons may be different from that of other carcinogens which produce cancer in both experimental
animals and humans (22,23). This possibility is further supported by the unusual situation that all carcinogens of this structural type have the liver as
their principal target organ. For carcinogens that
are activated to reactive metabolites, members of a
structural type almost always affect more than one
organ and often the principal organ affected varies
with the specific compound. For these and other reasons, we have suggested that chlorinated cyclic
hydrocarbons may be carcinogenic to rodents by
indirect mechanisms (22,23,24).

Mechanisms of Carcinogenesis

Chemical carcinogens are defined operationally
by their ability to induce tumors in exposed animals. A highly diverse collection of chemicals is
capable of producing this effect, including organic
and inorganic chemicals, solid state materials, hormones and immunosuppressants. The heterogeneity of
structures represented makes it improbable that all
chemicals would act through a single mechanism.
Therefore, Weisburger and Williams (23) have proposed
a classification that separates chemical carcinogens
into two major categories, genotoxic and epigenetic
(Table II).

Table II Classes of Carcinogenic chemicals

Type	Example
A. Genotoxic	
1. Direct-acting or primary carcinogen	Ethylene imine, bis- (chloromethyl)ether

2.	Procarcinogen or secondary carcinogen	Vinyl chloride, benzo-(a)pyrene, 2-naphtyl-amine, dimethylnitros-amine
3.	Inorganic carcinogen	Nickel, chromium
B.	**Epigenetic**	
4.	Solid-state carcinogen	Polymer or metal foils, asbestos
5.	Hormone	Estradiol, diethylstilbestrol
6.	Immunosuppressor	Azathioprine,
7.	Cocarcinogen	Phorbol esters, pyrene, catechol, ethanol, n-dodecane,
8.	Promoter	Phorbol esters, bile acids, saccharin

Carcinogens that interact with and alter DNA are classified as genotoxic. Thus, the genotoxic category contains the chemicals that function as electrophilic reactants as originally postulated by the Millers (25). Also, because some inorganic chemicals have displayed such effects they have tentatively been placed in this category. The second broad category designated as epigenetic carcinogens comprises those chemicals for which no evidence of direct interaction with genetic material exists. This category contains solid state carcinogens, hormones, immunosuppressants, cocarcinogens and promoters.

This classification and the underlying concepts, if ultimately validated, have major implications for risk extrapolation to humans of data on experimental carcinogenesis. Genotoxic carcinogens, as a consequence of their effects on genetic material, pose a clear qualitative hazard. These carcinogens are occasionally effective after a single exposure, are often carcinogenic at low doses, act in a cumulative manner, usually produce irreversible effects, and produce combined effects with other genotoxic carcinogens having the same target organ. In contrast, with some types of epigenetic carcinogens, it is known that the carcinogenic effects occur only with high and sustained levels of exposure that lead to prolonged physiologic abnormalities, hormonal imbalances, or tissue injury. In such cases, the effects are often entirely reversible upon cessation of exposure. Because of these features, the risk from exposure to epigenetic carcinogens seems to be of a quantitative nature.

Thus, a major element in assessing the potential hazard of a chemical is to evaluate its potential genotoxicity.

Lack of Genotoxicity of Organochlorine Pesticides

The genetic effects of organochlorine pesticides have been examined in a number of in vitro short-term tests (Table III).

Table III. Activity in Short-term Tests Measuring
DNA Interaction of Carcinogenic
Organochlorine Compounds

Compound	DNA Damage	DNA Repair	Mutagenesis Bacterial	Mutagenesis Mammalian
DDT	-a	-b,d	-f	-b
DDE	ND	-b	-g	ND
Dieldrin	-a	-b,+c	-e,f	+c
Chlordane	ND	+c,-d	ND	-b,+c
Heptachlor	ND	ND	-e,-h	-b
Kepone	ND	-b	-h	-b

a) Swenberg (26), b) Williams (24), c) Hart (27,28), d) Flamm (29), e) Marshall (30), f) Shirasu (31), g) Ames (32), h) Schoeny (33).

Although the results have been predominantly negative, their significance has been minimized by the frequent suggestion that lack of activity is simply a consequence of the absence of appropriate metabolism in the in vitro tests.

In our laboratory we have developed several tests for genotoxicity utilizing liver-derived cells (34,35). Since the organochlorine pesticides have the liver as their principal target organ, these tests represent the ideal system in which to evaluate the genotoxicity, as well as other effects, of these compounds.

The hepatocyte primary culture (HPC)/DNA repair test assesses the capability of chemicals to undergo

covalent interaction with DNA by measurement of autoradiographic DNA repair elicited as a result of the
DNA damage (36,37). The freshly isolated hepatocytes
used in this test retain a high level of activity for
biotransforming xenobiotics and thus the test re-
sponds to a wide spectrum of structural types of carcinogens requiring metabolic activation (34,35). Our
previous reports of lack of genotoxicity of organochlorine pesticides in the rat liver HPC/DNA repair
test (24,38) have been extensively confirmed (Table
IV).

Table IV.   HPC/DNA Repair Results

Compound	grains/nucleus[a]		
	Rat	Mouse	Hamster
2',3-Dimethyl-4-aminobiphenyl	60	25	>100
Biphenyl	-	-	-
Chlordane	-	-	-
DDT	ND	-	-
Mirex	ND	-	-
Kepone	-	ND	ND

[a] - = zero; ND = not done

In addition, since the organochlorine pesticides
are sometimes more active on mouse liver, these results were extended (38,39) to the mouse liver
derived HPC/DNA repair test, as well as the hamster
liver derived test (Table IV).

Another liver-derived test for genotoxicity is
the adult rat liver epithelial cell (ARL)/hypoxan-
thine-guanine phosphoribosyl transferase (HGPRT)
mutagenesis assay (40,41). This test assesses mutagenicity at the HGPRT locus through measurement of
conversion of liver epithelial cells to HGPRT-deficient mutants that are resistant to 6-thioguanine. As
with the HPC/DNA repair test, the cells in this assay
possess intrinsic metabolic capability for the biotransformation of activation-dependent carcinogens
(34). In spite of a mutagenic response to three
genotoxic carcinogens, the organochlorine pesticides
were all non-mutagenic in this assay (24) (Table V).

Table V.   ARL[a]/HGPRT Mutagenesis Assay Results

Compound	Concentration molar[b]	Induction of HGPRT deficient mutants
Aflatoxin B_1	10^{-6}	+
3-Methyl-4- dimethyl-aminoazobenzene	10^{-5}	+
2-Aminofluorene	10^{-4}	+
Chlordane	2.5×10^{-5}	−
Kepone	10^{-5}	−
Heptachlor	10^{-5}	−
Hexachloro-cyclopentadiene	10^{-6}	−
Endrin	3×10^{-3}	−
DDT	10^{-4}	−

[a] line ARL 6
[b] highest nontoxic dose that was negative or lowest dose that was positive.

The consistent lack of genotoxicity of organo-chlorine pesticides in liver derived tests strongly supports the negative data obtained in other tests. Thus, it appears that these chemicals are not geno-toxic carcinogens.

Epigenetic Mechanism of Action of Organochlorine Pesticides.

At least one organochlorine pesticide, DDT, has been shown to be a liver tumor promoter (42), enhanc-ing the carcinogenic effect of 2-acetylaminofluorene when given after the carcinogen.   Thus, we have pos-tulated that the organochlorine pesticides may be carcinogenic through a mechanism of tumor promotion (22,24,38).   All of the inbred strains of rats and mice used for carcinogen bioassay have a spontaneous incidence of liver tumors which in the case of some mouse strains is quite high (22).   As part of this condition, these animals also have a higher incidence of lesions regarded as preneoplastic or potentially neoplastic.   Thus, we postulated that the promoting effect of organochlorine pesticides would enable the pre-existing abnormal liver cells to progress to a higher frequency of tumor development than would occur under control conditions.

The mechanism of the promoting effect of chemicals when administered after a primary carcinogen is not yet known.   A compelling concept is that tumor promoters may act on the cell membrane.   Under normal conditions, the cells composing a tissue are in homeostasis in which the requirements for cell growth to balance cell loss are regulated throughout the tissue.   The regulation probably occurs through cell to cell communications.   Interruption of such communications could permit cells with an abnormal genotype to proliferate beyond the normal growth requirements, that is to form a neoplasm.   Recently, several groups (43,44) have reported in vitro studies which show that tumor promoters are capable of blocking intercellular communication.   We have extended these studies to the use of liver-derived cells to study liver tumor promoters (38).

The test system involves the measurement of inhibition of metabolic cooperation in mixed liver cell cultures.   Metabolic cooperation in cell culture involves the cell-to-cell transfer through gap junctions of a metabolic product from enzyme-competent to enzyme-deficient cells, as with the transfer of phosphoribosylated 6-thioguanine (TG) from HGPRT-competent cells to HGPRT-deficient cells.   In this case, HGPRT-deficient cells, such as those comprising an ARL-TG resistant strain, are not affected by the addition of TG to the medium because they lack the purine salvage pathway enzyme to convert TG to the mononucleotide, but are killed when cocultivated with HGPRT-competent cells as a result of transfer of the toxic metabolite.   As shown in Table VI, the colony forming efficiency of HGPRT-deficient ARL-TG[r] is comparable in control medium to that in TG-containing medium.

Table VI.   Inhibition of Metabolic Cooperation
between Hepatocytes and an ARL TG
Resistant Strain by the Liver Tumor
Promoter DDT

Condition	TG resistant colonies per flask[a]	
	- hepatocytes	+ hepatocytes
ARL 14-TG resistant cells	126[b]	-
+ TG	110	63
+ TG + DDT 10^{-7}	103	86
+ TG + DDT 10^{-6}	101	112
+ TG + DDT 10^{-5}	105	117
+ TG + DDT 10^{-4}	61	24

[a] 500 TG resistant cells were cocultured with 0.75×10^6 hepatocytes.
[b] Average of three flasks.

When HGPRT-competent cells, such as freshly isolated hepatocytes, are co-cultivated with TG resistant cells at ratios high enough to achieve significant cell to cell contacts, the HGPRT-competent cells metabolize the TG and transfer the mononucleotide to the TG resistant cells, thereby killing the TG resistant cells as well as themselves. Consequently, as shown in Table VI, the co-cultivation of hepatocytes with TG resistant cells in the presence of TG reduces the recovery of the colonies from TG resistant cells. The approach developed by Trosko and associates (44) and applied by us to liver (38) involves measurement of the ability of tumor promoters to inhibit this process and produce an increase in the recovery of TG resistant cells in the co-cultivation system. As shown in Table VI, the addition of DDT to co-cultivated hepatocytes and TG resistant cells exposed to TG restores the recovery of the mutant cells beginning at 10^{-7}M and reaching 100% at 10^{-6} and 10^{-5}M.

Conclusions

The studies described provide evidence for the

lack of genotoxicity of carcinogenic organochlorine
pesticides and demonstrate an effect on the intercellular lipid layer of the cell membrane. This process
may differ from that of other liver tumor promoters
such as phenobarbital. We have reported (45) that
phenobarbital alters the activity of certain membrane
associated enzymes such as gamma glutamyltranspepti-
dase and have suggested that phenobarbital modifies
gene expression to produce a biochemical change in
the composition of the cell membrane. Thus, both
types of tumor promoters may achieve the same inhibition of intercellular communication by different
processes.

The concept that the carcinogenicity of organochlorine pesticides is due to their promoting action
as a result of effects on the cell membrane has important implications. Inhibition of intercellular
communication presumably would not occur without substantial accumulation of the compounds in the cell
membrane. Thus, the carcinogenicity of these compounds only at high dose levels would be explained.
Furthermore, cessation of exposure would lead to elimination of the compounds and restoration of intercellular communication. This would suggest that the
carcinogenic effects, unlike those of genotoxic carcinogens, would be entirely reversible up to a point.

The absence of observable human carcinogenic effects following exposure to organochlorine pesticides
is interpretable in light of the proposed epigenetic
mechanisms of action. It could be that human exposures have been insufficient to achieve the cellular
levels required to effectively inhibit intercellular
communication. Certainly, this would seem to be the
case for exposures of the general population. It
could even be that human cells are more efficient in
intercellular communication and thus more resistant
to the effects of inhibitors. A third possibility is
that the exposed human populations lack the background of genetic alterations in the liver needed to
give rise to neoplasms in response to a promoting
agent.

These concepts and interpretations require
rigorous documentation. Nevertheless, sufficient
evidence is now available to suggest that projections
of the carcinogenic risks from organochlorine pesticide exposure require re-evaluation in light of newer
developments.

Acknowledgements

I wish to thank my collaborators Dr. Charles Tong, Dr. Shyla Telang and Ms. Carol Maslansky for their participation in these studies. Also, thanks to Mrs. Linda Stempel for preparing the manuscript.

Literature Cited

1. Davis, K.J.; Fitzhugh, O.G.; <u>Tox. Appl. Pharmacol.</u>, 1962, <u>4</u>, 187.
2. Division of Cancer Cause and Prevention, National Cancer Institute. "Bioassay of Chlordane for Possible Carcinogenicity" National Inst. Health, DHEW Publ. No (NIH) 77-808, Washington, D.C., 1977.
3. Innes, J.R.M.; Ulland, B.M.; Valerio, M.G.; Petrucelli, L.; Fishbein, L.; Hart, E.R.; Pallotta, A.J.; Bates, R.R.; Falk, H.L.; Gart, J.J.; Klein, M.; Mitchell, I; Peters, J. <u>J. Nat. Cancer Inst.</u>, 1969, 42, 1101.
4. Turusov, V.S.; Day, N.E.; Tomatis, L.; Gati, E.; Charles, R.T. <u>J. Nat. Cancer Inst.</u>, 1973, <u>51</u>, 983.
5. Reuber, M.D. Tumori, 1978, <u>64</u>, 571.
6. Walker, A.I.T.; Thorpe, E.; Stevenson, D.E.; <u>Fd. Cosmet. Toxicol.</u>, 1973, <u>11</u>, 415.
7. Deichman, W.B.; MacDonald, W.E.; Blum, E.; Bevilacqua, M.; Balkus, M. <u>Industr. Med. Surg.</u>, 1968, <u>37</u>, 837.
8. Division of Cancer Cause and Prevention, National Cancer Institute "Bioassay of Heptachlor for Possible Carcinogenicity". National Inst. Health, DHEW Publ. No (NIH) 77-809 Washington, D.C. 1977.
9. Cabral, J.R.P.; Mollner, T.; Raitano, F.; Shubik, P. <u>Int. J. Cancer</u>, 1979, <u>23</u>, 47.
10. Nagasaki, H.; Tomii, S.; Mega, T.; Marugami M.; Ito, N. <u>Gann</u>, 1971, <u>62</u>, 431.
11. Ito, H.; Nagasaki, H.; Aoe, H.; Sugihara, S.; Miyati, Y., Arai, M.; Shirai, T. <u>J. Natl. Cancer Inst.</u>, 1975, <u>54</u>, 801.
12. Division of Cancer Cause and Prevention, National Cancer Institute, "Report on Carcinogenesis of Technical Grade Chlordecone (Kepone)" National Cancer Institute, Bethesda, Md., 1976.
13. Ulland, B.M.; Page, N.P.; Squire, R.A.; Weisburger, E.K.; Cypher, R.L. <u>J. Nat. Cancer Inst.</u>, 1977, <u>58</u>, 133.

14. Kimbrough, R.D.; Linder, R.E. J. Natl. Cancer
     Inst. 1974, 53, 547.
15. Kimbrough, R.D.; Squire, R.A.; Linder, R.E.;
     Strandberg, J.D.; Montali, R.J.; Burse, V.W.
     J. Natl. Cancer Inst., 1975, 55, 1453.
16. Albert, R.E.; Train, R.E.; Anderson, E. J.
     Nat. Cancer Inst., 1977, 58. 1537.
17. Saffiotti, V. "Carcinogenic Risks/Strategies
     for intervention", IARC, Lyon, 1979, p. 151.
18. Epstein, S.S. In: "Origins of Human Cancer"
     Hiatt, H.H.; Watson, J.D.; Winsten, J.A., eds.,
     Cold Spring Harbor Laboratory, Cold Spring
     Harbor, 1977, pp. 243-266.
19. Laws, E.R., Jr., Maddrey, W.C.; Curley, A.;
     Burse, V.W. Arch. Environ. Health, 1973, 27,
     318.
20. Jager, K.W. "Aldrin, Dieldrin, Endrin & Telodrin
     -   An   Epidemiological   Study   of   Long-Term
     Occupational Exposure," Elsevier, Amsterdam,
     1970.
21. Deichmann, W.B.; MacDonald, W.E. Ecotoxicol and
     Environ. Safety, 1977, 1, 89.
22. Williams, G.M. Biochemica et Biophysica Acta
     Reviews on Cancer, 605:167-189, 1980.
23. Weisburger, J.H.;   Williams, G.M.   In:
     Toxicology The Basic Science of Poisons", 2nd
     Edition.   Doull, J.; Klasen, C.D., Amdur, M.O.
     Eds. Macmillan Publ. Co., Inc., NY, pp. 84-138,
     1980.
24. Williams, G.M.   In   "Advances in Medical
     Oncology Research and Education, Proceedings of
     the XIIth International Cancer Congress", Vol. I
     Carcinogenesis Margison, G.P., Ed., Pergamon
     Press, New York, 1979, pp. 273-280.
25. Miller, E.C., Miller, J.A.   In "Chemical
     Mutagens", Hollaender, A., ed., Plenum Press,
     N.Y., 1971, pp. 83-119.
26. Swenberg, J.A.; Petzold, G.L.; Harback, P.R.
     Biochem. Biophys. Res. Comm., 1976, 72, 732.
27. Ahmed, F.E.; Hart, R.W.; Lewis, N.J. Mutation
     Res., 1977, 42, 161.
28. Ahmed, F.E.; Lewis, N.J.; Hart, R.W.
     Chem.-Biol. Interactions, 1977, 19, 369.
29. Brandt, W.N.; Flamm, W.G.; Bernheim, N.J.
     Chem.-Biol. Interactions, 1972, 5, 327.
30. Marshall, T.C.; Dorough, H.W.; Swim, H.E. J.
     Agric. Food Chem., 1976, 24, 560.
31. Koda, T.; Moriya, M.; Shirasu, Y. Mutation Res,
     1974, 26, 243.

32. McCann, J.; Choi, E.; Yamasaki, E; Ames, B.N.
     Proc. Nat. Acad. Sci. (USA), 1975, 72, 5135.
     Marshall, T.C.; Dorough, H.W.; Swim, H.E.   J.
     Agric. Food Chem., 1976, 24, 560.
33. Schoeny,   R.S.;   Smith,   C.C.;   Loper,   J.C.
     Mutation Res., 1979, 68, 125.
34. Williams, G.M.   In: "Chemical Mutagens".   Vol.
     VI,   de Serres, F.J. and Hollaender, A.   eds.,
     Plenum Press, New York, 1980, pp. 61-79.
35. Williams, G.M.   In: "Short Term Tests for
     Chemical Carcinogens".   San, R.H.C. and Stich,
     H.F. eds.,   Springer-Verlag, NY, 1980, pp.
     581-609.
36. Williams, G.M. Cancer Letters, 1976, 1:231.
37. Williams, G.M. Cancer Research, 1977. 37:1845.
38. Williams, G.M.   Annals New York Acad. of Sci.,
     1980, in press.
39. Maslansky, C.J. and Williams, G.M.   J. Toxicol.
     Environ. Health. Submitted for publication.
40. Tong, C.; Williams, G.M. Mutation Res.
     1978, 58, 339.
41. Tong, C. and Williams, G.M.   Mutation Res.
     1980, 74, 1.
42. Peraino,   C.;   Fry,   R.M.J.;   Staffeldt,   E.;
     Christopher, J.P. Cancer Res., 1975, 35, 2338.
43. Murray,   A.W.;   Fitzgerald,   D.J.   Biochem.
     Biophs. Res. Comm., 1979, 28, 395.
44. Yotti, L.P.; Chang, C.C.; Trosko, J.E. Science,
     1979, 206, 1089.
45. Williams, G.M. Carcinogenesis, 1980, in press.

RECEIVED March 12, 1981.

The Role of Genetic Toxicology in a Scheme of Systematic Carcinogen Testing

GARY M. WILLIAMS and JOHN H. WEISBURGER

American Health Foundation, Naylor Dana Institute for Disease Prevention, Valhalla, NY 10595

DAVID BRUSICK

Director of Genetics and Cell Biology, Litton Bionetics, Inc., Kensington, MD 20795

Progress has been great in recent years in developing means for the evaluation of the hazardous potential of chemicals. A major achievement has been the introduction of in vitro techniques for the rapid identification of mutagenic or carcinogenic potential. However, no single test has been found to be sufficient for detection of all hazardous chemicals and therefore, agreement now exists that a battery of tests is required for toxicological evaluations in mass-screening programs (1-11). Over 100 short-term tests for detecting potential chemical carcinogens and mutagens are available (12) and the critical issue in developing a battery of tests is to formulate appropriate criteria for selecting the best combination of tests.

One of the first proposals for the systematic application of short-term tests for the detection of carcinogens and mutagens was that of Bridges (1), in which a three tiered protocol involving submammalian tests, whole mammal tests, and finally, in vivo tests for risk assessment was recommended. A similar approach was developed by Flamm (2), for mutagenicity testing, noting that no single genetic test could detect all genetic events of possible hazard to humans. Whereas Bridges favored tests for chromosomal damage, Flamm, and in a later modification, Green (4), recommended the dominant lethal test. It is interesting that none of these early proposals included a DNA damage test. A hierarchical approach described by Bora (3) did include a DNA damage test, but not Drosophila mutagenesis, which how

0097–6156/81/0160–0057$07.75/0

ever, was recommended by both Bridges and Flamm. Instead, Bora proposed the use of host-mediated systems. Thus, all these proposals include tests involving effects in whole organisms. The rationale for tests of this type has recently been reviewed by Rinkus and Legator (13).

With the awareness that testing schemes were becoming increasingly complex and expensive, recent efforts have been initiated to reduce the number of tests to an essential core (5,6,8,9-11,-14). Consideration of the various testing schemes that have been proposed reveals that most tiers or batteries are structured around seven tests (10): bacterial mutagenesis, eukaryote mutagenesis, Drosophila mutagenesis, mammalian cell mutagenesis, DNA damage, chromosome damage and malignant transformation. The emphasis on specific tests appears to be mainly a function of whether the proposed scheme is directed toward only mutagen testing, or carcinogen and mutagen detection. For example, all of the tests, except transformation, are listed by the International Commission for Protection against Environmental Mutagens and Carcinogens (15) as among the most widely used tests for mutagen screening.

Thus, there is substantial consensus regarding the most useful tests for screening for mutagenicity and carcinogenicity. As yet, however, such concepts are reflected in current testing requirements to only a limited degree (16).

Criteria for a Battery

The philosophy underlying a battery is that a group of tests should be performed before a decision is made regarding the potential hazard of the chemical. Thus, a battery corresponds to the initial "detection" phase that is part of most tiers (1,2). The crucial difference with a battery, however, is that it attempts to combine "detection" and the next step of a tier, "confirmation" in one stage. Implicit in this approach is the concept that all available short-term tests may yield false positive or false negative results that require parallel data for interpretation. The battery approach thus formally incorporates the concept that no decision on potential hazard should be made without the minimum data offered by the battery.

The choice of tests to comprise a battery

will vary depending upon whether the goal is to
define potential mutagens or carcinogens.   Little
is known about the validity of mutagenicity batteries because few chemicals have been shown to be
mutagenic to germ cells in experimental animals,
and no chemicals are known to produce human germinal mutations.   Thus, at present, mutagenicity
batteries must be constructed to identify the
broadest possible spectrum of genetic damage (17).
In contrast, carcinogenicity batteries can be
verified against in vivo data, albeit with an
important qualification.   Several lines of evidence now indicate that carcinogens may operated
by a variety of mechanisms (14,18,19).   Among
these, genotoxic effects can be readily detected
in short-term tests.   Other oncogenic mechanisms
of a presumed epigenetic nature are clearly not
detected in tests with a genetic end-point.   Some
tests such as malignant transformation and sister
chromatid exchange, which can be produced by
effects other than a direct attack on DNA, may be
capable of detecting non-genotoxic carcinogens.
In addition, efforts are being made to develop in
vitro tests for identifiying the promoting class
of epigenetic agents (20-23).   As yet, however,
none of these approaches can be recommended for
routine inclusion in a battery.   Therefore, in the
use of batteries for identification of carcinogens, it must be recognized that a whole category
of chemicals, containing such agents as saccharin,
hormones, certain organochlorine compounds and
pesticides, and several pharmaceuticals may not be
detected.
        Several other principles should guide the
construction of a battery.   Importantly, the tests
should be reliable and of clear biologic significance (24).   This means that they should truly
measure what they purport to measure and that the
end point should have conceptual relevance to
mutagenicity or carcinogenicity.   Secondly, a battery should seek to maximize the metabolic parameters provided by all tests.   As an example,
tests with intact cell metabolism should be
included to extend the metabolism obtained with
the commonly used exogenous subcellular preparations.   This may be of particular importance in
view of the artifactual enhancement of activation
over detoxification that is known to be characteristic of enzyme preparations (25,26).   Moreover,
the DNA adducts formed by activation through mic-

rosomes differ from those produced by intact cell
metabolism (27).

Adhering to these concepts, a battery of
short-term tests was proposed by Weisburger and
Williams (5,14,28) as part of a "decision point
approach.

"The Decision Point Approach"

The Decision Point Approach consists of five
sequential steps in the evaluation of the potential carcinogenicity of chemicals. (Table 1).
This approach was formulated to incorporate into
chemical evaluation several newer developments in
chemical carcinogenesis. Of prime importance
among these was the concept that chemicals could
produce an increase in the tumor incidence in exposed animals, i.e. be carcinogenic, by several
distinct mechanisms each having different theoretical and practical implications. One of these
mechanisms, proposed by Miller and Miller (29),
was through the generation of an electrophilic
reactant which would react covalently with cellular macromolecules. The work in several laboratories (see 14 and 30 for references) has strongly
indicated that DNA is the critical target. However, in addition to chemicals of this type,
others lacking this property were nevertheless
carcinogenic or oncogenic. Among chemicals of the
latter type were plastics, hormones, immunosuppressants, cytotoxic agents, co-carcinogens and
promoters. Thus, it was suggested that chemical
carcinogens could be divided into two main categories, based upon their capacity to damage DNA.
Carcinogens that reacted covalently with DNA were
categorized as genotoxic, while those lacking this
property and probably acting by other mechanisms
were categorized as epigenetic (14). The genotoxic category thereby contains the classic organic carcinogens that damage DNA either through
direct chemical reactivity or following metabolism
by enzyme systems (Table 2). In addition, the
availability of some evidence for DNA damage by
inorganic carcinogens led to their placement in
this category. The second category, epigenetic
carcinogens, is composed of those agents that have
not been found to damage DNA, but rather appear to
act through other indirect mechanisms (Table 2).

The decision point approach takes these two
categories or types of carcinogens into account in

Table I. Decision Point Approach to
Carcinogen Evaluation[a]

A. Structure of chemical

B. In vitro short-term tests
   1. Bacterial mutagenesis
   2. DNA repair
   3. Mammalian mutagenesis
   4. Sister chromatid exchange
   5. Cell transformation

B'. Decision point: tests under A and B.

C. Limited in vivo bioassays
   1. Skin tumor induction in mice
   2. Pulmonary tumor induction in mice
   3. Breast cancer induction in female
      Sprague-Dawley rats
   4. Altered foci induction in rodent
      liver

C'. Decision point. tests under A, B, and C

D. Chronic bioassay

E. Final evaluation: all tests.

[a] From Weisburger and Williams (14).

Macmillan Publishing Co., Inc.

Table II.   Classification of Carcinogenic
Chemicals[a]

Category and Class	Example
A. Genotoxic Carcinogens	
1. Activation-independent	Alkylating agent
2. Activation dependent aromatic	Polycyclic
	Hydrocarbon, Nitrosamine
3a Inorganic[b]	Metal
B. Epigenetic Carcinogens	
3b Inorganic[b]	Metal
4. Solid State	Plastics
5. Hormone	Estrogen
	Androgen
6. Immunosuppressor analog	Purine
7. Co-carcinogen	Phorbol ester
	Catechol
8. Promoter	Phorbol ester
	Bile acid

[a] after Weisburger and Williams (14)

[b] some are tentatively categorized as genotoxic because of evidence for damage of DNA; others may operate through epigenetic mechanisms such as alterations in fidelity of DNA polymerases.

two ways; (1) a battery of short term tests is constructed based upon an effort to include systems that may respond to epigenetic as well as genotoxic carcinogens; and (2) it is formally recognized that all types of subchronic testing may fail to detect chemicals that can induce tumors in animals under specific conditions upon chronic administration.

Other elements of the decision point approach are the use of a battery of short term tests that either may eliminate the need for further testing of the chemical or may enable the verification of carcinogenic potential in one of four limited in vivo bioassays for carcinogenicity. The battery also adds essential information for data evaluation when an already completed chronic test series has yielded ambiguous results.

The decision point approach therefore is a systematic approach to the reliable evaluation of the potential carcinogenicity of chemicals which provides a framework in which to minimize the necessary testing in chemical evaluation, and at the same time develop an understanding of the mechanism of action of a test chemical.

As shown in Table 1, the decision point approach involves a systematic step-wise progression of tests. A critical evaluation of information obtained and its significance in relation to the testing objective is performed at the end of each phase. A decision is made whether the data available are sufficient to reach a definitive conclusion or whether a higher level of tests is required. Attention is paid to qualitative - positive or negative - effects, and to quantitative - high, medium, or low - effects.

A. Structure of Chemical

For a number of reasons, the evaluation begins with a consideration of structure. Of principal importance is the fact that predictions as to whether or not a given chemical might be carcinogenic can be made with fair success within certain classes of chemicals (28). Structure, however, must always be considered in relationship to species metabolic parameters. The guinea pig, for example, differs from other rodents or humans, in that it has only limited amounts of the necessary enzymes to carry out N-hydroxylation, and thus metabolizes aromatic amines almost exclusive-

ly to detoxified metabolites. Consequently, the
arylamines so far tested in this species have not
been carcinogenic.   Other examples of species
selectivity   based upon metabolic capability are
well documented (30).
Information on structure and metabolism also provides a guide to the selection among limited bioassays at   stage   C   and,   as   more   information
accrues, may eventually contribute to selection of
specific short-term tests at stage B.

B.   In Vitro Short-term Tests

     No individual short-term test that has been
studied adequately has detected all carcinogens
tested, or even all genotoxic carcinogens.   Thus,
based on this fact alone, a battery of tests is
necessary.   However, the importance of a battery
becomes obvious upon consideration of the complexity of metabolism and mechanism of action of chemical carcinogens.   As indicated earlier, carcinogens can be classified as genotoxic or epigenetic.
Most in vitro tests identify genetic effects and
thus detect only genotoxic carcinogens.   If epigenetic carcinogens are to be detected, additional
tests will have to be developed. Known species
differences in response to carcinogens can be related to a large extent (but not exclusively) to
metabolism and thus, tests with different metabolic capabilities are important.
     A screening battery must include microbial
mutagenesis tests, because these have been the
most sensitive, effective, and readily performed
screening tests thus far (31-33).   In deciding
what other tests to include, it is important to
consider the contribution of the proposed test in
terms of metabolic capability, reliability, and
biological significance of the end point.   The
bacterial mutagenesis tests require a mammalian
enzyme preparation to provide for metabolism of
procarcinogens and hence, any test that is dependent upon an enzyme preparation does not expand the
metabolic capability of the battery since this
factor is the limiting part of a test series.
     Mutagenesis of mammalian cells is recommended
for inclusion in the battery because it has a definitive end point like bacterial mutagenesis, but
involves effects on the more highly organized
eukaryotic genome (34,35).   Moreover, differences
in the mutagenic response between microbial and
mammalian cells have been observed (36).

Tests for DNA damage or chromosome effects provide further evidence of the ability of a chemical to alter genetic material. Proposed indicators for DNA damage tests include DNA fragmentation (37), inhibition of DNA synthesis (38), and DNA repair (39). Measurement of DNA fragmentation, although showing an excellent correlation with carcinogenicity has a conceptual disadvantage in that DNA degredation can occur as a result of cell death. Inhibition of DNA synthesis has been suggested to offer an advantage in its ability to detect intercalating agents (38). This may not be a substantial advantage because pure intercalators are a limited group of chemicals comprising at best weak carcinogens and DNA synthesis is also inhibited by noncarcinogenic intercalators. Furthermore, the intercalating agents with reactive groupings do induce DNA repair.

DNA repair is a specific response to DNA damage and unlike DNA fragmentation and inhibition cannot be attributed to toxicity (40). Therefore, DNA repair tests offer an end point of high specificity biologic significance and are recommended in preference to these other assays.

A chromosomal test is included in the battery to provide detection of chemical effects at the highest level of genetic organization (13). In addition, such tests may detect nongenotoxic agents that operate through other effects involving processes such as DNA replication and chromosome separation.

Cell transformation is included because this alteration is potentially the most relevant to carcinogenesis. However, much more needs to be done to clarify the significance and limitations of this end point.

1. Bacterial Mutagenesis

Valuable bacterial screening tests have been developed in the laboratories of Ames (41) and Rosenkranz (42). The Ames test measures back mutation to histidine independence of histidine mutants of Salmonella typhymurium and can be conducted with strains that are also repair deficient, that possess abnormalities in the cell wall to make them permeable to carcinogens, or that carry an R factor enhancing mutagenesis. Hence, these organisms are highly susceptible to

mutagenesis making them sensitive indicators. The
test developed by Rosenkranz and associates utilizes DNA repair-deficient Escherichia coli and
measures their enhanced susceptibility to cell
killing by carcinogens. In this system, a chemical that interacts with DNA is more toxic to the
repair-deficient strain than to wild type E. coli
because the mutant strain cannot repair the damage. Thus, by measurement of relative toxicity an
indication of DNA interaction is obtained. These
tests are dependent upon mammalian enzyme preparations for metabolism of carcinogens. The capability of the Ames test to detect certain carcinogens
has been enhanced by application of preincubation
of the compound and the biochemical activation
system with the test organism (43).

2. Mammalian Mutagenicity Tests

The three mutational assays in mammalian
cells that have been most widely used for carcinogen screening are resistance to purine analogs
(44-46), bromodeoxyuridine (BUdR) (47) or ouabain
(48). Of these, purine analog resistance is the
most popular. In this assay, mutants lacking the
purine salvage pathway enzyme hypoxanthine-guanine
phosphoribosyl transferase are identified by their
resistance to toxic purine analogs such as
8-azaguanine or 6-thioguanine that kill cells that
utilize the analogs. This assay has the advantage
over ouabain resistance that it involves a nones-
sential function, unlike the membrane ATPase system involved in ouabain resistance, and hence
there are no lethal mutants. Its advantage over
the measurement of thymidine kinase-deficient
mutants by resistance to BUdR is that the gene for
hypoxanthine-guanine phosphoribosyl transferase is
on the X-chromosome rather than a somatic chromosome, as with thymidine kinase. Consequently, only
one functional copy is present in each cell and,
as a result, mutations in wild type cells can be
measured, whereas a heterozygous mutant is
required for measurable mutation to homozygous
thymidine kinase deficiency and BUdR resistance.
The target cells used in purine analog resistance assays have almost all been fibroblast-like,
such as the V79 and CHO lines, which have displayed little ability to activate carcinogens,
other than polycyclic aromatic hydrocarbons (45).
This deficiency has been overcome by providing

exogenous metabolism mediated by either cocultivated cells (45,49) or enzyme preparations (44). The latter again offers no extension in metabolic capability over that used for bacterial systems. However, the use of freshly isolated hepatocytes as a feeder system (49) offers additional possibilities since the metabolism of hepatocytes has been shown to be different than that of liver enzyme preparations (26,27). Another potentially useful development is the finding that liver epithelial cultures can be mutated by activation-dependent carcinogens (46) and may therefore provide another system with intact cell metabolism.

3. DNA Repair

The covalent interaction of chemicals with DNA provokes an enzymatic repair of the damaged regions of DNA known as excision repair. Two types of excision repair are recognized, base removal and nucleotide removal (50). The first step in each of these differs, but both processes result in incision of the strand of DNA near the point of damage and excision by an endonuclease of a stretch of DNA in the damaged region. The gap is then filled by resynthesis of a patch using the opposite strand as a template and the patch is closed by a ligase. Several of these steps could be measured as an indication of repair, but the resynthesis of the patch is most widely used to monitor repair in screening systems. Repair synthesis can be measured in a variety of ways (40, 50). Several of the definitive procedures are technically sufficiently demanding so that they have not received much attention for screening purposes. Of the simpler procedures available, autoradiographic measurement of repair synthesis has the advantage over liquid scintillation counting in that it excludes cells in replicative synthesis, whereas these are part of the background with liquid scintillation counting. In addition, with liquid scintillation counting, increases in incorporation can result from changes in uptake or the pool size of thymidine without any repair occurring. Furthermore, autoradiography permits a determination of the fraction of cells responding in the affected population. Two additional complications with most repair assays are that they require suppression of replicative DNA synthesis if continuously dividing lines are being used, and

that they are dependent upon enzymne preparations
for metabolic activation. Both of these complications are overcome in the hepatocyte primary cul-
ture/DNA repair assay of Williams (51/52) which
used freshly isolated non-dividing liver cells
that can metabolize carcinogens and respond with
DNA repair synthesis measured autoradiographical-
ly. This assay has demonstrated substantial sensitivity and reliability with activation-dependent
procarcinogens (51-53). It also offers the advantages of expanded metabolic capability in a battery and an end point of clear biological significance. Thus, it is a valuable addition to bacterial mutagenesis assays in a screening battery.

4.   Chromosome Tests

Chromosome tests are of conceptual importance
because they reveal damage at a higher level of
genetic organization than do mutagenesis assays.
There has been difficulty, however, in developing
means of objective analysis of many chromosomal
alterations. Measurement of sister chromatid exchanges (SCE) overcomes this problem and has shown
sensitivity to carcinogens not readily detected in
other in vitro assays (54,55). Therefore, determination of SCEs is presently recommended for a
chromosomal level test. The resulting extension
of the information base with SCE will be useful to
delineate further the value and limitations of
this relatively new test.

5.   Cell Transformation

The first reliable system for transformation
of cultured mammalian cells was introduced by
Sachs and associates (56). This system utilizing
hamster fibroblasts was subsequently developed into a colony assay for quantitative studies by
DiPaolo (57) and has been adapted as screening
test by Pienta et al (58). In addition, a quantitative focus assay for transformation using mouse
cells has been devised in the laboratory of
Heidelberger (59) and a quantitative assay for
growth of BHK cells in soft agar has been developed by Styles (6). The correlation between
transformation and malignancy appears to be good
in these systems, but a subject of concern is
their high frequency of induced transformation.
Nevertheless, they provide a useful indication of

the potential carcinogenicity of chemicals either through genotoxic or epigenetic mechanisms and will almost certainly assume a major role in screening in the future. Another approach under development is the use of cell systems carrying oncogenic viruses as a more sensitive means of detecting transforming chemicals. Also, because human cancers usually involve epithelial tissues, transformation in epithelial systems is actively being pursued.

Summary of Rapid In Vitro Tests

The five steps (A and B, 1-5) recommended thus far provide a basis for preliminary decision making. Survey of literature data on the application of the recommended test reveals a high degree of sensitivity and specificity for this battery (10,28).

Evidence of genotoxicity in only one test must be evaluated with caution. In particular, several types of chemicals such as intercalating agents are mutagenic to bacteria, but not reliably carcinogenic. Also positive results have been obtained with synthetic phenolic compounds or natural products with phenolic structures like flavones. In vivo, such compounds are likely to be conjugated and excreted readily. Their carcinogenicity, thus, would depend on in vivo splitting of such conjugates. Therefore, evidence of only bacterial mutagenesis must be evaluated with regard to the chemical structure and its metabolism.

If clear cut evidence of genotoxicity in more than one test has been obtained, the chemical is highly suspect. Confirmation of carcinogenicity may be sought in the limited in vivo bioassays without the necessity of resorting to the more costly and time-consuming chronic bioassay.

To facilitate the interpretation of results from a battery, Brusick (11) has developed a quantitative approach in which each assay is assigned a numerical value based on its contribution either positively or negatively to an assessment of genotoxicity. The assignment of a value takes into account the following:

1. The type of endpoint measured by the test and its presumed relationship to the development of chronic toxicity in vivo, including mutagenesis and carcinogenesis.

2.   The phylogenetic relationship of the test
      organism to mammalian species.
3.   The reported reproducibility of the test
      system within and between laboratories.
4.   The published data base supporting the util-
      ity of the test system to detect a broad
      range of chemical classes.
5.   The susceptibility of the test to incorrect
      designations of genotoxicity resulting from
      testing artifacts or anomalous nonspecific
      responses.
6.   The qualitative similarity between the test
      system metabolic or microsome bioactivation
      system and the in vivo bioactivation mechan-
      isms in mammals.
7.   The resolving power of the test system in-
      cluding the strength of the data analysis
      methods used with the assay.

      Applying these criteria, a set of assay
values was developed (Table 3). The tests listed
in Table 3 are among those which have been routinely proposed as screening methods for animal
mutagens and carcinogens and include the battery
proposed above. Values are assigned for positive
and negative responses ranging from -5 to +10.
The largest negative value represents the test and
test conditions providing the most powerful indication of a lack of genotoxicity. The values
between these extremes are weighted proportionally
according to the seven criteria listed above.

      Positive responses are obviously given signi-
ficantly greater weight than negative results
since negative results could mean either a lack of
potential or a lack of detectability by the assay.

      The differential weighting of results with or
without an S9 mix is predicated upon the assumption that a substance active without enzyme activation is unlikely to show species specificity;
whereas, an activation-dependent substance may be
species restricted and not amenable to generalized
extrapolation. For negative responses, however, a
greater negative value is assigned to tests employing an activation system. The presence of
such a system suggests that neither the parental
molecule nor microsomally-produced breakdown products have detectable activity. Assignment of the
specific values was arbitrary with +10 as a
maximum.

      A further attempt is made to bring a consid-
eration of potency into the scoring system.

Table III.   Score Table for Short-term Test Results Using the Weighed Contribution Method.

Test Procedure[a]	Positive Response[b]		Negative Response[b]	
	Without Activation	With Activation	Without Activation	With Activation
Salmonella, Ames method	+6	+5	-2	-4
Gene mutation cultured mammalian cells	+8	+7	-4	-5
DNA repair in cultured mammalian cells[c]	+3	+2	-1	-3
In vitro SCE induction	+3	+2	-1	-3
Morphological transformation in vitro	+10	+9	-4	-5
In vitro chromosome aberration induction	+4	+3	-2	-3
Sex-linked recessive lethal in Drosophila	NA	+6	NA	-4

a Study design must include dose selection criteria, suitable controls, and provisions for multiple doses.
b Based upon criteria and analysis consistent with those given in the Appendix.
c Only tests employing autoradiographic methods can be evaluated.

Because of the diversity in end points measured, the spontaneous rates for the detected events and the methods of scoring responses in the tests, a comparison of the absolute values is not feasible. A reasonable method to incorporate potency appears to be a measurement of the lowest test concentration producing a biologically significant increaseover the negative control. This value is designated as the Lowest Positive Concentration Reported (LPCR). Another value designated as the Highest Negative Concentration Reported (HNCR) defines the highest tested concentration which was negative. The HNCR may be limited by toxicity or a present maximum applicable concentration. The values listed in Table 3 under Positive and Negative responses may then be modulated by a potency factor as set forth in Table 4.

The product of the test values (TV) and the concentration score gives a Total Score (TS) for each test of the battery. The TS for each test will be either positive or negative. The algebraic sum represents the Activity Score (AS) for the compound in the battery of tests to which it has been subjected.

The next step in the scoring approach is to assign an effect definition to the AS for the test substance. The Effect Categories are shown in Table 5 and are calculated in the following manner.

1   A maximum (worst case) genotoxic effect, is calculated by taking each assay employed to evaluate the test material and calculating the TS products assuming it was positive without activation at a potency level equivalent to the maximum tested (not to exceed 1000 ug/unit).

2. This value, Maximum Positive Total (MP), is divided into the AS for the test material to obtain an index (expressed as percent) of what portion of the maximum genotoxic effect was obtained in the evaluations (% MP).

3. The percent value is then categorized using the Effect Table (Table 5).

4. Each category in the Effect Table defines the presumed genotoxic potential of the test substance and leads to an action response (11).

Most experimental carcinogens and/or mutagens fall into Category 1 (% MP > 60%). The only possible exception to this trend is benzene which is not identified as a genotoxic agent by in vitro

Table IV.   Test Concentration Factor for Short-
            Term Tests Score Table Using the
            Weighed Contribution Method.

LPCR or HNCR[a] Converted to uG/unit[c] Concentration		Concentration Score Point Factor[b]	
		Positive Response	Negative Response
$\leq$1.0	-	10	1
>1.0	- 5	9	1
>6	- 10	8	1
>11	- 25	7	1
>26	- 50	6	1
>51	- 100	5	2
>101	- 500	4	2
>501	- 1000	3	2
>1001	- 5000	2	2
>5000	-	1	2

[a]LPCR = Lowest Positive concentration reported.
 HNCR = Highest negative concentration reported.
[b]This factor is multiplied by the individual test
 score results   obtained from Table 1.
[c]uG/unit - Concentration in micrograms per
 milliliter or per plate, etc.

                Table V.   Effect Table

Percent Maximum Positive [MP]	Classification	Category
> 60%	Potent Genotoxic Agent	1
30 - 59%	Genotoxic Agent	2
10 - 29%	Suspect Genotoxic Agent	3
< 10%	Insufficient Response to Categorize the Agent as Genotoxic	4

tests.   Benzene has a % MP score of less than
10.   Two   presumed   nongenotoxic   carcinogens,
saccharin and nitrilotriacetic acid, also score at
a % MP of10 or less.   Both of these agents seem to
represent very little risk as carcinogens under
normal environmental exposure levels and are not
presently controlled as signficant human risks.
Several   compounds   such   as   ascorbate,
benzo(e)pyrene   and   lead   acetate   fall   into
categories 2 and 3.   Compounds in Category 2, are
candidates for limited in vivo bioassays.

Chemicals which fall into Category 3 are candidates for   further   genetic   evaluation.   These
include fluorene,   diphenylnitrosamine   and   Fyrol
FR-2.

This scoring scheme, then, takes the actual
test results from   a multitest battery through to
a specific action recommendation based on an index
of the maximum genotoxic effect for that group of
assays.   It   should   facilitate   the   interpretation
and decision making process which follow the actual testing program.

If the battery of test systems yields no
indication of genotoxicity and thus the chemical
falls   into   category   4   of   the   action table,   the
chemical may be given a low priority for further
testing depending on two criteria 1)   the structure and known physiological properties (e.g. hormone) of the material and 2)   the potential human
exposure.   If substantial human exposure is likely, careful consideration should be given to the
necessity for additional testing.   The chemical
structure and the properties of the material provide direct guidance on the proper relevant course
of action.   Organic chemicals with structures that
present possible sites for activation may reveal
their   carcinogenicity   in   limited   in   vivo   bioassays.   On the other hand chemicals such as solid
state materials,   hormones,   possibly   some   metal
ions and promoters which are negative in tests for
genotoxicity operate by complex and as yet poorly
understood mechanisms.   Thus, it is not certain
that the limited in vivo bioassays would yield
useful results with such materials.   Therefore the
standard chronic bioassay is, at this time, necessary to detect any potential activity with these
agents.   It is indeed urgent to develop reliable
means to detect such materials readily without
requiring the large investment associated with a
chronic bioassay.

The testing of metal ions in rapid bioassay tests may take advantage of the concept recently proposed by Loeb (60) that such ions affect the fidelity of enzymes concerned with DNA synthesis. Obviously, the nature of the metal ion, of which there are only a limited number, would provide the necessary insight as to the need for testing such a material further and what kind of assay would most likely reveal adverse effects.

Compounds with hormone-like properties exist outside of the strict androgen and estrogen types of hormones. Such chemicals are potential cancer risks mainly because they interfere with the normal physiological endocrine balance (14). More research on ways and means to test for such properties quickly is required. It is known, for example, that certain drugs lead to release of prolactin from the pituitary gland. Chronic intake of such drugs causing permanently higher serum and tissue prolactin levels might in turn alter the relative ratio of other hormones. With current understanding, any material with such properties needs to undergo a chronic bioassay with carefully and appropriately selected doses to evaluate whether endocrine sensitive tissues would be at higher risk. The interpretation of data must consider the normal diurnal, monthly and even seasonal cycles of the endocrine system and whether the test would have led to interference in this balanced, rythmic system.

The potential of halogenated polycyclic hydrocarbons to act as promoters in the production of liver tumors has been discussed in detail (14). As yet, the structural requirements for promoting activity are poorly understood outside the class of phorbol esters, and these agents can be identified only in initiation - promotion protocols in limited in vivo bioassays or in chronic bioassay.

The implications of the absence of convincing data for genotoxicity, but a positive response in chronic bioassays are discussed under the final evaluation.

C. Limited In Vivo Bioassays

This stage of evaluation employs tests that will provide further evidence of potential hazard of chemicals positive for genotoxicity without the necessity of undertaking chronic bioassay.

A number of tests for in vivo genotoxicity

have been developed; these include the dominant
lethal test, specific locus test, heritable trans-
location test, host-mediated mutagenicity, chromosomal damage, testicular DNA synthesis inhibition,
sebaceous gland suppression and DNA fragmentation
or repair in various organs. A chemical that is
negative in all the in vitro genotoxicity tests
is unlikely to be positive in any one of these in
vivo tests, with possible exception of chemicals
activated to genotoxic metabolites by host bacteria. Therefore, at present, little basis existsfor
recommending one of these. Furthermore, a positive result in one of these in vivo tests would
not be conclusive evidence of carcinogenicity and,
thus, would serve only as a further indication of
the need for chronic bioassay, which, as discussed, is already the only recourse for suspect
chemicals that are negative in the in vitro tests.
Such in vivo tests therefore serve primarily to
establish priorities for chronic bioassay of chemicals negative in in vitro tests.

    Thus, at this stage, the in vivo tests recom-
mended are those that will provide definitive evidence of carcinogenicity, including cocarcinogen-
icity and promotion, in a relatively short period
(i.e. 30 weeks or less). Unlike the in vitro
tests, these are not applied as a battery, but
rather used selectively according to the information available on the chemical. These tests which
have been described in detail by Williams and
Weisburger (28) include
    1. Skin tumor induction in mice.
    2. Pulmonary tumor induction in mice
    3. Breast cancer induction in female Sprague
Dawley rats
    4. Altered foci induction in rodent liver
    Each of these tests can be completed in 20-40
weeks and therefore provide a relatively rapid
means of assessing carcinogenicity.
    The classes of compounds active in limited in
vivo bioassays are shown in Table 6.
    Limited in vivo bioasays are recommended for
substances which yield equivocal results in the
battery of short-term tests or those positive, but
of such major economic significance that further
confirmation is desired. Also, in the absence of
genotoxicity, it is possible to test for promoting
activity on mouse skin initiated with small doses
of, for example, benzo(a)pyrene or 7,12-dimethyl-
benzo(a)anthracene.   A   material   exhibiting

Table VI.   Carcinogens Active in Limited
In Vivo Bioassays

1.   Skin Tumors in Mice
   A.   As Complete Carcinogens
        Polycyclic aromatic and heterocyclic
          hydrocarbons
        Direct acting alkylating agents
        Alkylnitrosoureas
   B.   As Initiators with a Promoter
        Polycyclic aromatic hydrocarbons
        Certain Arylamines
        Carbamic acid esters, Urethane
   C.   As Promoters or Cocarcinogens
          with Initiation
        Phorbol esters
        Anthralin
        Catechol

2.   Pulmonary Tumors in Mice
        Polycyclic aromatic hydrocarbons
        Carbamic acid esters and N-alkylated
          carbamates-urethan
        Alkylnitrosamides and alkylnitrosamines
        Alkylating agents
        Aziridines
        Hydrazines
        Arylamines (poor)

3.   Breast Cancer in Female Sprague-Dawley Rats
        Polycyclic aromatic hydrocarbons
        Arylamines
        Alkylnitrosoureas

4.   Altered Foci in Rodent Liver
   A.   Rats
        Polycyclic aromatic hydrocarbons
        Arylamines, certain aminoazo dyes,
          heterocyclic amines
        Nitrosamines
        Urethan
        Ethionine
        Aflatoxin
        Safrole
   B.   Mice
        Safrole
   C.   Hamsters
        Nitrosamines

endocrine properties likewise may show an effect in modifying breast cancer induction in animals given limited amounts of methylnitrosourea as an initiating dose. Similarly, promoters of urinary bladder cancer may be visulized by pretreatment with limited amounts of a genotoxic bladder carcinogen.

Summary of limited in vivo bioassays

The detection of two positive results, one in a battery of rapid in vitro bioassay tests reliably indicating genotoxicity and a definite positive result in the limited in vivo bioassays, makes a substance highly suspect as a potential carcinogenic risk to humans. This is true especially if the results were obtained with moderate dosages and more so if there was evidence of a good dose response, particularly as regards the multiplicity of the lung or mammary gland tumors.

Proven activity in more than one of the limited in vivo bioassays may be considered unequivocal qualitative evidence of carcinogenicity.

D.   Chronic Bioassay

Chronic bioassay is used in the decision point approach as a last resort for confirming questionable results in the more limited testing or in the case of compounds that are negative in the preceding stages of testing, but where extensive human exposure is likely, the development of data on possible carcinogenicity through epigenetic mechanisms. In the latter situation, multi-species and dose response data are important.

The conduct of chronic bioassay has been described in a number of review articles (61-63).

E.   Final Evaluation

If the decision point approach has led to chronic bioassays, then fairly definitive data on carcinogenicity would be obtained. Nevertheless, the results of the in vitro short-term tests are considered for evaluation of possible mechanisms of action and risk extrapolation to humans. Convincing positive results in the in vitro tests coupled with documented in vivo carcinogenicity permits classification of the chemical as a genotoxic carcinogen. It would, therefore, be antici-

pated that the chemical could display the properties characteristic of such carcinogens which include the ability under some circumstance to be
effective as a single dose cumulative effects,
and synergism or at least additive effects with
other genotoxic carcinogens. Genotoxic carcinogens, therefore, represent clear qualitative hazards to humans and the level of exposure permitted
must be rigorously evaluated and controlled.
Along those lines, no distinction should be made
between naturally occurring and synthetic carcinogens. In fact, there is growing evidence that the
majority of human cancers stem from exposure to
the former types of agents (64).

If, on the other hand, no convincing evidence
for genotoxicity is obtained, but the chemical is
carcinogenic in animal bioassays, then the possibility exists that the chemical is an epigenetic
carcinogen. The stregnth of this conclusion depends upon the relevance of the in vitro tests.
For example, the finding that certain stable organochlorine pesticides do not display genotoxic
effects in liver cell systems which are identical
to the in vivo target cell for these carcinogens,
strongly supports the interpretation that these
carcinogens may act by epigenetic mechanisms. The
nature of these mechanisms is poorly understood at
present and is probably quite different for different classes of carcinogens. They may involve
chronic tissue injury immunosuppressive effects,
hormonal imbalances, blocks in differentiation,
promotion of pre-existing altered cells, or processes not yet known. Regardless, most types of
epigenetic carcinogens share the characteristic of
being active only at high, sustained doses, and up
to a certain point, the lesion induced may be reversible. Thus, these types of carcinogens may
represent only quantitative hazards to humans and
safe levels of exposure may be established by
carrying out proper pharmacologic dose response
studies.

Quantitative Aspects

A number of distinct types of carcinogens and
mutagens differing in chemical structure are well
recognized. These agents differ from each other
as to effectiveness and target organ affected in
cancer causation in humans and in animal models.
In many cases, such differences are now understood

as a function of biochemical activation leading to
the ultimate carcinogen or mutagen, in contrast to
detoxification products. For example, benzo(a)-
pyrene is a much more powerful carcinogen than
benzo(a)anthracene, 2-fluoreneamine is more active
than 4-biphenylamine, and short chainlength aliphatic nitrosamines are more active than long
chainlength compounds. Because of distinct
ratios of activation over detoxification metabolites obtained in vitro compared to in vivo, these
qualitative and quantitative structure-activity
relationships do not always hold in studies involving in vitro effects such as mutagenicity
assays. In particular, most of the biochemical
activation systems used to convert promutagens to
the active metabolite are deficient in detoxification ability, thus accounting in part for the lack
of correlation in specific instances (65).

The primary objective of mutagenicity and
carcinogenicity testing is to provide a reliable,
sound data base for risk assessment of environmental chemicals and situations with respect to
somatic cell effects such as neoplastic disease or
germ cell effects such as genetic diseases. Thus,
the testing approach described should be used in a
manner such that data are generated which can indeed be used for objective definition of potential
adverse effects. This aspect necessarily needs to
consider quantitative potency, in addition to the
qualitative yes or no answers. It is quite evident that the protective measures needed for the
liver carcinogen aflatoxin B_1 (active at 1 ppb)
are distinct from those required for the liver
carcinogens safrole (active at 2000 ppm) or aceta-
mide (active at 12,500 ppm). The same is true
even for the complex environment in which the general public, as well as specific occupational
groups, is exposed to varied environmental influences and hazards. It is beyond the scope of this
review dealing with the role of genetic toxicology
in carcinogen and mutagen testing to review in detail the quantitative aspects of this field.
Nevertheless, it can be stated that the current
mathematical evaluations of dose response studies
have been based on very few carefully conducted
animal bioassays. In fact, a large number of
mathematical models have been based on a single
experimental series involving subcutaneous injection of polycyclic aromatic hydrocarbons, which
because of their relative insolubility, and slow

absorption from the injection site, exhibit a much broader dose response curve (active over 3-5 log units) than rapidly absorbed agents such as aromatic amines or nitrosamines. For example, even with the powerfully carcinogenic N2-fluorenyl-acetamide, a lowering of the dose by only one log unit, that is a factor of 10, converts a very powerful carcinogic stimulus (200 ppm) to a virtually inactive dose rate (20 ppm). On a larger scale, in the case of cigarette smoke, an individual smoking 40 standard cigarettes per day has a fairly high risk of disease, whereas 4 cigarettes per day would be a minimal risk. This again is only a factor of 10. Thus, quantitative aspects are most important if the goal of risk elimination and thus disease prevention is to be approached in a realistic manner.

In summary, the decision point approach provides a framework for systematic evaluation of the potential hazards of chemicals, which indicates the need for and can be integrated with other elements in toxicity testing (66). It is designed to yield a stepwise progression of data acquisition. An evaluation carefully conducted of this systematic program should provide sequentially a qualitative and a semi-quantitative data base, which need not necessarily terminate in an expensive and extensive long-term bioassay, and which provides an effective tool for the protection of the public against environmental cancer and mutagenic risks.

Literature Cited

1. Bridges, B.A. 1979. Short-term tests and human health--The central role of DNA repair. In Environmental Carcinogenesis, eds. P. Emmelot, E. Kriek, pp. 319-28. Amsterdam: Elsevier/No. Holland Press.
2. Flamm, W.G. 1974. A tier system approach to mutagen testing. Mutat. Res. 26:329-33.
3. Bora, K. C. 1976. A hierarchical approach to mutagenicty testing and regulatory control of environmental chemicals. Mutat. Res. 46:145.
4. Green, S. 1977. Present and future uses of mutagenicity tests for assessment of the safety of food additives. J. Environ. Pathol. Toxicol. 1:49.

5.   Weisburger, J.H., Williams, G.M. 1978.
     Decision point approach to carcinogen test-
     ing. In Structural Correlates of Carcino-
     genesis and Mutagenesis, ed. I.M. Asher C.
     Zervos, pp. 45-52. Rockville, Md.: FDA
     Office of Health Affairs.
6.   Purchase,I.F.H., Longstaff, E.,Ashby, J.,
     Styles,J.A.,Anderson, D., LeFevre, P.A.,
     Westwood,F.R. 1978. An evaluation of 6
     short-term tests for detecting organic
     chemical carcinogens. Br.J. Cancer
     37:873-959.
7.   Cramer, G.M., Ford, R.A., Hall, R.L. 1978.
     Estimation of toxic hazard--a decision tree
     approach. Food Cosmet. Toxicol. 16:255-76.
8.   Scientific Committee, Food Safety Council.
     1978. Food Cosmet. Toxicol. 16 (Suppl.2):35
9.   Ray, V. 1979. Application of microbial and
     mammalian cells to the assessment of muta-
     genicity. Pharmacolo. Revs. 30:537-46.
10. Williams, G., Kroes, R., van de Poll, K.W.
     and Waaijers, H.W eds. 1980. The Predic-
     tive Value of In Vitro Short-Term Screening
     Tests in Carcinogenicity Evaluation.,
     Amsterdam: Elsevier/No. Holland Biomed.
     Press.
11. Brusick, D. Unified scoring system and
     activity definitions for results from in
     vitro and submammalian mutagenesis test
     batteries. In: Proceedings of the Third
     Life Sciences Symposium on Health Risk
     Analysis, in press.
12. Hollstein, M., McCann, J., Angelssanto,
     F.A., Nichols, W.W. 1979. Short-term test
     for carcinogens and mutagens. Mutat.Res.
     46:145.
13. Rinkus, S.J., Legator, M.S. 1980. The need
     for both in vitro and in vivo systems in
     mutagenicity screening. In Chemical Muta-
     gens. Principles and Methods for Their De-
     tection. Vol. 6. de Serres, F.J. and
     Hollaender, A. pp. 365-473 New York: Plenum
     Press.
14. Weisburger, J.H., Williams, G.M. 1980.
     Chemical carcinogens. In Casarett and
     Doull's Toxicology, eds. J. Doull C.D.
     Klaassen, M.O. Amdur, 2nd ed., pp. 84-138.
     New York. Macmillan.

15. International Commission for Protection Against Environmental Mutagens and Carcinogens 1979. Advice on screening of chemicals for mutagenicity. Mutat. Res. 64:155-58.

16. Prival, M.J. 1979. Genetic toxicology: Regulatory aspects. J. Environ. Pathol. Toxicol.

17. DeSerres, F.J. 1979. Problems associated with the application of short-term tests for mutagenicity in mass-screening programs. Environmental Mutagenesis 1:203-8.

18. Kroes, R. 1979. Animal Data, Interpretation and Consequences. In: Environmental Carcinogenesis: Occurrence, Risk Evaluation and Mechanisms, Emmelot, P., Kriek, E., eds. pp. 287-299. Amsterdam: Elsevier/No. Holland.

19. Kolbye, A.C., 1981. The application of fundamentals in risk assessment. American Chemical Society Symposium Series, in press.

20. Mondal, S., Brankow, D.W., and Heidelberger, C. 1978. Enhancement of Oncogenesis in C3H/10T $1/2$ mouse embryo cell cultures by saccharin. Science 201:1141-1142.

21. Chang, C.C., Trosko, J.E. and Warren, S.T. 1978. In vitro assay for tumor promoters and antipromoters. J. Environ. Path. & Toxicol. 2:43-64.

22. Weinstein, I.B. 1980. Evaluating substances for promotion, cofactor effects and synergy in the carcinogenic process. J. Environ. Pathol. and Toxicol. 3:89-102.

23. Williams, G.M. 1980. Classification of genotoxic and epigenetic hepatocarcinogens using liver culture assays. Annals New York Academy of Sciences, 349:273-282.

24. Dunkel, V.C. and Williams, G.M. Biological significance of endpoints in short-term tests for carcinogens and mutagens. In: Proceedings of the Third Life Sciences Symposium on Health Risk Analysis, in press.

25. Selkirk, J. 1977. Divergence of metabolic activation systems for short-term mutagenesis assays. Nature 270:604-7.

26. Schmeltz, I., Tosk, J., Williams, G.M. 1978. Comparison of the metabolic profiles of benzo(a)pyrene obtained from primary cell cultures and subcellular fractions derived from normal and methylcholanthrene-induced rat liver. Cancer Letters 5:81-89.

27.   Bigger, C.A.H., Tomaszewski, J.E., Dipple,
       A. 1980. Limitations of metabolic activa-
       tion systems used with in vitro tests for
       carcinogens. Science 209:503-4.
28.   Williams, G.M. and Weisburger, J.H. System-
       atic carcinogen testing through the decision
       point approach. Annual Review of Pharmacol-
       ogy and Toxicology, in press.
29.   Miller, E.C. and Miller, J.A. 1971. The
       mutagenicity of chemicals. In Chemical
       Mutagens: Principles and Methods for Their
       Detection Vol. 1. A. Hollaender, ed. pp.
       83-119 New York: Plenum Press.
30.   Weisburger, J.H. and Williams, G.M. 1981.
       Metabolism of chemical carcinogens. In:
       Cancer: A Comprehensive Treatise., 2nd Edi-
       tion, Ed. F.F. Becker, Plenum Press, N.Y.,
       pp. 185-234.
31.   Ames, B.N. 1979. The identification of
       chemicals in the environment causing
       mutations and cancer. In Naturally Occur-
       ring carcinogens-Mutagens and Modulators of
       Carcinogenesis, eds. E. C. Miller, J.A.
       Miller, I. Hirono, T. Sugimura, S. Takayama,
       pp.345-58. Baltimore: University Park
       Press.
32.   Nagao, M., Sugimura, T., Matsushima, T.
       1978. Environmental mutagens and carcino-
       gens. Ann. Rev. Genet. 12:117-59.
33.   Rosenkranz, H.S., Poirier, L.A. 1979.
       Evaluation of the mutagenicity and DNA-
       modifying activity of carcinogens and non-
       carcinogens in microbial systems. J. Natl.
       Cancer Inst. 62:873-92.
34.   Chu, E.H.Y. Induction and analysis of gene
       mutations in mammalian cells in culture. In
       Chemical Mutagens: Principles and Methods
       for Their Detection, Vol. 1. Hollaender,
       A., ed. pp. 411-444. New York: Plenum
       Press.
35.   Radman, M., Caillet-Fauquet, P., Defais, M.,
       Villani, G. 1980.The molecular mechanism of
       induced mutations and in vitro biochemical
       assay for mutagenesis. IARC Sci. Pub.
       12:537-46.
36.   Bartsch, H., Malaveille, C., Camus, A.M.,
       Martel-Planche, G., Brun, G., Hautefeville,
       A., Sabadie, N., Barbin, A., Kuroki, T.,
       Drevon, C., Piccoli, C. and Montesano, R.
       Bacterial and mammalian mutagenicity tests:

validation and comparative studies on 180 chemicals. In Molecular and Cellular Aspects of Carcinogen Screening Tests. R. Montesano, H. Bartsch and L. Tomatis eds. pp. 179-241. Lyon, France: IARC.

37. Swenberg, J.A. and Petzold, G.L. 1979. The usefulness of DNA damage and repair assays for predicting carcinogenic potential of chemicals. In: Strategies for Short-Term Testing for Mutagens/Carcinogens. B.E. Butterworth, ed. West Palm Beach, Fla.: CRC Press.

38. Painter, R.B. 1978. DNA synthesis inhibition in Hela cells as a simple test for agency that damage human DNA. J. Environ. Pathol. & Toxicol. 2:65-78.

39. Stich, H.F., San, R.H.C., Lam, P., Koropatnick and Lo, L. 1977. Unscheduled DNA synthesis of human cells as a short-term assay for chemical carcinogens. In: Origins of Human Cancer. Hiatt, H.H., Watson, J.D., and Winston, J.A. eds. pp. 1499-1512 Cold Spring Harbor, N.Y.: Cold Spring Harbor Laboratory.

40. Williams, G.M. 1979. The status of in vitro test systems utilizing DNA damage and repair for the screening of chemical carcinogens. J. Assoc. Official Analytical Chemists 62:857-63.

41. Ames, B.N., Haroun, L. 1980. An overview of the Salmonella mutagenicity test. In Microsomes, Drug Oxidations, and Chemical Carcinogenesis, vol. 2, eds. M.J. Coon, A.H.Conney R.W. Estabrook, H.V. Gelboin, J.R. Gillette, P.J. O'Brien, pp. 1025-1040. Academic Press. New York.

42. Hyman, J., Leifer, Z., Rosenkranz, H.S. 1980. The E. coli Pol A_1 assay. A quantitative procedure for diffusible and nondiffusible chemicals. Mutat. Res. 74:107-11.

43. Sugimura, T., Yahagi, T., Nagao, M., Takeuchi, M., Kawachi, T., Hara, K., Yamasaki, E., Matsushima, T., Hashimoto, Y., Okada, M. 1976. Validity of mutagenicity tests using microbes as a rapid screening method for environmental carcinogens. Lyon, France: IARC Sci. Pub. 12:81-104.

44.   O'Neill, J.P., Brimer, P.A., Machanoff, R.,
      Hirsch, G.P., Hsie, A.W. 1977. A quantita-
      tive assay of mutation induction at the
      hypoxanthine-guanine phosphoribosyl trans-
      ferase locus in Chinese hamster ovary cells
      (CHO/HGPRT System): Development and defini-
      tion of the system. Mutat. Res. 45:91-101.

45.   Huberman, E. 1976. Cell-mediated mutagen-
      icity of different genetic loci in mammalian
      cells by carcinogenic polycyclic hydrocar-
      bons. In Screening Tests in Chemical Car-
      cinogenesis, eds. R. Montesano, H. Bartsch,
      L. Tomatis. IARC Sci. Pub. 12, pp.521-36.
      Lyon, France: IARC.

46.   Tong, C., Williams, G.M. 1980. Definition
      of conditions for the detection of genotoxic
      chemicals in the adult rat-liver epithelial
      cell/hypoxanthine-guanine   phosphoribosyl
      transferase (ARL/HGRPT) mutagenesis assay.
      Mutat. Res. 74:1-9.

47.   Clive, D., Johnson, K.O., Specter, J.F.S.,
      Batson, A.G., Brown, M.M. 1979. Validation
      and characterization of the L5178Y/TK mouse
      lymphoma mutagen assay system. Mutat. Res.
      59:61-108.

48.   Davies, P.J., Parry, J. 1974. The induction
      of oabain-resistant mutants by N-methyl-N'-
      nitro-N-nitrosoguanidine in Chinese hamster
      cells. Genet. Res. 24:311-14.

49.   San, R.H.C., Williams, G.M.   1977.   Rat
      hepatocyte primary cell culture-mediated
      mutagenesis of adult rat liver epithelial
      cells by procarcinogens. Proc. Society
      Experimental Biology Medicine 156:534-38.

50.   Roberts, J.J. 1980. Cellular responses to
      carcinogen-induced DNA damage and the role
      of DNA repair. British Medical Bulletin,
      36:25-31.

51.   Williams, G.M. 1980. The detection of chem-
      ical mutagens/carcinogens by DNA repair and
      mutagenesis in liver cultures In Chemical
      Mutagens, Principles and Methods for Their
      Detection vol. 6, eds. F.J. deSerres, A.
      Hollaender, pp. 61-79. New York Plenum.

52.   Williams, G.M. 1980. Liver culture indi-
      cators for the detection of chemical carcin-
      ogens. In: Short-Term Tests for Chemical
      Carcinogens, eds. R.H.C. San and H.F. Stich,
      pp. 581-609. New York: Springer-Verlag.

53. Probst, G.S., Hill, L.E., Brewsey, B.J. 1980. Comparison of three in vitro assays for carcinogen-induced DNA damage. J. Toxicol. Environ. Hlth. 6:333-49.
54. Wolff, S. 1977. Sister chromatid exchange. Annu. Rev. Genet. 11:183-201.
55. Perry, P.E. 1980. Chemical mutagens and sister chromatid exchange. In Chemical Mutagens. Principles and Methods for Their Detection Vol. 6. de Serres, F.J. and Hollaender, A. eds. pp. 1-39 New York: Plenum Press.
56. Berwald, Y., Sachs, L. 1963. In vitro cell transformation with chemical carcinogens. Nature 200:1182.
57. DiPaolo, J.A. 1979. Quantitative transformation by carcinogens of cells in early passage. In Environmental Carcinogenesis: Occurrence, Risk Evaluation and Mechanisms, eds., P. Emmelot, E. Kriek, pp. 365-380, Amsterdam: Elsevier/North-Holland.
58. Pienta, R. 1980. Transformation of Syrian hamster embryo cells by diverse chemicals and correlation with their reported carcinogenic and mutagenic activities. In Chemical Mutagens: Principles and Methods for their Detection, Vol. 6. de Serres, F.J. and Hollaender A. eds. pp. 175-202. New York: Plenum Press.
59. Heidelberger, C. 1980. Mammalian cell transformation and mammalian cell mutagenesis. J. Environ. Pathol. Toxicol. 3:69-88.
60. Sirover, M.A. aand Loeb, L.A. 1976. Metal-induced infidelity during DNA synthesis. Proc. Natl. Acad. Sci., 73:2331-2335.
61. Weisburger, J.H. In: Searle, C.E. ed. 1976. Chemical Carcinogenesis. American Chemical Society Monograph 173. New York: American Chemical Society.
62. Sontag, J., ed. 1980. Carcinogens in Industry and Environment. New York: Marcel Dekker.
63. Page, N.P. 1977. Current concepts of a bioassay program in environmental carcinogenesis. In Advances in Modern Toxicology, vol. 3, ed. H.F. Kraybill, M.A. Mehlman, pp. 87-172. New York: John Wiley.

RECEIVED March 2, 1981.

Pesticides: Mutagenic and Carcinogenic Potential

MICHAEL D. WATERS and STEPHEN NESNOW—Genetic Toxicology Division,
Health Effects Research Laboratory, U.S. Environmental Protection Agency,
Research Triangle Park, NC 27711

VINCENT F. SIMMON[1], ANN D. MITCHELL, and TED A. JORGENSON—
SRI International, Menlo Park, CA 94025

RUBY VALENCIA[2]—WARF Institute, Inc., Madison, WI 53704

In recent years, major advances have been made in the
methodology of evaluating chemical compounds for their mutagenic
and carcinogenic potential. By correlating genetic and related
effects in short-term studies with mutagenic and carcinogenic
effects in whole animals, these tests can also be used to screen
chemicals for mutagenic as well as presumptive carcinogenic
activity.

There are three major classes of genetic damage: gene or
point mutation, chromosomal alteration, and primary DNA damage
that can be detected by short-term bioassays.

Alterations affecting single genes are termed point
mutations. This category includes base pair substitutions and
frame-shift mutations, in addition to other small deletions and
insertions. For point mutations, the in vitro test systems are
forward and reverse mutation assays in bacteria ([1,2,3]),
yeast ([4]), fungi ([5]), and mammalian cell cultures ([6-11]).

Metabolic activation has been incorporated into most short-
term in vitro assays, usually by use of a mammalian liver
microsomal preparation. Some genotoxicants have to be converted
into reactive forms before producing observable effects.
Metabolism by oxidative enzymes and formation of electrophilic
metabolites that bind covalently to deoxyribonucleic acid (DNA)
([12,13]) are the presumed mechanisms for most genetic activity.
Gene mutagens may be screened in short-term in vivo assays

[1] Current address: Genex Laboratories, Rockville, MD 20852.
[2] Current address: Zoology Department, University of Wisconsin, Madison, WI
53706.

including tests in insects, plants, and mammals, the most utilized being the sex-linked recessive lethal test in the fruit fly Drosophila melanogaster ([14]). A few short-term assays for gene mutation in rodents are available. One of these, the "spot test" in mice, has recently been described ([15]). Unfortunately, this test detects somatic rather than germinal mutations.

Chromosomal alterations refer to changes in number and structure of chromosomes. They may involve loss or gain of entire chromosomes, chromosome breaks, nondisjunctions, and translocations. These abnormalities are detected by searching for chromosomal changes either in somatic or germinal cells. When chromosomal aberrations are observed in the germinal tissues of intact animals, they produce important evidence as to whether reproductive organs can be affected by the test chemical ([16]). Positive findings of specific gene mutations in in vitro tests and of chemically induced chromosomal damage in germinal tissues of rodents constitute strong evidence for a chemical's ability to produce heritable effects.

Assays of DNA damage and repair, rather than measuring mutation per se, measure direct damage caused by a chemical to the DNA and ensuing repair. Detection of DNA damage and repair can be accomplished with bioassays using bacterial ([17,18]), yeast([19,20]),and mammalian cells ([21,22]), and whole animals ([23,24,25]).

The whole animal bioassays that detect DNA damage and repair in germinal tissues are valuable in mutagenesis testing to indicate potential reproductive and heritable mutagenic effects.

Certain other short-term tests focus more specifically on carcinogenesis as an end point. The process by which normal cells grown in culture are converted into malignant cells after treatment with a carcinogen is termed oncogenic cellular transformation. By injecting transformed cells into intact animals, malignancy can be confirmed. However, for purposes of short-term bioassay, this procedure is not requisite. The most common means of distinguishing transformed cells in culture is altered morphology and growth in agar. Several mammalian oncogenic transformation bioassays are available that use cells derived from different rodent species ([26,27,28,29]).

In the biological analysis of an environmental chemical, the function of short-term tests is predictive--they examine the potential, in qualitative terms, for producing carcinogenesis, mutagenesis, and related toxic effects. To appropriately utilize this capacity of short-term test systems, it is necessary to gain an understanding of the way their results relate to corresponding biological phenomena.

Good correlation exists between test results of point mutations in microorganisms and carcinogenesis bioassay results ([30-37]). In studies in Salmonella typhimurium and

Escherichia coli, for example, a range of 80 to 90 percent of chemical carcinogens can cause increased mutation or DNA damage (1,33,34,37,38).   It has been established that most carcinogens are mutagenic in short-term tests, when appropriate metabolic activation is included. Given the inadequacies of metabolic activation systems and inherent limitations in the indicator organisms themselves, however, no single in vitro system can be universally applied. The fact that the Salmonella mutagenesis assay fails to demonstrate mutagenicity in certain carcinogenic halogenated organic and metallic compounds(34) should advise caution in using the Ames test alone for screening.

The potential for genetic effects to show up in later generations is the main concern in mutagenesis testing. Even if it is several generations before a trait appears, the genetic burden to the offspring of the exposed population is increased. For measuring mutational effects in germinal tissues, the test requires an intact animal, such as Drosophila or rodents.

The changes produced by environmental mutagens in chromosomal germ cell structures are identical to the aberrations in somatic cells produced by these same compounds (39,40). Somatic cell damage is not transmitted to offspring, but suggests at least a potential for heritable effects.

In programs designed to evaluate large numbers of potentially hazardous environmental agents, inexpensive short-term bioassays are useful to set priorities, to be followed by long-term whole animal procedures, for more in-depth evaluation. Using the approach in biological testing that proceeds from simple short-term detection systems to long-term whole animal bioassays in sequential steps or "tiers" decreases the number of substances requiring complete evaluation, and is, thus, most cost-effective. Several variations of this "tiered" approach are discussed in the literature of carcinogenicity testing (41,42) with concensus on key points of emphasis in the evaluation process: detection at the first tier, confirmation at the second tier, and risk assessment at the third tier (43,44).

In mutagenicity testing where several kinds of genetic damage must be evaluated, a "battery" of tests is recommended. The battery should include tests for point or gene mutations, chromosomal effects, and primary damage to DNA. Ideally, tests for genetic damage in germinal cells are included as well. The simultaneous performance of such a battery of tests is not as cost effective as one would desire.

The phased approach recommended by the authors (45,46,47) combines the tiered approach to carcinogenicity testing and the "battery" approach to mutagenicity testing (48,49) in one useful framework, a three-level test matrix which organizes genetic and related bioassays according to: (1) end point examined and degree of selectivity of test, (2) sensitivity and statistical power of test (potential number of respondents per total number

treated), and (3) complexity and cost of test. Phase 1 is the
detection group. Phase 2 confirms or refutes results in Phase 1,
and its tests are generally more expensive and complex. Phase 3,
hazard or risk assessment, has the greatest relevance for humans.

Figure 1 illustrates the phased approach, showing the three
step matrix with a battery of tests at each step. Some of these
tests apply to mutagenesis only, while others apply to
carcinogenesis only. However, certain tests have a broad enough
data base to permit correlation of results with either mutagenesis
or carcinogenesis. The Phase 1 test battery emphasizes detection
of point mutations and primary DNA damage in microbial species,
and detection of chromosomal changes in mammalian cells (in vivo
exposure preferred). The Phase 2 battery uses more narrow and
definitive genetic and related bioassays of corresponding endpoints in mammalian cells in culture, plants, insects, and mammals

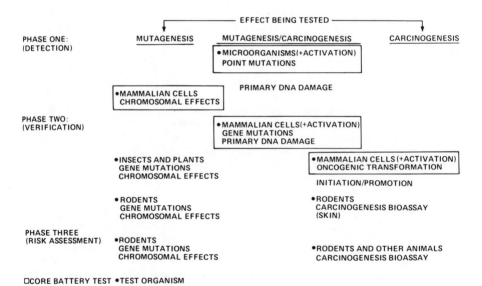

*Figure 1.   A phased approach for evaluating mutagenesis and presumptive carcino-
genesis of environmental chemicals*

to verify results of Phase 1. In addition, Phase 2 bioassays contain tests for mutagenicity as well as carcinogenicity. Phase 3 testing concerns quantitative hazard assessment and includes appropriate whole animal bioassays. Thus, the pursuit of positive responses from Phase 1 continues tier-wise through tests involving similar end points in Phases 2 and 3, with greater focus and relevance at each stage.

The special value of the "core battery" of short-term tests is its ability to delineate a probable negative result for mutagenicity and presumptive carcinogenicity. Included in this battery are tests for point mutation in microorganisms and gene mutation in mammalian cell cultures; a test (preferably in vivo) for chromosomal alterations; a test for primary damage to DNA using mammalian (preferably human) cells in culture; and an in vitro test for oncogenic transformation. This battery, selected from Phase 1 and Phase 2 tests, contains the most essential short-term tests in the phased evaluation system.

Experimental Methods

Pesticides used for these studies were procured from the manufacturers by Battelle Memorial Laboratories, Columbus, Ohio, and the EPA Office of Pesticide Programs, Washington, D.C. The quality of each chemical was a technical grade or its equivalent. Information about manufacturers, lot numbers, and purity of these compounds may be obtained from the first author. Descriptions of the experiments and the numerical data from Phase 1 and Phase 2 assays are available in contract reports ([50,51,52,53]). Qualitative results are reported in the remainder of the paper. Oncogenic transformation assays were performed according to the procedures of Reznikoff et al ([27,28]).

For all compounds, the following in vitro procedures were employed. The tests are designated in subsequent tables by the abbreviations shown in parentheses.

(1) Reverse mutation in Salmonella typhimurium strains TA1535, TA1537, TA1538, and TA100 (Ames).

(2) Reverse mutation in Escherichia coli WP2 (uvrA⁻) (WP2).

(3) Mitotic recombination in the yeast Saccharomyces cerevisiae D3 (D3).

(4) Differential toxicity assays in DNA repair-proficient and deficient strains of E. coli (strains W3110 and P3478, respectively) (POL A) and Bacillus subtilis (strains H17 and M45, respectively) (REC).

(5) Unscheduled DNA synthesis in human fetal lung fibroblasts (WI-38 cells) (UDS).

(6) The Drosophila melanogaster sex-linked recessive lethal test (DRL).

(7) The mouse dominant lethal test (MDL).
(8) Oncogenic transformation in C3H10T1/2CL8 cells
(OT).

Results

The findings from a selection of both Phase 1 detection systems and Phase 2 confirmatory bioassays are summarized in Table 1.

TABLE I

Pesticide Mutagenesis/Carcinogenesis Evaluation

Summary of Results

Group A Probable Positive Chemicals (6/38)[a]
         Positive for Point/Gene Mutation and DNA Damage
         in Pro- and/or Eukaryotic Systems

Group B Low Priority for Further Evaluation (6/38)
         Positive Only for DNA Damage in Prokaryotic Systems

         Medium Priority for Further Evaluation (5/38)
         Positive Only for DNA Damage in Eukaryotic Systems

         High Priority for Further Evaluation (2/38)
         Positive Only for Gene Mutation in Insects

Group C Probable Negative Chemicals (19/38)
         Negative in All Tests Performed

[a]The number in the parentheses represents the number of chemicals found positive over the number evaluated.

They have been grouped into three classes: Group A, pesticides eliciting a positive response in point or gene mutation and DNA damage in prokaryotic and eukaryotic systems; Group C, pesticides negative in all tests; and Group B, agents requiring further evaluation. Group B has been divided into three subgroups: pesticides producing positive responses only for gene mutations in

insects (high priority for further evaluation); pesticides
inducing positive responses only for DNA damage in eukaryotic
systems (medium priority for further evaluation); and pesticides
evoking positive responses only for DNA damage in prokaryotic
systems (low priority for further evaluation).

What constitutes a positive response for the in vitro assays
is a reproducible, dose-related increase in the observed effect.
For 19 of the 38 pesticides tested in the bioassays (grouped in
Tables III-VI), a mutagenic or related effect was found. Of the 19,
however, 11 were positive for primary damage to DNA only.

Of the 19 pesticides grouped in Table II, all were negative in
five Phase 1 bioassays and the Phase 2 bioassays performed. These
compounds included insecticides (I), fungicides (F), and
herbicides (H). Malathion, parathion, pentachloronitrobenzene
(PCNB), and phorate were also negative for heritable chromosomal
effects in the mouse dominant lethal test. The six compounds
grouped in Table III that were positive in three or more bioassays
were acephate, captan, demeton, folpet, monocrotophos, and
trichlorfon. Positive results were seen for demeton in all in
vitro tests in Phase 1 and Phase 2. Folpet and captan were
positive in all Phase 1 and all Phase 2 in vitro assays except the
test for unscheduled DNA synthesis in WI-38 cells. Trichlorfon
was positive in all Phase 1 and Phase 2 in vitro tests, with the
exclusion of relative toxicity tests with E. coli and B.
subtilis.

Acephate and monocrotophos produced mutagenic effects in S.
typhimurium, an increase in mitotic recombination in S. cerevisiae
D3 and unscheduled DNA synthesis (UDS) in WI-38 cells. Acephate
and monocrotophos produced no effects in E. coli or B. subtilis
relative toxicity assays. The negative findings for acephate,
monocrotophos, and trichlorfon in bacterial relative toxicity
assays may mean that these pesticides did not diffuse into the
agar. Both acephate and trichlorfon were tested for oncogenic
transformation in C3H10T1/2 CL8 cells; only the latter was
positive.

The pesticides chlorpyrifos, 2,4-D acid, 2,4-DB acid,
dicamba, dinoseb, and propanil were positive in the bacterial
relative toxicity assays (propanil in B. subtilis only), but in
all other assays, produced no activity (Table IV). Increased
mitotic recombination was induced by azinphos-methyl, crotoxyphos,
cacodylic acid, and parathion-methyl in S. cerevisiae D3;
disulfoton enhanced UDS in WI-38 cells. As shown in Table V,
however, these pesticides produced no other effects. Bromacil
and simazine were positive only in the Drosophila sex-linked
recessive lethal test (Table VI).

Discussion

There have been numerous proposals on methods of evaluating
the mutagenic and presumptive carcinogenic hazards of

TABLE II

Pesticides Having a Negative Response
In Six or More Genetic Bioassays†
Group C

Aspon (I)	Fensulfothion (I)	MSMA (H)
Carbofuran (I)	Fenthion (I)	Parathion (I)‡
Diazinon (I)*	Fonofos (I)*	PCNB (F)‡
DSMA (H)	Malathion (I)‡	Phorate (I)‡
Endrin (I)*	Methomyl (I)	Siduron (H)
Ethion (I)*	Methoxychlor (I)	Trifluralin (H)
	Monuron (H)	

†Negative results in Ames, WP2, D3, POL A, REC, UDS, and DRL
bioassays (I, insecticide; H, herbicide; F, fungicide).
*Not tested in DRL bioassay.
‡Negative result in MDL bioassay.

TABLE III

Pesticides Positive for Point/Gene Mutation
and DNA Damage in Pro and/or Eukaryotic Systems (Group A)

Pesticide	Phase 1 Detection (Microbial)					Phase 2 Confirmation			
	Ames	WP2	D3	POL A	REC	UDS	DRL	MDL	OT
Acephate (I)	Pos	Neg	Pos	Neg	Neg	Pos	Neg	NT*	Neg
Captan (F)	Pos	Pos	Pos	Pos	Pos	Neg	Pos	Neg	NT
Demeton (I)	Pos	Pos	Pos	Pos	Pos	Pos	Neg	NT	NT
Folpet (F)	Pos	Pos	Pos	Pos	Pos	Neg	Pos	Neg	NT
Monocrotophos (I)	Pos	Neg	Pos	Neg	Neg	Pos	Neg	Neg	NT
Trichlorfon (I)	Pos	Pos	Pos	Neg	Neg	Pos	Neg	NT	Pos

*Not tested.

TABLE IV

Pesticides Positive Only for DNA Damage in Prokaryotic Systems
Group B Low Priority

Pesticide	Phase 1 Detection (Microbial)					Phase 2 Confirmation		
	Ames	WP2	D3	POL A	REC	UDS	DRL	MDL
Chlorpyrifos (I)	Neg	Neg	Neg	Pos	Pos	Neg	Neg	NT
2,4 D Acid (H)	Neg	Neg	Neg	Pos	Pos	Neg	NT	NT
2,4 DB Acid (H)	Neg	Neg	Neg	Pos	Pos	Neg	NT	NT
Dicamba (H)	Neg	Neg	Neg	Pos	Pos	Neg	Neg	NT
Dinoseb (H)	Neg	Neg	Neg	Pos	Pos	Neg	Neg	NT
Propanil (H)	Neg	Neg	Neg	Neg	Pos	Neg	NT	NT

TABLE V

Pesticides Positive Only for DNA Damage in Eukaryotic Systems
(Group B Medium Priority)

Pesticide	Phase 1 Detection (Microbial)					Phase 2 Confirmation		
	Ames	WP2	D3	POL A	REC	UDS	DRL	MDL
Azinphos-Methyl (I)	Neg	Neg	Pos	Neg	Neg	Neg	Neg	Neg
Cacodylic Acid (H)	Neg	Neg	Pos	Neg	Neg	Neg	Neg	NT*
Crotoxyphos (I)	Neg	Neg	Pos	Neg	Neg	Neg	Neg	NT
Disulfoton (I)	Neg	Neg	Neg	Neg	Neg	Pos†	Neg	NT
Parathion-Me (I)	Neg	Neg	Pos‡	Neg	Neg	Neg	Neg	Neg

*Not tested.
†Not detected in Phase 1 bioassays.
‡Weak positive.

TABLE VI

Pesticides Positive Only for Gene Mutation in Insects
(Group B High Priority)

Pesticide	Phase 1 Detection (Microbial)					Phase 2 Confirmation		
	Ames	WP2	D3	POL A	REC	UDS	DRL	MDL
Bromacil (H)	Neg	Neg	Neg	Neg	Neg	Neg	Pos*	Neg
Simazine (H)	Neg	Neg	Neg	Neg	Neg	Neg	Pos*	NT†

*Not detected in Phase I bioassays.
†Not tested.

environmental chemicals ([41,42,44,48,54,55,56,57]). An effective way of setting priorities for use of the whole animal testing resources is offered by short-term tests for genetic effects. Whole animal testing resources are too limited to be applied to large numbers of chemicals and they are expensive and time consuming as well. It is recommended that current bioassay methodology be systematized to evaluate the many chemicals now in use or being developed.

The application of a stepwise or phased approach to evaluation of mutagenicity and presumptive carcinogenicity of pesticides described in this paper has produced some intriguing results. There are a number of compounds that give a positive response in two to five Phase 1 detection systems: acephate, captan, demeton, folpet, monocrotophos, and trichlorfon (Table III). These also were positive in the Phase 2 in vitro test or in one of the Phase 2 in vivo tests now completed. In addition, trichlorfon was found positive in producing oncogenic transformation of C3H10T1/2 mouse embryo fibroblasts ([58]). This compound is being studied in the carcinogenesis bioassay program at the National Cancer Institute.

In Phase 2 tests for unscheduled DNA synthesis in WI-38 cells, a positive response with disulfoton was found. When examined in Phase 1 microbial tests, this compound had not been detected. Positive responses were also induced by bromacil and simazine in the Drosophila sex-linked recessive lethal test in Phase 2. As these results may represent false negatives for the Phase 1 tests, they are a matter for concern. Under ordinary circumstances, these three pesticides would not have been evaluated in Phase 2. Further testing will explore the possibility of false negatives and if there is additional evidence, it may be necessary to modify the present Phase 1 test battery.

There are interesting structural similarities observed within groups of pesticides which exert similar biological effects. Captan, folpet, and trichlorfon all contain a trichloromethyl substituent and all were found to be gene mutagens and to damage DNA. Bromacil and simazine, which were positive only in the Drosophila sex-linked recessive lethal test, are related diazine and triazine heterocylics. However, minor changes in chemical structure can also alter biological activity. Crotoxyphos, positive only in S. cerevisiae and monocrotophos, positive in S. typhimurium, S. cerevisiae, and unscheduled DNA synthesis in WI-38 cells differ only in carbonyl substituent. These structural and biological relationships indicate that structure-activity studies may be useful in defining mechanisms of action of mutagenic or carcinogenic pesticides and potentially in the identification of hazardous environmental agents.

Present mutagenicity results are compared to some available results from whole animal carcinogenesis bioassays in Table VII.

TABLE VII

Comparison of Mutagenicity and Carcinogenicity Results

Pesticide	Present Mutagenicity Test Results	NCI or IARC Carcinogenicity Test Results
Azinphos-methyl (59)*	Positive for mitotic recombination in yeast. Negative in all other tests.	Equivocal evidence of positive response in male Osborne-Mendel rats. Negative in female Osborne-Mendel rats and in B6C3F1 mice of both sexes.**
Captan (60)	Positive in all tests except UDS in human cells and the mouse dominant lethal test.	Positive in B6C3F1 mice of both sexes. Negative in Osborne-Mendel rats of both sexes.**
Diazinon (61)	Negative in all tests.	Negative in B6CF1 mice and in F344 rats of both sexes.**
2,4-D and esters (62)	Positive only in microbial relative toxicity tests.	Negative or inconclusive data.***

TABLE VII (continued)

Comparison of Mutagenicity and Carcinogenicity Results

Pesticide	Present Mutagenicity Test Results	NCI or IARC Carcinogenicity Test Results
Endrin (63)	Negative in all tests.	Negative or inconclusive data.***
Fenthion (64)	Negative in all tests	Equivocal in male B6C3F1 mice. Negative in female B6C3F1 mice and in male and female F344 rats.
Malathion (65)	Negative in all tests.	Negative in B6C3F1 mice and in Osborne-Mendel rats of both sexes.**

TABLE VII (continued)

Comparison of Mutagenicity and Carcinogenicity Results

Pesticide	Present Mutagenicity Test Results	NCI or IARC Carcinogenicity Test Results
Methoxychlor (66)	Negative in all tests	Negative in B6C3F1 mice and in Osborne-Mendel rats of both sexes.**
Monuron (67)	Negative in all tests.	Equivocal in cs mice and rats suggestive evidence.***
Parathion (68)	Negative in all tests.	Equivocal in Osborne-Mendel rats of both sexes. Negative in B6C3F1 mice of both sexes.**

TABLE VII (continued)

Comparison of Mutagenicity and Carcinogenicity Results

Pesticide	Present Mutagenicity Test Results	NCI or IARC Carcinogenicity Test Results
Parathion-methyl (69)	Negative (marginally positive for mitotic recombination in yeast).	Negative in B6C3F1 mice and in F344 rats of both sexes.**
Pentachloro-nitrobenzene (PCNB) (70)	Negative in all tests.	Negative in B6C3F1 mice and in Osborne-Mendel rats of both sexes.**
Trifluralin (71)	Negative in all tests.	Positive in female B6C3F1 mice. Negative in male B6C3F1 mice and in Osborne-Mendel rats of both sexes.**

*Numbers in parentheses are references to published data.
**Results obtained from the National Cancer Institute.
***Results obtained from the International Agency for Research on Cancer.

There is substantial agreement between the mutagenicity and carcinogenicity test results.

The compound trifluralin is an exceptional case; it proved negative in mutagenicity tests, but positive in carcinogenesis bioassays at the National Cancer Institute. However, technical grade trifluralin containing 84 to 88 ppm dipropylnitrosamine was used in the NCI studies. The findings of liver tumors in the treated animal may indicate that nitrosamine was involved in the carcinogenic activity.

On the basis of preliminary data presented here, a phased testing strategy appears useful for identifying carcinogens and mutagens, particularly when large numbers of chemicals must be investigated efficiently, accurately, and quickly.

Abstract

Methodology has been developed to evaluate the mutagenic and carcinogenic potential of pesticide chemicals. Short-term bioassays for gene mutation, chromosomal effects, and primary damage to DNA permit evaluation of genetic and related biological effects that may be correlated with mutagenic and carcinogenic activity in whole animals. Other bioassay techniques allow observation of chemical transformation of normal cells in culture to cells that can induce tumors in animals.

This paper presents a phased approach to evaluating chemicals for mutagenic and carcinogenic effects. This method allows cost-effective utilization of limited testing resources and protection of human health according to anticipated hazards. Relevant tests are described and the rationale for the approach is explained using results from tests on 38 pesticides. The sequence of bioassay groups emphasizes first, detection, then, confirmation, then, hazard assessment. Whole animal tests are used to pursue investigation of chemicals testing positive in short-term detection systems and confirmatory bioassays. A "core battery" of tests is proposed for delineation of probable negative results in short-term bioassays.

Literature Cited

1. Ames, B.N., McCann, J., and Yamasaki, E., "Methods for Detecting Carcinogens and Mutagens with the Salmonella/ Mammalian-Microsome Mutagenicity Test", Mutat. Res. (1975), 31:347-364.

2. Bridges, B.A., Dennis, R.E., and Munson, R.J., "Mutation in Escherichia coli B/r WP2 Try-reversion or Suppression of a Chain Termination Codon", Mutat. Res. (1967), 4:502-504.

3. Mohn, G.J., Ellenberger, J., and McGregor, D., "Development of Mutagenicity Testing Using *Escherichia coli* K-12 as Indicator Organism", Mutat. Res. (1974), 5:187-196.

4. Zimmerman, F.K., "Detection of Genetically Active Chemicals Using Various Yeast Systems", in: "Chemical Mutagens: Principles and Methods for Their Detection", vol. 3, Hollaender, A., ed., pp. 209-239, Plenum Press, New York, N.Y., 1973.

5. de Serres, F.J., and Malling, H.V., "Measurement of Recessive Lethal Damage Over the Entire Genome and at Two Specific Loci in the ad-3 Region of a Two-Component Heterokaryon of *Neurospora Crassa*", in: "Chemical Mutagens: Principles and Methods for Their Detection", vol. 2, Hollaender, A., ed., pp. 311-342, Plenum Press, New York, N.Y., 1971a.

6. Chu, E.H.Y., "Induction of Gene Mutations in Mammalian Cells in Culture", in: "Chemical Mutagens: Principles and Methods for Their Detection", vol. 2, Hollaender, A., ed., pp. 411-444, Plenum Press, New York, N.Y., 1971.

7. Huberman, E., and Sachs, L., "Mutability of Different Genetic Loci in Mammalian Cells by Metabolically Activated Carcinogenic Polycyclic Hydrocarbons", Proc. Natl. Acad. Sci., USA (1976), 73:188-192.

8. Krahn, D.F., and Heidelberger, C., "Liver Homogenate-Mediated Mutagenesis in Chinese Hamster V79 Cells by Polycyclic Aromatic Hydrocarbons and Aflatoxins", Mutat. Res. (1977), 46:27-44.

9. Clive, D., and Spector, J-A.F.S., "Laboratory Procedure for Assessing Specific Locus Mutations at the TK Locus in Cultured L5178Y Mouse Lymphoma Cells", Mutat. Res. (1975), 31:17-29.

10. Hsie, A.W., Brimer, P.A., Mitchell, T.J., and Gosslee, D.G., "The Dose Response for Ethyl Methanesulfonate-Induced Mutations at the Hypoxanthine-Guanine-Phosphoribosyltransferase Locus in Chinese Hamster Ovary Cells", Somatic Cell Genet. (1975a), 1:247-261.

11. Hsie, A.W., Brimer, P.A., Mitchell, T.J., and Gosslee, D.G., "The Dose Response Relationship for Ultraviolet-Light-Induced Mutations at the Hypoxanthine-Guanine-Phosphoribosyltransferase Loci in Chinese Hamster Ovary Cells", Somatic Cell Genet. (1975b), 1:383-389.

12. Miller, E.C., and Miller, J.A., "The Mutagenicity of Chemical Carcinogens: Correlations, Problems, and Interpretations", in: "Chemical Carcinogens: Principles and Methods for Their Detection", vol. 1, Hollaender, A., ed., pp. 83-120, Plenum Press, New York, N.Y., 1971.

108 THE PESTICIDE CHEMIST AND MODERN TOXICOLOGY

13. Miller, E.C., and Miller, J.A., "The Metabolism of
    Chemical Carcinogens to Reactive Electrophiles and Their
    Possible Mechanisms of Action in Carcinogenesis", in:
    "Chemical Carcinogens", ACS Monograph 173, Searle, C.,
    ed., pp. 737-762, American Chemical Society, Washington,
    D.C., 1976.
14. Wurgler, F.E., Sobels, F.H., and Vogel, E., "Drosophila
    as Assay System for Detecting Genetic Changes", in:
    "Handbook of Mutagenicity Test Procedures", Kilbey, B.J.,
    Legator, M., Nichols, W., and Romel, C., eds., pp. 335-
    373, Elsevier North Holland Biomedical Press, Amsterdam,
    The Netherlands, 1977.
15. Fahrig, R., "The Mammalian Spot Test (Fellfleckentest)
    with Mice", Arch. Toxicol. (1977), 38:87-98.
16. Cohen, M.M., and Hirschhorn, K., "Cytogenetic Studies
    in Animals", in: "Chemical Mutagens: Principles and
    Methods for Their Detection", vol.2, Hollaender, A.,
    ed., pp. 515-534, Plenum Press, New York, N.Y., 1971.
17. Slater, E.E., Anderson, M.D., and Rosenkranz, H.S.,
    "Rapid Detection of Mutagens and Carcinogens", Cancer
    Res. (1971), 31:970-973.
18. Kada, T., Tutikana, K., and Sadaie, Y., "In vitro and
    Host-Mediated 'Rec-assay' Procedures for Screening
    Chemical Mutagens and Phioxine, a Mutagenic Red Dye
    Detected", Mutat. Res. (1972), 16:165-174.
19. Brusick, D.J., and Mayer, V.W., "New Developments in
    Mutagenicity Screening Techniques Using Yeast", Environ.
    Health Perspec. (1973), 6:83-96.
20. Marguardt, H., "Mutation and Recombination Experiments
    with Yeast as Prescreening Tests for Carcinogenic
    Effects", Z. Krebsforsch (1974), 81:333-346.
21. Perry, P., and Evans, H.J., "Cytological Detection of
    Mutagen-Carcinogen Exposure by Sister Chromatid
    Exchange", Nature (London) (1975), 258:121-125.
22. San, R.H.C., and Stich, H.F., "DNA Repair Synthesis of
    Cultured Human Cells as a Rapid Bioassay for Chemical
    Carcinogens", Int. J. Cancer (1975), 16:284-291.
23. Stich, H.F., and Kieser, D., "Use of DNA Repair Synthesis
    in Detecting Organotropic Actions of Chemical
    Carcinogens" (38009), Proc. Soc. Exp. Biol. Med. (1974),
    145:1339-1342.
24. Sega, G., Owens, J., and Cumming, R., "Studies on DNA
    Repair in Early Spermatid Stages of Male Mice after in
    vivo Treatment with Methyl-, Ethyl-, Propyl-, and
    Isopropyl-Methane-Sulfonate", Mutat. Res. (1976),
    36:193-212.
25. Stetka, D.G., and Wolff, S., "Sister Chromatic Exchanges
    as an Assay for Genetic Damage Induced by Mutagen-
    Carcinogens, I, in vivo Test for Compounds Requiring
    Metabolic Activation", Mutat. Res. (1976), 41:333-342.

26. DiPaolo, J.A., Nelson, R.L., and Donovan, P.J., "In vitro Transformation of Syrian Hamster Embryo Cells by Diverse Chemical Carcinogens", Nature (London) (1972), 235:278-280.
27. Reznikoff, C.A., Brankow, D.W., and Heidelberger, C., "Establishment and Characterization of a Cloned Line of C3H Mouse Embryo Cells Sensitive to Postconfluence Inhibition of Division", Cancer Res. (1973), 33:3231-3238.
28. Reznikoff, C.A., Bertram, J.S., Brankow, G.W., and Heidelberger, C., "Quantitative and Qualitative Studies of Chemical Transformation of Cloned C3H Mouse Embryo Cells Sensitive to Postconfluence Inhibition of Cell Division", Cancer Res. (1973), 33:3239-3249.
29. Kakunaga, T., "Requirement for Cell Replication in the Fixation and Expression of the Transformed State in Mouse Cells Treated with 4-nitroquinoline-1-oxide", Int. J. Cancer (1974), 14:736-742.
30. Bartsch, H., "Predictive Value of Mutagenicity Tests in Chemical Carcinogenesis", Mutat. Res. (1976), 38:177-190.
31. de Serres, F.J., "The Utility of Short-term Tests for Mutagenicity", Mutat. Res. (1976), 38:1-2.
32. de Serres, F.J., "Prospects for a Revolution in the Methods of Toxicological Evaluation", Mutat. Res. (1976), 38:165-176.
33. McCann, J., and Ames, B.N., "Detection of Carcinogens as Mutagens in the Salmonella/Microsome Test: Assay of 300 Chemicals: Discussion", Proc. Natl. Acad. Sci. USA (1976), 73:950-954.
34. McCann, J., Choi, E., Yamasaki, E., and Ames, B.N., "Detection of Carcinogens as Mutagens in the Salmonella/ Microsome Test: Assay of 300 Chemicals", Proc. Natl. Acad. Sci. USA (1975), 72:5135-5139.
35. Meselson, M., and Russell, K., "Comparisons of Carcinogenic and Mutagenic Potency", in: "Origins of Human Cancer", Hiatt, H., Watson, J.D., and Winsten, J.A., eds., pp. 1473-1481, Cold Spring Harbor Laboratory, Cold Spring Harbor, N.Y., 1977.
36. Purchase, I.F.H., Longstaff, F., Ashby, J., Styles, J.A., Anderson, D., Lefevre, D.A., and Westwood, F.R., "Evaluation of Six Short-term Tests for Detecting Organic Chemical Carcinogens and Recommendations for Their Use", Nature (1976), 264:624-627.
37. Sugimura, T., Sato, S., Nagao, M., Yahagi, T., Matsushima, T., Seino, Y., Takeuchi, M., and Kawachi, T., "Overlapping of Carcinogens and Mutagens", in: "Fundamentals in Cancer Prevention", Magee, P.N., Takayama, S., Sugimura, T., and Matsushima, T., eds., pp. 191-213, University Park Press, Baltimore, Md., 1976.

38. Rosenkranz, H.S., Gutter, B., and Speck, W.T.,
     "Mutagenicity and DNA Modifying Activity: A Comparison
     of Two Microbial Assays", Mutat. Res. (1976), 41:61-70.
39. Evans, H.J., "Cytological Methods for Detecting Chemical
     Mutagens", in: "Chemical Mutagens: Principles and
     Methods for Their Detection", vol. I, Hollaender, A.,
     ed., pp. 1-29, Plenum Press, New York, N.Y., 1976.
40. Vogel, E., "Identification of Carcinogens by Mutagen
     Testing in Drosophila: The Relative Reliability for the
     Kinds of Genetic Damage Measured", in: "Origins of
     Human Cancer", vol. 4, Hiatt, H., Watson, J.D., and
     Winsten, J.A., eds., pp. 1483-1497, Cold Spring Harbor
     Laboratory, Cold Spring Harbor, N.Y., 1977.
41. Flamm, W.G., "A Tier System Approach to Mutagen
     Testing", Mutat. Res. (1974), 26:329-333.
42. Dean, B.J., "A Predictive Testing Scheme for
     Carcinogenicity and Mutagenicity of Industrial
     Chemicals", Mutat. Res. (1976), 41:83-88.
43. Mayer, V.W., and Flamm, W.G., "Legislative and
     Technical Aspects of Mutagenicity Testing", Mutat. Res.
     (1975), 29:295-300.
44. Sobels, F.H., "Some Problems Associated with the Testing
     for Environmental Mutagens and a Perspective for Studies
     in Comparative Mutagenesis", Mutat. Res. (1978),
     46:245:260.
45. Waters, M.D., and Epler, J.L., "Status of Bioscreening
     of Emissions and Effluents from Energy Technologies",
     in: "Energy/Environment III, Proceedings of the Third
     National Conference on the Interagency Energy/
     Environment Research and Development Program",
     Washington, D.C., June 1-2, 1978, National Technical
     Information Center, Springfield, Va., publication
     EPA-600/9-78-022, U.S. Environmental Protection Agency,
     Washington, D.C., 1978.
46. Waters, M.D., "Monitoring the Environment", in:
     "Toxicity Testing in vitro", Nardone, R.M., ed.,
     Academic Press, New York, N.Y., in press.
47. Waters, M.D., Simmons, V.F., Mitchell, A.D., Jorgenson,
     T.A., and Valencia, R., "An Overview of Short-term
     Tests for the Mutagenic and Carcinogenic Potential of
     Pesticides", J. Environ. Sci. and Health Part B Food
     Contaminants and Agricultural Wastes, in press.
48. Drake, J.W., "Environmental Mutagenic Hazards",
     publication prepared by Committee 17 of the
     Environmental Mutagen Society, Drake, J.W., Chairman,
     Science (1975), 187:503-514.
49. Flamm, W.G., "Approaches to Determining the Mutagenic
     Properties of Chemicals: Risk to Future Generations",
     J. Environ. Pathol. Toxicol. (1977), 1:301-352.

50. Simmon, V.F., "In vitro Microbiological Mutagenicity and Unscheduled DNA Synthesis Studies of Eighteen Pesticides", National Technical Information Center, Springfield, Va., publication EPA-600/1-79-049, Research Triangle Park, N.C., 1979.
51. Simmon, V.F., "In vivo and in vitro Mutagenicity Assays of Selected Pesticides", in: "Proceedings of the Conference: A Rational Evaluation of Pesticidal vs. Mutagenic/Carcinogenic Action", Bethesda, Md., September 15, 1976, DHEW Publication, (NIH) 78-1306, 1978.
52. Simmon, V.F., Mitchell, A.D., and Jorgenson, T.A., "Evaluation of Selected Pesticides as Chemical Mutagens, in vitro and in vivo Studies", National Technical Information Center, Springfield, Virginia, May 1977, publication EPA-600/1-77-028, Research Triangle Park, N.C., 1977.
53. Valencia, R., "Mutagenesis Screening of Pesticides Using Drosophila", (Final Report) WARF Institute, Inc., Madison, Wisc., U.S. Environmental Protection Agency Contract No. 68-01-2474, 1977.
54. Bridges, B.A., "Use of a Three-Tier Protocol for Evaluation of Long-term Toxic Hazards Particularly Mutagenicity and Carcinogenicity", in: "Screening Tests in Chemical Carcinogenesis", Montesano, R., Bartsch, H., and Tomatis, L., eds., World Health Organization, International Agency for Research on Cancer, Lyon, France, WHO/IARC Publ. No. 12, pp. 549-568, 1976.
55. Green, S., "Present and Future Uses of Mutagenicity Tests for Assessment of the Safety of Food Additives", J. Environ. Pathol. Toxicol. (1977), 1:49-54.
56. U.S. Environmental Protection Agency, "Mutagenicity Testing Requirements Section of the FIFRA Registration Guidelines for Hazard Evaluation of Humans and Domestic Animals", (Draft) Office of Pesticides Programs, U.S. Environmental Protection Agency, July 12, 1977.
57. National Academy of Sciences, "Principles and Procedures for Evaluating the Toxicity of Household Substances", publication prepared for the Consumer Product Safety Commission by the Committee for the Revision of NAS Publication 1138, pp. 86-98, National Academy of Sciences, Washington, D.C., 1977.
58. Nesnow, S., unpublished results.
59. U.S. Department of Health, Education, and Welfare, "Bioassay of Azinphos-methyl for Possible Carcinogenicity", National Cancer Institute, Bethesda, Md., NCI-CG-TR-69, vol. 69, 1978.

60.   U.S. Department of Health, Education, and Welfare,
      "Bioassay of Captan for Possible Carcinogenicity",
      National Cancer Institute, Bethesda, Md.,
      NCI-CG-TR-15, vol. 15, 1977.
61.   U.S. Department of Health, Education, and Welfare,
      "Bioassay of Diazinon for Possible Carcinogenicity",
      National Cancer Institute, Bethesda, Maryland,
      NCI-CG-TR-137, vol. 137, 1979.
62.   International Agency for Research on Cancer, 2,4-D
      and esters, in: "IARC Monographs on the Evaluation
      of the Carcinogenic Risk of Chemicals to Man:
      Some Fumigants, the Herbicides 2,4-D and 2, 4,5-T,
      Chlorinated Dibenzodioxins and Miscellaneous
      Industrial Chemicals", IARC, Lyon, France, vol. 15,
      p. 111, 1977.
63.   International Agency for Research on Cancer, Endrin,
      in: "IARC Monographs on the Evaluation of the
      Carcinogenic Risk of Chemicals to Man: Some
      Organochlorine Pesticides", IARC, Lyon, France, vol. 5,
      p. 157, 1974.
64.   U.S. Department of Health, Education, and Welfare,
      "Bioassay of Fenthion for Possible Carcinogenicity",
      National Cancer Institute, Bethesda, Md.,
      NCI-CG-TR-103, vol. 103, 1979.
65.   U.S. Department of Health, Education, and Welfare,
      "Bioassay of Malathion for Possible Carcinogenicity",
      National Cancer Institute, Bethesda, Md.,
      NCI-CG-TR-24, vol. 24, 1979.
66.   U.S. Department of Health, Education, and Welfare,
      "Bioassay of Methoxychlor for Possible Carcinogenicity",
      National Cancer Institute, Bethesda, Md.,
      NCI-CG-TR-35, vol.35, 1979.
67.   International Agency for Research on Cancer, Monuron,
      in: "IARC Monographs on the Evaluation of the
      Carcinogenic Risk of Chemicals to Man: Some
      Carbamates, Thiocarbamates, and Carbazides", IARC,
      Lyon, France, vol. 12, p. 167, 1976.
68.   U.S. Department of Health, Education, and Welfare,
      "Bioassay of Parathion for Possible Carcinogenicity",
      National Cancer Institute, Bethesda, Md.,
      NCI-CG-TR-70, vol. 70, 1979.
69.   U.S. Department of health, Education, and Welfare,
      "Bioassay of Methyl Parathion for Possible
      Carcinogenicity", National Cancer Institute, Bethesda,
      Md., NCI-CG-TR-157, vol. 157, 1979.

70. U.S. Department of Health, Education, and Welfare, "Bioassay of Pentachloronitrobenzene for Possible Carcinogenicity", National Cancer Institute, Bethesda, Md., NCI-CG-TR-61, vol. 61, 1978.
71. U.S. Department of Health, Education, and Welfare, "Bioassay of Trifluralin for Possible Carcinogenicity", National Cancer Institute, Bethesda, Md., NCI-CG-TR-34, vol. 34, 1978.

RECEIVED February 19, 1981.

Reproductive and Teratogenic Effects: No More Thalidomides?

ROCHELLE WOLKOWSKI–TYL

Chemical Industry Institute of Toxicology, P.O. Box 12137, Research Triangle Park, NC 27709

The term teratology was first coined by the father and son team Etienne and Isidore Geoffrey Saint-Hilaire for their investigations of malformations, or monsters (teras, plural terata, from the Greek monster), mostly in the chick embryo (books published 1822 and 1832). However, human concern with congenital malformations is as ancient as human awareness, mentioned in Biblical references and discussed by Aristotle (cited in 136). Earliest views were that embryos and fetuses were affected structurally by maternal experience during pregnancy; the concept that "maternal impressions" directly affected the unborn gained widespread credence for centuries. With the advent of the enlightened scientific atmosphere in western Europe in the nineteenth century, attitudes shifted to the opinion that the embryo and fetus were inviolate in the uterus, untouchable by the environment. The presentation and subsequent appreciation of Mendel's Laws of genetics provided the apparent explanation for observed abnormal births: all flaws arose from genetic mishaps during gametogenesis and the zygote developed based solely on the incoming genetic information.

The twentieth century brought with it the first experimental evidence for the role of the environment in production of abnormal offspring. Early experiments on pregnant mammals involved studies with ionizing radiation (43, 66) and sex hormones (33, 34, 69). Studies with dietary deficiencies, drugs and chemicals followed rapidly. In 1933, Hale reared pregnant pigs on a vitamin A deficient diet and produced anophthalmic piglets (37, 38). Data were also presented in 1948 for the effects of trypan blue and nitrogen mustard on developing rat embryos (28. 44). The supposed safety of the human fetus in utero was directly challenged by Gregg in 1941 (35) who reported that a German measles epidemic in Australia resulted in offspring with cataracts, deafness and congenital heart disease. These results were confirmed with the birth of almost 20,000 defective children following a rubella epidemic in

the United States in 1964 (121). The thalidomide disaster occurring worldwide in 1955-1965, ultimately involving over 8,000
children in 28 countries, was first reported by Lenz (67, 68) and
McBride (75). These events triggered awareness of the vulnerability of the intrauterine occupant to outside influences.
However, anecdotal evidence for reproductive effects of substances
such as lead or mercury on women in industrial exposures has been
accumulating for centuries.

The current view is that embryos and fetuses may be especially vulnerable to environmental insult because of qualitative
and/or quantitative differences from adults. These factors
include:

1. Small cell number
2. Rapid rates of proliferation
3. High proportion of undifferentiated cells
4. Requirement for precise temporal and spatial sequencing
   of specific cells and cell products at the appropriate
   place and time for normal differentiation, including
   programmed cell death
5. Unique metabolism: presence or absence of inducible
   and/or constitutive repair enzymes, activating and
   detoxifying enzymes, eg. DNA repair enzymes, the mixed
   function oxidases, etc.
6. Unique tissue sensitivities which may be transient
7. Immaturity of immunosurveillance mechanisms, of special
   concern for the induction of transplacental carcinogenesis

There is also the awareness that sensitivity to environmental
insult, and subsequent expression of that insult, does not cease
with birth. The mammal at term is not a miniature adult; a partial list of systems still undergoing differentiation include:
the nervous, endocrine, urogenital, digestive and immune systems.
Expression of an insult incurred in utero may not develop until
after birth, in the human up to ten years of age for most detected
anomalies, but with a latency of 15-30 years for carcinogenic
events.

A current working definition of teratology, taking into
account the above considerations and concerns has been generated
by Wilson (136): Teratology is the study of adverse effects of
the environment on developing systems; that is, on germ cells,
embryos, fetuses and immature postnatal individuals. More com-
prehensively, it deals with the causes, mechanisms and manifestations of developmental deviations of either structural or functional nature. Agents which alter the rate of growth of the fetus
or are lethal to the fetus without producing specific anatomic or
functional anomalies are thought by some to be better termed
developmental toxins than teratogens (36).

Increasing concern is being raised as to reproductive and
teratogenic risks for a number of reasons: the increase of women
especially of childbearing age in the workforce in non-traditional

jobs, the increasingly rapid introduction of new chemicals (1500-2000 new chemicals synthesized or otherwise produced each year), and the awareness that relatively little is known concerning the reproductive and teratological risk involved with exposure to chemicals already in the workplace; TSCA lists over 60,000 chemicals in current usage. There is also increasing evidence from both human and laboratory animal data that the male may mediate teratogenic effects on the developing fetus.

Categories of Teratogenic Agents. Many substances are known to be teratogenic in one or more species of mammals (Table I). The emphasis has been primarily on drugs, with data generated by drug research companies adhering to FDA Guidelines (31) for reproductive testing of drugs, and the awareness that in our drug-permissive society women consume an average of four drugs, both by prescription and over-the-counter administration, during pregnancy (76, 84, 88). Schardein (98) has listed over 1200 drugs evaluated as teratogens; Shepard's catalog (102) lists 600 teratological agents, only 20 of which are documented as human teratogens. Meyers and Meyers (77) list 527 teratogenic substances but their list is based on human or animal data.

Human teratogenic agents have been discovered initially from anecdotal observations, and then more rigorously examined in epidemiological studies. Suspect human teratogens have been defined one of two ways: clinicians use anecdotal data, animal model researchers have suggested that any agent positive in two or more mammalian species must be considered a suspect human teratogen. Some examples of agents in both categories are presented in Table II.

According to the National Foundation (3) about 7% of all liveborn humans will have birth defects. This value may be as high as 10% if children are evaluated to age 10 years to include subtle structural, functional deficits such as minimal brain dysfunction. More than 560,000 lives out of approximately three million births per year in the United States are lost through infant death, spontaneous abortion, stillbirths and miscarriage due presumably to defective fetal development. The relative contributions to human teratogenesis have been estimated by Wilson (136, 140) as follows: known germinal mutations: 20%; chromosomal and gene aberrations: 3-5%; environmental causes such as radiation: < 1%; infections: 2-3%, maternal metabolic imbalance: 1-2%; drugs and environmental chemicals: 4-5%; contributions from maternal dietary deficiencies or excesses and combinations or interactions of drugs and environmental chemicals are unknown. Wilson (136, 140) estimates the contribution from unknown sources as 65-70%. The estimated 20-25% pregnancy loss due to chromosomal aberrations may be even higher due to early losses diagnosed as late menstrual bleeding. Recovered tissues from spontaneous abortions prior to the thirteenth week of gestation exhibit chromosomal anomalies on the order of 560 per 1000 abortions; the value

TABLE I. SOME TYPES OF DRUGS AND ENVIRONMENTAL CHEMICALS THAT
HAVE BEEN SHOWN TO BE TERATOGENIC IN ONE OR MORE
SPECIES OF MAMMALS[a] (137)

Salicylates (e.g., aspirin, oil of wintergreen)
Certain alkaloids (e.g., caffeine, nicotine, colchicine)
Tranquilizers (e.g., meprobamate, chlorpromazine, reserpine, diazepam)
Antihistamines (e.g., buclizine, meclizine, cyclizine)
Antibiotics (e.g., chloramphenacol, streptonigrin, penicillin)
Hypoglycemics (e.g., carbutamide, tolbutamide, hypoglycins)
Cortoids (e.g., triamcinolone, cortisone)
Alkylating agents (e.g., busulfan, chlorambucil, cyclophosphamide, TEM)
Antimalarials (e.g., chloroquine, quinacrine, pyrimethamine)
Anesthetics (e.g., halothane, urethan, nitrous oxide, pentobarbital)
Antimetabolites (e.g., folic acid, purine and pyrimidine analogs)
Solvents (e.g., benzene, dimethylsulfoxide, propylene glycol)
Pesticides (e.g., aldrin, malathion, carbaryl, 2,4,5-T, captan, folpet)
Industrial effluents (e.g., some compounds of Hg, Pb, As, Li, Cd)
Plants (e.g., locoweed, lupins, jimsonweed, sweet peas, tobacco stalks)
Miscellaneous (e.g., trypan blue, triparanol, diamox, etc.)

[a]Teratogenic effects were usually seen only at doses well above therapeutic levels
for the drugs, or above likely exposure levels for the environmental chemicals.

Teratology

TABLE II.   HUMAN TERATOGENS

A.   Known:

    alcohol
    antibiotics (tetracycline, sulfonamides, chloramphenicol)
    anticonvulsants (diphenylhydantoin / barbiturates)
    folate antagonists (aminopterin, methotrexate)
    lead
    methylmercury (Minimata disease)
    smoking
    steroid hormones (oral progestins, androgens, estrogens)
    Thalidomide
    Vitamin D (excess)

B.   Suspect

    amphetamines
    anticonvulsants (paramethadione, trimethadione)
    antihistamines
    antimalarials (quinine, chloroquine)
    antithyroid drugs, iodides and iodine lack (temporary?)
    aspirin
    barbiturates
    blighted potatoes (solanine)
    folate deficiency
    hormonal pregnancy tests and contraceptives
    hypoglycemic agents (oral)
    lysergic acid diethylamide (LSD)?
    operating room environment – probable
    organic solvents
    pesticides, fungicides, herbicides
    Polychlorinated biphenyls (PCBs) (Yusho disease)
    Warfarin (anticoagulant) – probable
    Vitamin D (deficiency)

Taken from references 25, 70, 76, 83,
105 and 137

at term is 5 per 1000. Of the children born live who subsequently die in the first year of life, approximately 20% of the deaths are associated with or caused by birth defects, more than any other single factor (136).

There is one final, almost plaintive maxim sometimes termed Karnofsky's law (87) that almost any substance may be teratogenic if given in appropriate dose regimens to a genetically susceptible organism at susceptible stage or stages of embryonic or fetal development.

Determinants of Teratogenic Susceptibility. Factors which influence the teratogenic response are listed in Table III. Genetic susceptibility varies among species, for example: aspirin is teratogenic in rodents but not in primates, imipramine is teratogenic in rabbits, but not in humans, thalidomide is teratogenic in primates but not in rodents. Differences also exist among strains. Inbred mouse strains differ radically in their response to many teratogenic agents, for example to cortisone-induction of cleft palate (54) and cadmium-induced testicular and embryotoxicity (144, 145). Individuals also vary in their response to teratogenic agents in outbred strains and heterogeneous human populations. The current interpretation is that teratogens act on a susceptible genetic locus or loci which may control disposition of the agent including absorption, metabolism, transport or excretion and/or direct susceptibility of the target tissue or organ. The teratogen therefore increases the incidence of previously existing malformations; its action must be viewed against the "background noise" of spontaneous malformation rates, which also vary among species, strains and individuals. The phocomelic syndrome, induced by thalidomide, occurs at a low rate spontaneously in human populations; approximately 20-80% of the human fetuses exposed, presumably to the appropriate dose at the appropriate time, developed the malformations (20). This concept of environmentally induced imitations of genetic anomalies was presented first by Landauer (63) with experimental evidence of "phenocopies".

There is some specificity of agent on the teratological response (Table IV) with acetazolamide causing perhaps the most specific lesion (74). However, there are almost always effects on other systems derived in many cases from different primary embryonic germ layers. The gestational stage of the embryo or fetus at the time of environmental insult appears to be the most critical determining factor. Figure 1 examines time periods of embryonic and fetal development in humans, mice, and rats. The predifferentiation period, from fertilization to establishment of the three primary embryonic germ layers, is considered refractory to teratogenic agents (although there are some exceptions such as hypoxia, 125 hypothermia, 104; and actinomycin D, 135). This resistance has been explained as due to the small, omnipotent cell population of the pre- and immediately post-implantation embryo. Cell damage

TABLE III. DETERMINANTS OF TERATOGENIC SUSCEPTIBILITY

1. Genotype: species, strain, individual
2. Specificity of agent
3. Gestational stage at exposure
4. Dose
5. Route of administration: inhalation, percutaneous absorption, po, iv, ip, sc, gavage
6. Duration of administration: short-term, chronic, intermittent
7. Disposition of agent (maternal, placental and fetal pharmacokinetics)
   a. Absorption
   b. Equilibrium
      (1) maternal compartments: blood, organs
      (2) Placental (maternal/placental, placental/fetal)
      (3) Fetal and fetal compartments
   c. Metabolism: activation/inactivation
   d. Transport, especially transplacental
   e. Excretion
8. Animal status
   a. Age
   b. Health
   c. Interactions: synergisms, antagonisms, protections

TABLE IV. SPECIFICITY OF AGENT ON TERATOGENIC RESPONSE

Agent	Characteristic Anomaly	Other Effects	References
Excess vitamin A (d 9–16 gestation, rat)	central nervous system (brain, eye, skull)	cleft palate, syndactyly, genitourinary defects	15
Thalidomide (d 34–50 postmenstruation, human)	musculoskeletal system Phocomelia (d 39–45) face	anorectal stenosis (d. 49–51)	67, 68, 75
Androgenic hormones	masculinization of female fetuses		98
Alcoholism–human (poor nutrition? contaminants?)	"Fetal Alcohol Syndrome" small size, shortened palpebral fissures, stub nose, hirsutism	cleft palate, cardiac anomalies	98
Aspirin (salicylic acid) (d 9–11, rats)	brain (exencephaly, hydrocephaly) facial clefts, eye defects, vertebrae and ribs, spina bifida	heart defects	120
Acetazolamide (and other carbonic anhy- drase inhibitors)	post axial defects in right forepaw— fourth and fifth digits and corres— ponding metacarpals		74
Cortisone (d 11–14 in mouse, 23 agents last count)	cleft palate		27
Trypan blue, (d 6–9, rat)	hydrocephalus, spina bifida, ear, eye, cardiovascular defects	cleft palate, skull, tail defects, umbilical hernia	28
Myleran (d 12, rat)	syndactylous fore (86%) and rear (80%) paws, short kinky tail	cleft palate	81
TEM (d 12, rat)	syndactylous fore (78%) and rear (18%) paws	encephalocele, edema	81

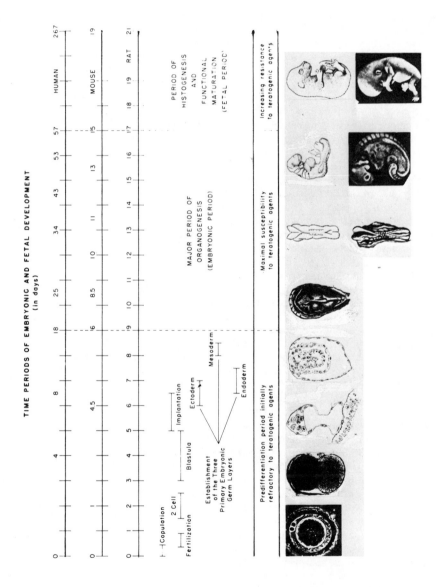

Figure 1. Time periods of embryonic and fetal development in the human, mouse, and rat

or death is either corrected for by the surviving cells, which
regulate to produce a normal, albeit small term fetus, or the cell
loss is so devastating that the embryo dies. Once implantation
and establishment of the primary germ layers have occurred, the
major period of organogenesis begins, a period of 8-9 days in
rodents and approximately 40 days in humans. This is the period
of maximal susceptibility to teratogenic agents causing structural
anomalies. Using data generated from studies with actinomycin D
and other chemicals, Wilson (133, 134, 135, 136) has described the
differential susceptibilities of embryonic organ systems to teratogenic agents during organogenesis in the rat (Table V and VI,
Figure 2). In the data from Table VI, fatalities parallel malformation rate. An increase in deaths may also obscure the detection
of the abnormality generated which caused the fatality, so that
the relationship of deaths to anomalous fetuses becomes inverse.
Administration of a lower dose of the test agent may be useful to
detect the anomalies responsible for the fetal wastage. From
Figure 2, it is apparent that administration of an agent on gestation day 10 would affect eye, brain, heart and anterior axial
skeletal development. The same agent, administered on day 15
would affect palate, urogenital and posterior axial skeletal
development. These times of specific sensitivity need not correspond to the morphological appearance of the organ or organ
system, but to the time of cell biochemical commitment: the shift
of cells from presumptive to determined status.

Once histogenesis has begun: the differentiation of tissue-
specific biochemical and morphological characteristics, the con-
ceptus is termed a fetus and is viewed as increasingly refractory
to teratogenic agents. However, this is true only of most morphological or structural manifestations. Increasing evidence indicates susceptibility of the fetus to agents causing functional
deficits which presumably have a biochemical or micro-structural
basis. Those systems not yet complete, especially the nervous
system, are most vulnerable. For example: Vitamin A (118), lead
(85), methyl mercury (109, 110) and methyl azoxymethanol (53, 103)
all cause neurofunctional lesions when administered during this
period. In addition, transplacental carcinogens, such as diethylstilbesterol, ethyl or methylnitrosourea, 7,12-dimethylbenzanthra-
cene and nitrosomethylurethan, act during this period in humans,
rodents and rabbits (91). The lesion is expressed as a system
specific tumor after a long latency in the postnatal mature animal
but the only exposure and therefore the initiation of the later
carcinogenic event occurs in utero.

The route and duration of administration of the agent is also
critical for the development of the teratogenic anomaly. Human
industrial exposure is almost always by inhalation or percutaneous
absorption of fumes, aerosols or vapors. Consumer or other secondary exposure would be by more varied routes. Experimental teratology endeavors to duplicate the human route of exposure for
experimental animal models. Inhalation presents problems of

TABLE V
SOME RAT TERATOGENS THAT HAVE LITTLE EMBRYOTOXIC EFFECT
ON THE SIXTH DAY OF GESTATION BUT ARE HIGHLY EFFECTIVE
3 OR 4 DAYS LATER (136)

Agent	Treatment			20-day fetuses	
	Dose (mg/kg)	Day	Total implants	% dead resorbed	% survivors malformed
5-Fluorodeoxy-uridine	20	6	114	6	1
		9	209	10	38
Retinoic acid	20	6	95	5	0
		9	79	44	84
Actinomycin D	0.3	6	207	7	5
		10	88	48	65
Controls	(vehicle)		558	7	1

Academic Press, Inc.

TABLE VI
RELATIONSHIP BETWEEN DEATH AND MALFORMATION FOR
ACTINOMYCIN D IN THE RAT

Dose, mg/kg	Day treated	Fetuses	
		% dead	% malformed
0.2	7	11.5	1.9
	8	4.2	16.0
	9	32.5	28.1
	10	12.3	4.4
	11	7.7	0
0.3	6	10.3	2.8
	7	13.0	11.2
	8	84.8	26.6
	9	99.2	100.0
	10	57.9	65.2
	11	12.1	0.9

Harper Hospital, Bulletin

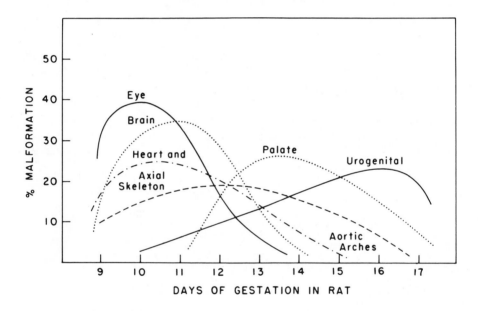

Figure 2. Differential sensitivities of embryonic organ systems to teratogens (136)

concentration monitoring and dose assessment for the pregnant female since pulmonary parameters such as ventilation, tidal volume, expiration reserve volume, respiratory tract capillary dilation and anatomic considerations change radically during pregnancy (122). The effect of stress due to the inhalation exposure in confined, wire-rack cages is also a confounding factor. However, classical teratological evaluations are being done on chemicals exposed by inhalation, most notably by the group at Dow Chemical Company (99, 100, 101). Exposure by skin absorption is difficult to quantify and requires a complex application regimen. In addition, opinion is mixed as to whether route of entry of the agent makes a difference in the ultimate distribution and metabolism of the agent under examination. First pass organ absorption and metabolism may differ if the exposure is by inhalation to the lung, or orally to digestive system and liver, although subsequent transport and organ exposure may yield equivalent metabolite patterns. Most teratology studies usually employ administration of the test compound in the feed, by oral intubation or injection into the dam.

Timing is important. Experimental exposure before implantation or during early organogenesis may result in interference with implantation or in early embryonic death, resulting in no term fetuses. Exposure before peak susceptibility or repeated exposure may induce activating and/or detoxifying enzymes in dam, placenta and/or fetus. This may result in increased or decreased blood levels of the active metabolite in the dam, and therefore altered exposure to the fetus. Conversely, these enzymes may be inhibited by accumulation of metabolite(s) again altering blood levels of parent compound and metabolite(s). Other effects of repeated or early exposure may be to alter liver or kidney function, for example, as well as to induce pathological changes in these organs which will affect quantity and quality of compound reaching the fetus. Saturation of protein-binding sites may also occur in the dam to alter transport. All of these effects may alter the disposition parameters listed in Table III and obscure or change any teratological effects of the agent being examined (136).

Dose range and schedule are also critical. Three to four dose levels are usually employed: high dose: toxic to the maternal organism, perhaps lethal to 10-15% of dams, essentially to obtain an effect, and to establish target organ(s); mid dose(s): embryotoxic or embryolethal and a slightly lower dose to obtain teratogenic level with overlap between these two dose levels; and low dose: comparable on a body weight basis to possible human exposure levels or small multiples thereof.

Teratological Testing. Following the reports of the effects of thalidomide on fetuses exposed during the first trimester which appeared in 1961-1965, (67, 68, 75), the United States Food and Drug Administration (FDA) established Guidelines for Reproductive Studies for Safety Evaluation of Drugs for Human Use (31). These

are presented diagrammatically in Figure 3. These guidelines were
promulgated "as a routine screen for the appraisal of safety of
new drugs for use during pregnancy and in women of childbearing
potential." (31). Phase I designated Study of Fertility and
General Reproductive Performance involves exposure of the males
for 10 weeks prior to mating to include exposure during all phases
of spermatogenesis, estimated as 8 weeks duration in rodents, and
of the females for 2 weeks to include oogenesis, a 5 day cycle in
rodents. Exposure is continued in the females through pregnancy,
parturition and lactation. One-half the dams are sacrificed on
gestation day 13 for examination of number and distribution of
embryos in uterine horns, empty implantation sites and resorptions. The dams allowed to litter are examined for litter size,
stillborn and live births. Dead pups are examined for skeletal
anomalies. Live pups are examined for gross anomalies and individually weighed at delivery, postnatal day 4 and 21. Phase II,
entitled Teratological Study, involves treatment during organogenesis gestational days 6-15 (mouse) or 7-16 (rat). Since evaluation in two species, one other than rodent, is called for,
parameters for the rabbit are also indicated. Dams are sacrificed
1-2 days before the anticipated date of parturition and fetuses
are delivered by cesarean section. Data to be collected include
number of ovarian corpora lutea, live and dead fetuses, and early
and late resorptions. Live fetuses are to be weighed and examined
for external malformations. In rats, one-third of each litter
will be examined for soft tissue deficits by dissection or the
Wilson technique (133, 136), two-thirds preserved and stained for
examination for skeletal anomalies. Rabbit fetuses are to be
incubated for 24 hours to assess viability, then all fetuses are
examined for external, visceral and skeletal anomalies.

    Phase III, entitled Perinatal and Postnatal Study, involves
exposure of the dam during the final one-third of gestation and
continuing through parturition, and lactation to weaning. This
segment "should delineate effects of the drug on late fetal
development, labor and delivery, lactation, neonatal viability,
and growth of the newborn." (31). Cross-fostering is suggested
for this phase if survival of test-pups is impaired. Rearing of
pups from this phase and phase I to evaluate reproductive and
fertility performance in these F_1 animals is also suggested as a
possibility.

    These guidelines have survived essentially intact and are now
incorporated into proposed U. S. Environmental Protection Agency
(EPA) guidelines (22) as well as recent Interagency Regulatory
Liaison Group (IRLG) draft guidelines (79). FDA has further
proposed a three generation reproductive study (24) currently in
use to evaluate long-term effects on reproduction and fertility
including effects on the germinal cell line developing in utero
during exposure to the test compound in the P_o generation and
subsequent generations under continuous exposure to the test
substances (Figure 4).

(FDA, 1966)

Phase I: Study of Fertility and General Reproduction Performance

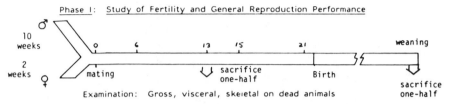

Examination: Gross, visceral, skeletal on dead animals

Information on: Breeding, fertility, nidation, parturition, lactation,
neonatal effects

Phase II: Teratological Study

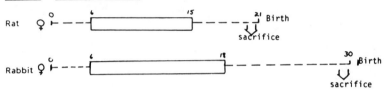

Examination: Gross, one-third visceral (Wilson sections)
two-thirds skeletal (alizarin staining)
(rabbit, 24 hr incubation, all fetuses gross, visceral
and skeletal)

Information on: Embryotoxicity, teratogenicity

Phase III: Perinatal and Postnatal Study

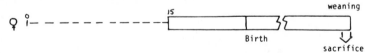

Examination: litter size, pup weight, etc., possible cross-fostering

Information on: Parturition, lactation, neonatal effects

*Figure 3.   Guidelines for reproductive studies for safety evaluation of drugs for
human use (FDA, 1966)*

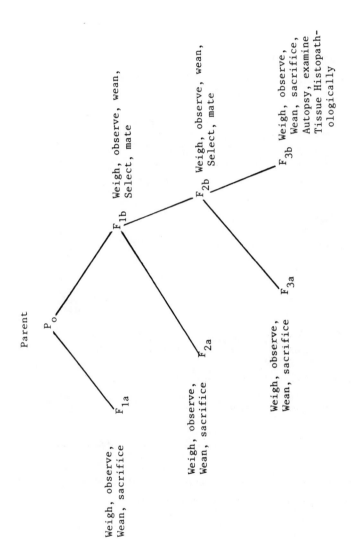

Figure 4. *Three-generation reproduction study (FDA, 1970)*

The FDA Guidelines (24, 31) have been the protocol used in almost all drug and chemical tests done in industry. There has been growing concern that these tests are inadequate to identify subtle morphological and functional deficits expressed pre- or postnatally. Research to improve the established tests and to develop new tests and approaches is gaining momentum.

The term sacrifice, the basis for the phase II Teratology Study test involves counting implantations, resorptions, dead and live fetuses. Live fetuses are then examined for soft tissue and skeletal anomalies. Detection of early implantation sites, not discernible by placental remains, may be visualized by staining fresh uteri with ammonium sulfide (61). Examination for live fetus soft tissue anomalies by the Wilson technique (133, 136) involves fixation of fetuses in Bouin's solution for decalcification, subsequent free hand sections through the head and 1 mm free hand razor cross sections of the trunk. Disadvantages include: difficulties in recognizing cardiac malformations in serial cross-sections, difficulties in duplicating the sections from fetus to fetus and litter to litter, and the inability to examine skeleton of same fetus. A number of modifications have been suggested (6, 23, 113). The Staples technique entails microdissection of decapitated fetuses immediately after cesarean section at term. Advantages include ease of examination for functional heart anomalies such as septal wall defects with no distortion due to fixation, with the eviscerated carcass available for skeletal examination, and the head preserved in Bouin's for later sectioning by the Wilson technique. A modification of Staples' technique has also been suggested (26). Examples of anomalies demonstrable by the Wilson technique in the fetal head region are presented in Figure 5 (trypan blue: unilateral anophthalmia, hydroxyurea: cleft palate).

The skeletal examination involves preservation of the eviscerated fetus in 70–95% ethanol, maceration and clearing in potassium hydroxide and staining with Alizarin Red S, specific for calcium and therefore bone (18, 19). Many versions exist (46, 113). This procedure stains areas of ossification but the researcher cannot distinguish between sites that would have ossified if the fetus had continued development, hence delayed ossification, from totally absent cartilaginous anlagen, hence missing bone. Counterstaining with a cartilage-specific stain such as alcian blue (49) allows distinction between delayed and absent ossification. Skeletal anomalies demonstrable by the classic Alizarin technique are presented in Figure 6 (hydroxyurea: fused ribs, doubled centra).

The basic techniques enumerated above will detect missing, ectopic or grossly abnormal organs, malposition of or missing major blood vessels, retarded, abnormal or absent ossification. Embedding, sectioning and staining each fetus would allow detection of microscopic lesions but this would require an extended period of time and many person-hours and so is not applicable to

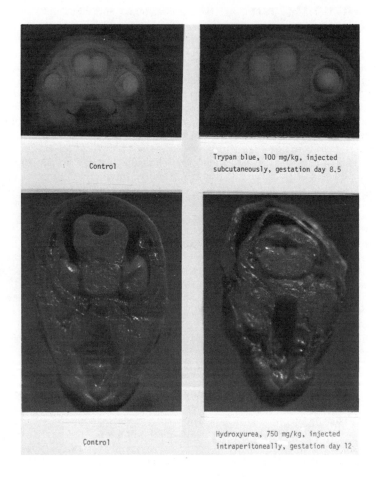

Figure 5.   Representative soft tissue anomalies detected in Fischer-344 rat fetuses, gestation day 20, by Wilson sections

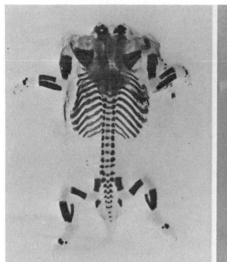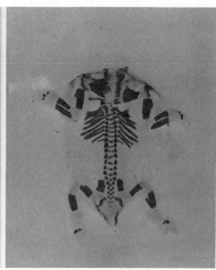

Figure 6.   Representative skeletal anomalies detected in Fischer-344 rat fetuses, gestation day 20, by Alizarin Red S staining: control (left); hydroxyurea, 200 mg/ kg/d administered by gavage on gestation days 7–20 (right).

rapid screening techniques. In addition, functional deficits in
terms of biochemical alterations and effects demonstrable only in
the postnatal period would not be detected. Detection of these
lesions would be very important for human risk assessment. In
addition, the animal model utilized, usually rat, mouse and rabbit
may not be the most suitable test system for particular agents if
metabolism of these substances differs between the test animal and
the human. Differences in metabolism may render results in a test
animal system misleading or irrelevant to evaluation of drug or
chemical risk in humans.

At the Chemical Industry Institute of Toxicology, (CIIT), a
chemical is first examined to evaluate parameters of toxicokinetics, disposition and metabolism in the pregnant rat and feto-
placental unit including evaluation of placental transport of the
parent compound and/or identified metabolites to characterize the
system prior to any teratological testing. Once the characteristics and limits of the test system are defined, teratological
studies or evaluation of reproductive performance are then performed. Whole body autoradiography (WBAR) of the pregnant animal
after exposure to a radiolabelled test chemical is valuable to
assess disposition and target organ specificity with minimum of
person-hours expended. This technique compares favorably with
classical disposition studies done by radioisotopic analysis on
dissected maternal and fetal organs (50, 51). Figure 7 presents
WBAR results for three CIIT priority chemicals illustrating distribution in the dam and fetuses.

During term sacrifice, relative fetal organ weights may be
determined as part of the soft tissue examination at sacrifice
(112). Data generated on a CIIT priority chemical shown to be
toxic to the adult spleen are shown in Table VII. Hematological
parameters may also be evaluated such as complete blood count
using an automated counting system, and examination of blood
smears for evaluation of nucleated red blood cells and reticulo-
cytes. These latter cell types are very numerous in the fetus and
neonate. Touch preparations may also be generated from cut surfaces of fresh maternal and fetal organs as a rapid alternative to
fixing, embedding and sectioning these tissues, to evaluate cell
integrity, differentiation and function.

In vitro test systems are being considered for teratogenicity
screening (7, 59). These systems include unicellular organisms,
somatic cell tissue or organ culture, and culturing of intact
invertebrate, lower vertebrate, mammalian and avian embryos.
Systems using mammalian embryos include culture of pre-implanta-
tion or post-implantation embryos and specific organ cultures such
as palate or limb bud (73). Preimplantation embryos from mouse
(126) or man (114) have been grown successfully in culture up to
the blastocyst stage. Using techniques developed by D.A.T. New
and co-workers (82), postimplantation embryos from rat (9) or mouse
(96) have been cultured in vitro for up to four to five days, with
best results obtained from young postimplantation embryos at

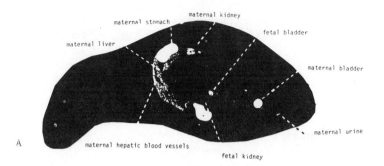

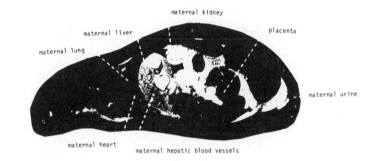

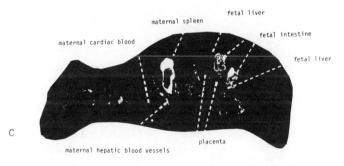

50 micron sections

Figure 7.   WBARs of pregnant Fischer-344 rats, gestation day 20–21 exposed to various ^{14}C priority chemicals. (A) WBAR of F-344 dam (gestation day 21) given ^{14}C-terephthalic acid by gavage; dose = 12.5 mg/kg (tracer dose), 30 μCi/dose, sacrificed 5.5 h after dose. (B) WBAR of F-344 dam (gestation day 20) given ^{14}C-2,4-dinitrotoluene by gavage; dose = 35 mg/kg, 100 μCi/dose, sacrificed 6 h after dose. (C) WBAR of F-344 dam (gestation day 21) given ^{14}C-aniline HCl by gavage for 5 d (gestation day 17–21); dose = 100 mg/kg/d, 4 μCi/dose, sacrificed 6 h after last dose.

TABLE VII. DATA FROM PREGNANT F-344 FEMALES EXPOSED TO
^{14}C-ANILINE-HCL (100 MG/KG) FOR 1 OR 5 DAYS

A. DAM PARAMETERS	1 day (n = 14 dams)[1]	5 day (n = 11 dams)[1]
Hematocrit ± SE	36.5 ± 0.4	32.6 ± 0.7* (10.68%↓)
Body weight ± SE	214.02 ± 3.47	215.65 ± 2.83
Total implantations/dam ± SE	7.9 ± 0.6	8.0 ± 0.8
Live fetuses/dam ± SE	7.6 ± 0.6	7.5 ± 0.7
% Organ/BW ratios ± SE		
Liver	3.541 ± 0.074	3.348 ± 0.075
Kidneys (2)	0.585 ± 0.010	0.570 ± 0.011
Spleen	0.171 ± 0.006	0.247 ± 0.010* (44.44%↑)
B. FETAL PARAMETERS		
Hematocrit ± SE[2]	30.2 ± 0.6	29.2 ± 0.6
Placenta weight/fetus ± SE[3]	0.415 ± 0.009	0.430 ± 0.033
Body weight/live fetus ± SE[4]	3.485 ± 0.249	4.157 ± 0.111 (19.28%↑)
% Organ/BW ratios ± SE[5]		
Liver	6.665 ± 0.282	6.301 ± 0.641
Kidneys (2)	0.698 ± 0.033	0.722 ± 0.022
Bladder	0.245 ± 0.014	0.260 ± 0.018
Spleen	0.109 ± 0.014	0.148 ± 0.009* (35.78%↑)

[1]Dams sacrificed 1-12 hours after (last) gavage and data pooled

[2]Hematocrits done on blood pooled from fetuses of one uterine horn.

[3]Placenta weight/fetus is determined by dividing total placental weight/litter by number of fetuses/litter.

[4]Body weight/fetus is determined by dividing total weight of entire litter by number of fetuses in the litter.

[5]Organs were pooled from each litter to obtain enough tissue for weight and radioisotope determinations. Hence the organ/BW ratios represent litter total weight of each organ divided by the total weight of entire litter (sum of individual organs plus carcasses) for each dam.

*Significantly different from 1 day exposed dams at p < 0.05.

primitive streak or early head-fold stage. Preliminary reports indicate that this explant system is sensitive to known teratogenic agents when they are administered to rats whose serum is then collected for use as part of the culture system for test embryos (56). Further work (13) indicates that serum from human subjects treated with cancer therapeutic or anticonvulsive agents causes lethality or teratogenicity in cultured rat embryos. For a review of this promising technique, see Wilson (141).

Postnatal testing is also becoming an important component of teratological testing. It is now recognized that in utero administration of many classic structural teratogens, at lower dosages and/or later times than usually administered for production of structural effects, results in neurofunctional and endocrine deficits. These lesions may be of a permanent nature and are detectable only in postnatal life. Agents so tested include methyl mercury (32, 108, 111), cadmium (32), Vitamin A (118), lead (60), and 5-azacytidine (94). These findings agree with human data on children exposed to methyl mercury (2, 39) and lead (5) in utero. Some agents that have been shown to cause neurological deficits had not been considered teratogens at all.

A few protocols for evaluating postnatal development are currently in use on a limited basis both undergoing and awaiting validation. One such profile in use at CIIT is presented in Figure 8. Except for open field, which is run at CIIT on postnatal day 29 or 31, one of the most widely used behavioral tests, the other parameters listed are not volitional behaviors and are perhaps better termed developmental landmarks. More sophisticated behavioral tests are also being utilized (12, 48) with preliminary attempts to develop and standardize screening methods for behavioral teratology (10). Problems exist in experimental design, statistical analysis and interpretation (17). But, this aspect of teratological testing will become an increasingly important area of research and required addendum to teratological testing in the future, especially in the light of increased regulatory agency concern in this area in the United States and other countries world wide. A number of thought-provoking essays have emerged on the evolution (142), current methodology and suggested improvements of teratological testing (7, 16, 116, 117, 139) and reproduction and fertility studies (86, 93) which are fruitful reading beyond the scope of this paper.

Extrapolation to Human Risk Assessment. The difficulty in extrapolating from animal toxicity data to man is compounded in reproduction and teratology risk assessment. Even before the experiment is begun the choice of animal test system is important. Is the rodent the best model for this chemical? Which genus and species should be used: rat or mouse? The rabbit is commonly used in addition to a rodent since it is related to Rodentia but belongs to the order Lagomorpha and therefore satisfies governmental regulatory agency requirements for two mammalian species,

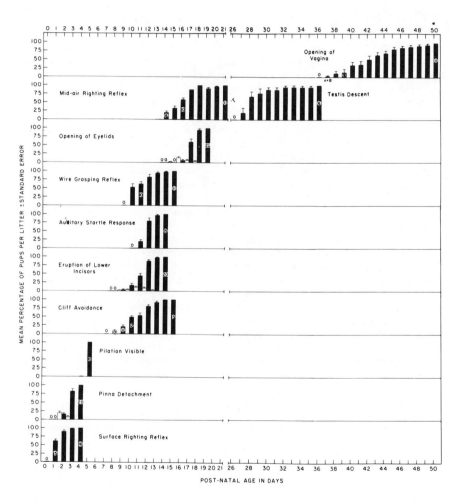

Figure 8.   Acquisition of postnatal developmental landmarks in the Fischer-344 rat

one preferably non-rodent. Which strain should be chosen?
Should it be inbred or outbred? Use of an inbred strain with
genetically uniform animals allows observed variations (anomalies,
abnormalities) to be ascribed to the environment, i.e., the agent
under test. Use of genetically heterogeneous animals, an outbred
strain, allows detection of teratogenic effects involving chemi-
cal-gene interactions and approximates the human genetic situation
with a multiplicity of genotypes (heterozygosity). However, use
of an outbred strain increases variability, and therefore the
number of animals to be used, and complicates interpretation of
results including causality of observed anomaly. Based on toxico-
kinetic, disposition and metabolism data, the rodent may not be
the test animal of choice. Teratological researchers are also
using, or contemplating using, the rabbit, hamster, guinea pig,
armadillo, ferret, dog, miniature swine, cat, or non-human primates for drugs prescribed specifically for pregnant women.
   Prenatal development in the Rodentia and Lagomorpha differs
in significant ways from that in humans. All three have a chorioallantoic placenta but that of humans is hemochorial, where the
chorionic villi of the fetus are bathed in maternal blood and one
layer of syncitial trophoblast separates the maternal blood from
the fetal capillaries (21). The chorioallantoic placenta of
rodents and lagomorphs is a complex hemoendothelial type composed
of intimately juxtaposed and modified fetal and maternal cells,
bathed by a labyrinth of blood sinuses (55) with three (rat, mouse
and hamster) or two (rabbit) trophoblastic layers separating
maternal blood from fetal capillaries (21). The human and rat
placenta also differ functionally with secretory patterns of
placental lactogen differing and with the presence in primates of
chorionic gonadotropin (80). What effect if any these differences
have on placental transport is not fully understood. In addition,
rodents and lagomorphs also form a yolk sac placenta immediately
after implantation, which is the major (only) mechanism for nutrient processing and transport until gestation day 11-11½, and
persists as functional, even when the chorioallantoic placenta
forms, almost to parturition. Again, what effect this has on
embryo and fetal vulnerability is not yet known, although at least
one teratogenic agent, trypan blue, appears to act solely on the
yolk sac placenta (8). In multifetal pregnancies there are
differences in blood flow to left and right uterine horns and to
implants at ovarian versus cervical ends of the uterine horns.
Different fetuses within the same dam have been shown to be at
differential risk (119, 145). In addition, fetal loss is handled
differently: dead implants are not expelled in a spontaneous
abortion as in single-birth mammals but are resorbed in situ. It
is not uncommon to recover healthy, viable fetuses side-by-side
with large numbers of resorption sites. Maternal, placental and
fetal metabolism of xenobiotics may also differ hence the need for
prior characterization, at least, of the test organism's metabolic
capabilities of the substance to be tested.

The placenta is both a transport and metabolizing organ. Transport is accomplished by simple diffusion, facilitated diffusion, active transport across membranes and by special processes such as pinocytosis, phagocytosis and breaks in the "barrier" (29). Characteristics of chemicals showing high transfer from maternal blood to placenta include: low molecular weight (< 500 daltons optimal), high lipid/water partition coefficient (lipophilic), low ionization at blood pH (pKa) and low binding to plasma proteins (4). The placenta contains a full complement of mixed function oxidases located in the microsomal and mitochondrial subcellular fractions capable of induction (eg. benzo(a)-pyrene hydroxylase, 24).

Metabolism in the test dam and/or fetus and its relevance to the human gravida is also critical. For example, the parent compound may be teratogenic and is metabolized to innocuous products as with diphenylhydantoin, an anti-seizure drug used in the treatment of epilepsy (41). In contrast, the parent compound may be harmless and must be metabolized to the proximal teratogenic agent as in chlorcyclizine, an antihistamine metabolized in vivo to the active teratogen norchlorcyclizine (57, 89). One of the current hypotheses concerning mechanism of thalidomide-induced teratogenesis suggests that thalidomide is transmitted to the human fetus and metabolized to more polar metabolite(s), the putative proximal teratogenic agent(s), which cannot cross the placenta back to the maternal organism for further metabolism and excretion (58, 129). This sequence may be qualitatively or quantitatively different in the insensitive pregnant rodent. In contrast, imipramine, an antidepressant, is teratogenic in rabbits where blood levels of the parent compound stay high. In the human, imipramine is rapidly metabolized by demethylases and is not teratogenic (42, 92).

Mechanisms of Teratogenesis. Most toxicologists have viewed the experimental animal as a "black box" wherein one inserts test chemicals usually at high dose and observes effects out. There has recently been a call for low-dose exposure, examination of the effects and the mechanisms by which they arise and assessment of human risk of the effects seen in the test systems (30). Similarly, in teratology, the pregnant mammal has been considered a "black box" whereon exposures are done and resulting fetuses examined with little or no attention to mechanisms.

Without elucidation of mechanisms, teratologists are doomed to an endless succession of empirical testing screens. Teratology must be concerned with anticipation of teratogenic risk which requires knowledge of mechanisms. Researchers must be able to extrapolate results from a test compound at (relatively) high dose on a test animal to risk assessment to the human at (usually) lower doses. Given the huge number of potential teratogens in use today and new ones entering the environment yearly, teratologists must be able to generalize from known agents with known mechanisms to new agents with similar functional groups, for example, with projected similar effects and mechanisms. Landauer, just before his death, was beginning to examine the role of functional groups on the teratological response in the chick embryo (63). The elucidation of structure–activity relationships (SARs) has been one of the major tools in pharmacology and toxicology. Teratological screens can be improved if mechanisms are known, by focussing on specific gestational times of exposure or close examination of certain organs expected to be targets. One can evaluate the appropriate test model if the mechanism of action is known. For example, if the specific biochemical pathways involved were identified, the experimental animal with such pathways most similar to humans would be the test system of choice. Longer range goals made possible by understanding mechanisms of action would include prevention or amelioration of the developmental defect prior to final manifestation by diverting the initiating mechanism or intervention at some point(s) in the process of pathogenesis. This could be done by supplementation of deficient enzyme or substrate, restriction of diet to avoid excess accumulation of a deleterious metabolite, or augmentation of inadequate transport. One might, in the future, prevent the "spontaneous" birth defects which result from interactive or multiple causes yet unidentified if the components of the effect, mechanisms and interactions can be identified and the sequence from cause to manifestation interrupted (136). These interactions, based on information already known, may be with nutritional status (47), with other teratogens or with a non–teratogen, a so–called "pro–teratogen" (14, 95, 130).

Wilson (140) suggests a number of mechanisms of teratogenesis including: 1. mutations (somatic, that is non–heritable); 2. chromosomal non–disjunctions and breaks; that is clastogenic events; 3. mitotic interference; 4. altered nucleic acid integrity or function; 5. lack of precursors and substrates required for biosynthesis; 6. altered energy sources; 7. enzyme inhibitions; 8. osmolar imbalance; 9. altered membrane characteristics. A current estimation is that 70% of all mutagens are teratogens (40) but not all teratogens are mutagens. Thalidomide is perhaps the best example of a human epigenetic teratogenic agent.

These alterations induced by the teratogen may occur in the intracellular compartment in the nucleus and cytoplasm, at the cell surface, in the extracellular matrix and/or at the level of the fetal environment: fetal organism, placental or maternal interactions (97).

CONCLUSIONS

This review has so far focussed on the maternal organism as the source to the fetus of the teratogenic agent. However, the male has been implicated as the cause of a teratogenic event in animal studies for example, with methadone (106), thalidomide

(71), lead, narcotics, alcohol and caffeine (cited in Science 202:733, 1978). Human male mediation has been statistically confirmed in studies indicating increased incidence of spontaneous abortions, stillbirths and congenital defects from male operating room personnel exposed to waste anesthetic gases and vapors (1) and been implicated in congenital heart defects from male production-worker exposure to Oryzalin (90). Apparently 1,2-dibromo-3-chloropropane (DBCP) is a human male sterilant (127, 128). High caffeine consumption by the male has also been implicated in spontaneous abortion, stillbirths, and premature births (123). The putative mechanisms may include damage to the sperm, presence of the agent or its metabolite(s) in the semen which may affect the embryo directly or act on the gravid uterus (72), or an indirect action on the male affecting hormone levels and perhaps libido (52, 107). These results have grave implications for production workers of both sexes under risk of exposure during child-siring or child-bearing years.

Wilson (136) has suggested criteria for recognizing a new teratogenic agent in humans. These include an abrupt increase in the incidence of a particular defect or association of defects (syndrome) and coincidence of this increase with a known environmental change such as introduction of a new drug or environmental exposure to other chemicals. The appearance of characteristically malformed offspring should be correlated with known exposure to the environmental change early in pregnancy, and there should be absence of other factors common to those pregnancies yielding infants with the characteristic defect(s).

Hunt (45) has made a number of recommendations to increase the data available on births to include maternal and paternal work experience, to encourage analysis of data already collected to identify possible relationships between occupational history of the mother (and father) and pregnancy outcome, and to encourage and support research on fetal development and maternal physiology in relation to exposure and handling of toxic substances. She suggests promotion of information exchange with other countries especially in Eastern Europe. The importance is stressed of health education programs especially in the workplace, consideration of the pregnant worker in all investigations and analyses of occupational safety and health standards, and a concerted effort to impress clinicians and epidemiologists with the importance of occupational history for any study on reproduction from both males and females. The U. S. Department of Health, Education and Welfare has established a Congenital Malformations Surveillance published yearly to monitor birth defects in the United States divided into four regions to attempt to detect any alterations in frequency of a dozen major malformation syndromes.

Teratological research has made great advances in the last decade, with recognition of the need for new approaches, with refinement of analytical tools, and with the awareness of the importance of subtle structural and functional alterations pre-

and postnatally. With input from clinicians and epidemiologists, toxicologists, pharmacologists, analytical chemists and behavioral scientists, teratologists are working to meet the challenge to guarantee the birthright of health to children of this chemical age.

ABSTRACT

The discipline of Teratology is introduced, including an historical perspective and current definition. Categories of teratogenic agents are discussed including human and animal teratogens. Determinants of teratogenic susceptibility are detailed: specificity of agent, specificity of target and disposition of agent. State-of-the-art teratological testing is presented as well as new trends evolving such as toxicokinetics and metabolism on maternal and fetal tissues, in vitro test systems and postnatal testing procedures. Extrapolation of animal data to human risk assessment is discussed. Possible mechanisms of teratogenesis and site(s) of action are suggested and conclusions tentatively drawn as to the sensitivity of current teratological testing, new methodology developing and the limitations of available techniques to guarantee the birthright of health to children of this chemical age.

Literature Cited

1. Ad Hoc Committee on the Effect of Trace Anesthetics on the Health of Operating Room Personnel: Occupational disease among operating room personnel: A National Study. Anesthesiology, 1974, 41:321–340.

2. Amin–Zaki, L., M. A. Majeed, S. B. Elhassani, T. W. Clarkson, M. R. Greenwood, R. A. Doherty: Prenatal methyl mercury poisoning: Clinical observations over five years. Am. J. Dis. Child, 1979, 133:172–177.

3. Anon., National Foundation/March of Dimes: Facts. National Foundation, New York, 1975.

4. Asling, J. and E. L. Way, Placental transfer of drugs. In Fundamentals of Drug Metabolism and Drug Disposition (B. N. LaDu, H. G. Mandel and E. L. Way, eds.) Williams and Wilkins; Baltimore, MD, 1971, pp. 88–105.

5. Baloh, R., B. Sturm, B. Green and G. Gleser, Neuropsychological effects of chronic asymptomatic increased lead absorption. Arch. Neurol., 1975, 32:326–330.

6. Barrow, M. V. and W. J. Taylor, A rapid method for detecting malformations in rat fetuses. J. Morphol., 1969, 127:291-306.

7. Beck, F., Model systems in teratology. Brit. Med. Bull., 1976, 32(1):53-59.

8. Beck, F., J. B. Lloyd and A. Griffiths, Lysosomal enzyme inhibition by trypan blue. A theory of teratogenesis. Science, 1967, 157:1180-1182.

9. Buckley, S. K. L., C. E. Steele and D. A. T. New, In vitro development of early postimplantation rat embryos. Develop. Biology, 1978, 65:396-403.

10. Buelke-Sam, J. and C. A. Kimmel, Development and standardization of screening methods for behavioral teratology. Teratology, 1979, 20(1):17-30.

11. Butcher, R. E., R. L. Brunner, T. Roth and C. A. Kimmel, A learning impairment associated with maternal hypervitaminosis-A in rats. Life Sciences, 1972, 11(1):141-145.

12. Butcher, R. E., C. V. Vorhees and C. A. Kimmel, Learning impairment from maternal salicylate treatment in rats. Nature New Biol., 1972, 236:211-212.

13. Chatot, C. L., N. W. Klein, J. Piatek and L. J. Pierro, Successful culture of rat embryos on human serum: Use in the detection of teratogens. Science, 1980, 207:1471-1473.

14. Clegg, D. J. Teratology. Annual Rev. Pharmacol., 1971, 11:409-424.

15. Cohlan, S. Q., Congenital anomalies in the rat produced by excessive intake of vitamin A during pregnancy. Pediatrics, 1954, 13:556-567.

16. Collins, T. F. X. and E. V. Collins, Chapter 6. Current Methodology in Teratology Research. In Advances in Modern Toxicology Volume 1, part 1: New Concepts in Safety Evaluation (Mehlman, M. A., R. E. Shapiro and H. Blumenthal, eds.) John Wiley and Sons; New York, 1976, pp. 155-175.

17. Coyle, I., M. J. Wayner and G. Singer, Behavioral teratogenesis: A critical evaluation. Pharmacology, Biochemistry and Behavior, 1976, 4:191-200.

18. Crary, D. D., Modified benzyl alcohol clearing on alizarin-stained specimens without loss of flexibility. Stain Technology, 1962, 37:124–125.

19. Dawson, A. B., Note on the staining of the skeleton of cleared specimens with alizarin red S. Stain Tech., 1926, 1:123–124.

20. Degenhardt, K., Thalidomide effects in children. Presented at the Symposium: Thalidomide and the Embryo, Third Annual Meeting of the Teratology Society, The Chantecler, Ste. Adele, Quebec, 1963.

21. Enders, A. C., A comparative study of the fine structure of the trophoblast in several hemochorial placentas. Am. J. Anat., 1965, 116:29–65.

22. Environmental Protection Agency, Part IV. Proposed Health Effects Test Standards for Toxic Substances Control Act Test Rules and Proposed Good Laboratory Practice Standards for Health Effects. Federal Register 44(145):44059–44060, 44087–44092, July 26, 1979.

23. Faherty, J. F., B. A. Jackson and M. F. Greene: Surface staining of 1 mm (Wilson) slices of fetuses for internal visceral examination. Stain Technology, 1972, 47(2):53–58.

24. FDA Advisory Committee on Protocols for Safety Evaluations: Panel on Reproduction Report on Reproduction Studies in the Safety Evaluation of Food Additives and Pesticide Residues. Toxicol. and Appl. Pharmacol., 1970, 16:264–296.

25. Forfar, J. O., What drugs are unsafe?. Symposium in Advances in Medicine, 1974, 10:34–50.

26. Fox, M. H. and C. M. Goss, Experimentally produced malformations of the heart and great vessels in rat fetuses. Am. J. Anat., 1958, 102:65–92.

27. Fraser, F. C. and T. D. Fainstat, Production of congenital defects in the offspring of pregnant mice treated with cortisone. Pediatrics, 1951, 8:527–533.

28. Gilman, J., C. Gilbert and G. C. Gilman, Preliminary report on hydrocephalus, spina bifida and other congenital anomalies in rats produced by trypan blue. S. Afr. J. Med., 1948, Sci. 13:47–90.

29. Ginsburg, J. Placental drug transfer. Ann. Reviews Pharma-col., 1971, 11:387-476.

30. Golberg, L., Toxicology: Has a new era dawned? Pharmacol. Reviews, 1979, 30(4):351-370.

31. Goldenthal, E. I. (Chief, Drug Review Branch - Division of Toxicological Evaluation, Bureau of Scientific Standards and Evaluation) Guidelines for Reproduction Studies for Safety Evaluation of Drugs for Human Use, letter dated March 1, 1966.

32. Grady, R. R., J. I. Kitay, J. M. Spyker and D. L. Avery, Postnatal endocrine dysfunction induced by prenatal methyl-mercury or cadmium exposure in mice. J. Environ. Pathol. and Toxicol., 1978, 1:187-197.

33. Greene, R. R., M. W. Burrill and A. C. Ivy, Experimental intersexuality: the effect of antenatal androgens on sexual development of female rats. Amer. J. Anat., 1939, 65:415-469.

34. Greene, R. R., W. W. Burrill and A. C. Ivy, Experimental intersexuality: the effects of estrogens on the antenatal sexual development of the rat. Amer. J. Anat., 1940, 67:305-345.

35. Gregg, N. M., Congenital cataract following German measles in the mother. Tr. Ophth. Soc. Australia, 1941, 3:35-46.

36. Haas, J. F. and D. Schottenfeld, Risks to the offspring from parental occupational exposure. J. Occupat. Medicine, 1979, 21(9):607-613.

37. Hale, F., Pigs born without eyeballs. J. Hered., 1933, 24:105-106.

38. Hale, F., The relation of vitamin A to anopthalmos in pigs. Amer. J. Opth., 1935, 18:1087-1093.

39. Harada, M., Congenital Minamata Disease: Intrauterine Methylmercury Poisoning. Teratology, 1978, 18:285-288.

40. Harbison, R. D., Chemical-biological reactions common to teratogenesis and mutagenesis. Environmental Health Per-spectives, 1978, 24:87-100.

41. Harbison, R. D. and B. A. Becker, Effect of phenobarbital and SKF 525A pretreatment on diphenylhydantoin teratogenicity in mice. J. Pharmacol. Exp. Therapeutics, 1970, 175(2):283–288.

42. Hendrickx, A. G., Teratologic evaluation of imipramine hydrochloride in bonnet (Macaca radiata) and rhesus monkeys (Macaca mulatta). Teratology, 1975, 111:219–222.

43. Hipple, V. and H. Pagenstrecher, Über den Einfluss des Cholins und der Röntgenstrahlen auf den Ablauf der Gravidität. Munch. Med. Wochenschr., 1907, 54:452–456.

44. Hoskins, D., Some effects of nitrogen mustard on the development of external body form in the rat. Anat. Rec., 1978, 102:493–512.

45. Hunt, V. R., Occupational Health Problems of Pregnant Women. A Report and Recommendations for the Office of the Secretary of DHEW Order No. SA-5304-75, April 30, 1975.

46. Hurley, L. S., Demonstration "A" Alizarin staining of bone. Supplement to Teratology Workshop Manual, p. 121, 1965.

47. Hurley, L. S., Chapter 7, Nutritional Deficiencies and Excesses. In Handbook of Teratology volume 1 (Wilson, J. G. and F. C. Fraser, eds.). Plenum Press:N.Y., 1977, pp. 261-308.

48. Hutchings, D. E. and J. Gaston, The effects of vitamin A excess administered during the mid-fetal period on learning and development in rat offspring. Dev. Psychobiol., 1974, 7:225–233.

49. Inouye, M.: Differential staining of cartilage and bone in fetal mouse skeleton by alcian blue and alizarin red S. Cong. Anom., 1976, 16:171–173.

50. Irons, R. D. and E. A. Gross, Standardization and calibration of whole-body autoradiography for routine semi-quantitative analysis of the distribution of ^{14}C-labelled compounds in animal tissues. Toxicol. Appl. Pharm., 1980, in press.

51. Irons, R. D., E. A. Gross, R. M. Long and D. E. Rickert, Absorption and initial distribution of ^{14}C-2,4-dinitrotoluene in pregnant Fischer-344 rats. Comparison of whole body autoradiography with conventional tracer methods. Toxicol. Appl. Pharmacol., 1980, in press.

52. Joffe, J. M., Influence of drug exposure of the father on perinatal outcome. Clinics in Perinatology, 1979, 6(1):21-36.

53. Johnston, M. V., R. Grzanna and J. T. Coyle, Methylazoxy-methanol treatment of fetal rats results in abnormally dense noradrenergic innervation of neocortex. Science, 1979, 203:369-371.

54. Kalter, H., The inheritance of susceptibility to the terato-genic action of cortisone in mice. Genetics, 1954, 39:185-196.

55. Kaye, M. D., The evolution of placentation. Aust. N. Z. J. Obstet. Gynaec., 1971, 11:197-207.

56. Klein, N. W., M. A. Volger, C. L. Chatot and L. J. Pierro. The use of cultured rat embryos to evaluate the teratogenic activity of serum: cadmium and cyclophosphamide. Tera-tology, 1979, 19(2):35A.

57. King, C. T. G., S. A. Wenver and S. A. Narrod, Antihistamines and teratogenicity in the rat. J. Pharmac. Exp. Ther., 1965, 147:391-398.

58. Knightley, P., H. Evans, E. Potter, and M. Wallace, Suffer the Children: The Story of Thalidomide. The Viking Press:N. Y., 1979.

59. Kochhar, D. M., The use of in vitro procedures in teratology. Teratology, 1975, 11(3):273-288.

60. Konat, G. and J. Clausen, Triethyl lead-induced hypomyelina-tion in the developing rat forebrain. Exper. Neurology, 1976, 50:124-133.

61. Kopf, R., D. Lorenz and E. Salewski, Der Einfluss von Thali-domide auf die Fertilität von Ratten in Generationversuch über zwei Generationen. Naunyn-Schmiedebergs Arch. exp. Path. a. Pharmak., 1964, 247:121-135.

62. Landauer, W., On the chemical production of developmental abnormalities and of phenocopies in chicken embryos. J. Cell Physiol., 1954, 43:261-305.

63. Landauer, W., Molecular shape and teratogenic specificity. Tox. Res. Proj. Direct., 1977, 2(4):I-159.

64. Layton, W. M., An analysis of teratogenic testing procedures in Congenital Defects. New Directions to Research (D. T. Janevich, ed.), 1974, pp. 205–217.

65. Lendon, R. G., A synergistic interaction between the teratogenic effect of trypan blue and dietary deficiency in the rat. (Specialia) Experientia, 1978, 34(4):510–511.

66. Lengfellner, K. Über Versuche von Einwirkung der Röntgenstahlen auf Ovarien und den schwangeren Uterus von Meerschweinchen. Munch. Med. Wochenschr., 1906, 53:2147–2148.

67. Lenz, W., Kindliche Missbildungen nach Medikament-Einnahme wahrend der Graviditat? Deutsch. Med. Wochenschr., 1961, 86:2555–2556.

68. Lenz, W., Thalidomide and congenital abnormalities Lancet, 1962, 1:45.

69. Lillie, F. R., The free-martin: A study of the action of sex hormones in the fetal life of cattle. J. exp. Zool., 1917, 23:371–452.

70. Lucey, J. F., Hazards to the newborn infant from drugs administered to the mother. Ped. Clin. N. America, 1961, 8:413–420.

71. Lutwak-Mann, C., Observations on progeny of thalidomide-treated male rabbits. Brit. Med. J., 1964, 1:1090–1091.

72. Lutwak-Mann, C., K. Schmid and H. Keberle, Thalidomide in rabbit semen. Nature, 1967, 214(5092):1018–1020.

73. Manson, J. M. and R. Simons, In vitro metabolism of cyclophosphamide in limb bud culture. Teratology, 1979, 19(2):149–158.

74. Maren, T. H., Editorial: Teratology and carbonic anhydrase inhibition. Arch. Ophthal., 1971, 85:1–2.

75. McBride, W. G., Thalidomide and congenital abnormalities. Lancet, 1961, 2:1358.

76. McKay, R. J. and J. F. Lucey, Neonatology, New England J. of Medicine, 1964, 270:1231–1236.

77. Meyers, V. K. and C. V. Meyers, Chemicals which cause birth defects: Teratogens. A Brief Guide. Southern Illinois Univ.:Carbondale, IL, 1980, p. 1–37.

78. Miller, R. W., Transplacental Chemical Carcinogenesis in Man.
     Nat. Can. I. M, 1979, M52:13-16.

79. Morgenroth, III, V. (Chairman, Testing Standards and Guide-
     line's Work Group of the Interagency Regulatory Liaison Group
     OI.R.L.G.o), Draft I.R.L.G. Guidelines for Selected Acute
     Toxicity Tests. August 1979, pp. 113-141.

80. Munro, H. N., Placenta in relation to nutrition. Introduc-
     tion. Fed. Proc., 1980, 39(2):236-238.

81. Murphy, L. M. A comparison of the teratogenic effects of
     fine polyfunctional alkylating agents on the rat fetus.
     Pediat., 1959, 23:231-244.

82. New, D. A. T., Techniques for assessment of teratologic
     effects: Embryo culture. Environ. Health Perspectives,
     1976, 18:105-110.

83. Nishimura, H. and T. Tanimura, Ch. VII Risk of environmental
     chemicals to human embryos. In Clinical Aspects of the
     Teratogenicity of Drugs, Excerpta Medica:Amsterdam, 1976,
     pp. 271-278.

84. Nora, J. J., A. H. Nora, R. J. Sommerville, R. M. Hill and D.
     G. McNamara, Maternal exposure to potential teratogens.
     J. Am. Med. Assoc., 1967, 202:1065-1069.

85. Overmann, S. R., Behavioral effects of asymptomatic lead
     exposure during neonatal development in rats. Toxicol.
     Appl. Pharmacol., 1977, 41:459-471.

86. Palmer, A. K., Some thoughts on reproductive studies for
     safety evaluation. Proc. Europ. Soc. Study of Drug Toxicity,
     1973, XIV:79-90.

87. Palmer, A. K., Assessment of current test procedures.
     Environ. Health Perspec., 1976, 18:97-104.

88. Peckham, C. H. and R. W. King, A study of intercurrent con-
     ditions observed during pregnancy. Am. J. Obstet. Gynecol.,
     1963, 87:609-624.

89. Posner, H. S., Significance of cleft palate induced by
     chemicals in the rat and mouse. Fd. Cosmet. Toxicol., 1972,
     10:839-855.

90. Rawls, R. L., Reproductive hazards in the workplace. Chem.
     and Eng. News, 1980, 58(6):28-31.

91. Rice, J. M., Carcinogenesis: A late effect of irreversible toxic damage during development. Env. Health Perspec., 1976, 18:133–139.

92. Robson, J. M. and F. M. Sullivan, The production of foetal abnormalities in rabbits by imipramine. Lancet, 1963, (1963)i:638–639.

93. Robson, J. M., Testing drugs for teratogenicity and their effects on fertility. Brit. Med. Bull., 1970, 26(3):212–216.

94. Rodier, P. M., S. S. Reynolds and W. N. Roberts, Behavioral consequences of interference with CNS development in the early fetal period. Teratology, 1979, 19:327–336.

95. Runner, M. N., Inheritance of susceptibility to congenital deformity. Metabolic clues provided by experiments with teratogenic agents. Pediatrics, 1959, 23:245–251.

96. Sadler, T. W., Culture of early somite mouse embryos during organogenesis. J. Embryol. Exp. Morph., 49:17–25.

97. Saxén, L., Mechanisms of Teratogenesis. J. Embryol. Exp. Morph., 1976, 36(1):1–12.

98. Schardein, J. L. Drugs as Teratogens, CRC Press, Inc.: Cleveland, OH 44128, 1976.

99. Schwetz, B. A., B. K. J. Leong and P. J. Gehring, Embryo and fetotoxicity of inhaled chloroform in rats. Toxicol. Appl. Pharmacol., 1974, 28:442–451.

100. Schwetz, B. A., B. K. J. Leong and P. J. Gehring, Embryo and fetotoxicity of inhaled carbon tetrachloride, 1,1-dichloro-ethane and methyl ethyl ketone in rats. Toxicol. Appl. Pharmacol., 1974, 28:452–464.

101. Schwetz, B. A., B. K. J. Leong and P. J. Gehring, The effect of maternally inhaled trichloroethylene, perchloroethylene, methyl chloroform and methylene chloride on embryonal and fetal development in mice and rats. Toxicol. Appl. Pharmacol., 1975, 32:84–96.

102. Shepard, T. H., Catalog of Teratogenic Agents, Second Edition, Johns Hopkins University Press:Baltimore, 1976.

103. Singh, S. C., Ectopic neurones in the hippocampus of the postnatal rat exposed to methylazoxymethanol during foetal development. Acta Neuropath. (Berl.), 1977, 40:111–116.

104. Smith, A. U., The effects on fetal development of freezing pregnant hamsters (Mesocricetus auratus). J. Embryol. Exp. Morphol., 1957, 5:311-323.

105. Smithells, R. W., Environmental teratogens of man. Brit. Med. Bull., 1976, 32:27-33.

106. Soyka, L. F., J. M. Peterson and J. M. Joffe, Lethal and sublethal effects on the progeny of male rats treated with methadone. Toxicol. Appl. Pharmacol., 1978, 45:797-807.

107. Soyka, L. F. and J. M. Joffe, Male mediated drug effects on offspring. In Progress in Clinical and Biological Research: Drugs and Chemical Risk to the Fetus and Newborn (R. H. Schwarz and S. J. Yaffe, eds) Alan R. Liss, Inc.:New York, 1980; 36:49-66.

108. Spyker, J. M., S. B. Sparber and A. M. Goldberg. Subtle consequences of methylmercury exposure: behavioral deviations in offspring of treated mothers. Science, 1972, 177:621-623.

109. Spyker, J. M. and M. Smithberg, Effects of methylmercury on prenatal development in mice. Teratology, 1972, 5:181-190.

110. Spyker, J. M. and Chang, Delayed effects of prenatal exposure to methylmercury: brain ultrastructure and behavior. Teratology, 1974, 9:37A.

111. Spyker, J. M., Ch. 12, Behavioral Teratology and Toxicology. In Behavioral Toxicology (Weiss, B. and V. G. Laties, eds.). Plenum Press:New York, 1975, pp. 311-349.

112. Staples, R. E., Detection of visceral alterations in mammalian fetuses. Teratology, 1974, 9(3):A37-38.

113. Staples, R. E. and V. L. Schnell, Refinements in rapid clearing technic in the KOH-Alizarin Red S method for fetal bone. Stain Technology, 1964, 39(1):62-63.

114. Steptoe, P. C., R. G. Edwards and J. M. Purdy, Human blastocysts grown in culture. Nature (London), 1971, 229:132-133.

115. Tuchmann-Duplessis, H. Teratogenic drug screening. Present procedures and requirements. Teratology, 1972, 5:271-286.

116. Tuchmann-Duplessis, H., Teratogenic Action of Drugs. Pergamon Press:Oxford, 1965.

117. Tuchmann-Duplessis, H., Design and interpretation of terato-
     genic tests. In Embryopathic Activity of Drugs (Robson, J.
     M., F. M. Sullivan and R. L. Smith, eds.). Little, Brown and
     Co.:Boston, 1965, pp. 56–93.

118. Vorhees, C. V., R. L. Brunner, C. R. McDaniel and R. E.
     Butcher, The relationship of gestational age to vitamin A
     induced postnatal dysfunction. Teratology, 1978, 17:271–276.

119. Warkany, J. and E. Schraffenberger, Congenital malformations
     induced in rats by roentgen rays. Amer. J. Roentgenol.
     Radium Ther., 1947, 57:455–463.

120. Warkany J. and E. Takacs, Experimental production of con-
     genital malformations in rats by salicylate poisoning.
     Amer. J. Pathol., 1959, 35:315–331.

121. Warkany, J., Congenital Malformations - Notes and Comments.
     Year Book Medical Publications:Chicago, 1971.

122. Warshaw, L. J. (chairman) Guidelines on Pregnancy and Work
     NIOSH Research Report (The American College of Obstetricians
     and Gynecologists) U.S. Dept. HEW, Rockville, Maryland
     (Contract No.210-76-0159), 1977.

123. Weathersbee, P. S. and J. R. Lodge, Caffeine: Its direct and
     indirect influence on reproduction. J. Reprod. Med., 1977,
     19:55–63.

124. Welch, R. M., Y. E. Harrison, B. W. Gommi, P. J. Poppers, M.
     Finster and A. H. Conney. Stimulatory effect of cigarette
     smoking on the hydroxylation of 3,4-benzpyrene and the N-
     demethylation of 3-methyl-4-monomethylaminoazobenzene by
     enzymes in human placenta. Clin. Pharmacol. Ther., 1969,
     10(1):100–109.

125. Werthemann, A. and M. Reiniger, Uber Augenentwicklungssto-
     rungen bei Rattenembryonen durch Sauerstoffmangel in der
     Fruhschwangerschaft. Acta Anat., 1950, 11:329–347.

126. Whittingham, D. G., Fertilization, early development and
     storage of mammalian ova in vitro. In The Early Develop-
     ment of Mammals (M. Balls and A. E. Wild, eds.) Cambridge
     University Press:Cambridge, 1975, pp. 1–24.

127. Whorton, D., R. M. Krauss, S. Marshall and T. H. Milby,
     Infertility in male pesticide workers. Lancet, 1977, ii:
     1259–1261.

128. Whorton, D., T. H. Milby, R. M. Krauss and H. A. Stubbs,
Testicular function in DBCP exposed pesticide workers.
J. Occup. Medicine, 1979, 21(3):161-166.

129. Williams, R. T., Thalidomide: A study of biochemical teratology. Arch. Environ. Health, 1978, 16:493-502.

130. Wilson, J. G., Teratogenic interaction of chemical agents in
the rat. J. Pharmacol. Exp. Ther., 1967, 144:429-436.

131. Wilson, J. G., Experimental teratology. Amer. J. Obstet.
Gynecol., 1964, 90:1181-1192.

132. Wilson, J. G. and J. Warkany (eds.). Teratology: Principles and Techniques, U. of Chicago Press:Chicago, IL, 1965.

133. Wilson, J. G., Methods for administering agents and detecting
malformations in experimental animals. In Teratology:
Principles and Techniques (Wilson, J. G. and J. Warkany,
eds.) U. of Chicago Press:Chicago, IL, 1965, pp. 262-277.

134. Wilson, J. G., Embryological considerations in teratology.
Ann. N.Y. Acad. Sci., 1965, 123:219-227.

135. Wilson, J. G., Effects of acute and chronic treatment with
actinomycin D on pregnancy and the fetus in the rat.
Harper Hosp. Bull., 1966, 24:109-118.

136. Wilson, J. G., Environment and Birth Defects, Academic Press,
Inc., NY (Environmental Sciences An Interdisciplinary Monograph Series), 1973.

137. Wilson, J. G., Present status of drugs as teratogens in man.
Teratology, 1973, 7:3-16.

138. Wilson, J. G., Reproduction and teratogenesis: Current
methods and suggested improvements. Journal Assoc. Off.
Anal. Chem., 1975, 58:657-667.

139. Wilson, J. G., Critique of current methods for teratogenicity
testing and suggestions for their improvement. In Methods
for detection of environmental agents that produce
congenital defects (T. Shepard, J. R. Miller and M. Marois,
eds.) American Elsevier, New York, 1975, pp. 29-48.

140. Wilson, J. G. and F. C. Fraser, eds. Handbook of Teratology
Volume 1 General Principles and Etiology, Plenum Press:NY and
London, 1977.

141. Wilson, J. G., Review of in vitro systems with potential for use in teratogenicity screening. J. Environ. Pathol. and Toxicol., 1978, 2:149–167.

142. Wilson, J. G., The evolution of teratological testing. Teratology, 1979, 20(2):205–212.

143. Wolkowski, R. M., Differential cadmium–induced embryotoxicity in two inbred mouse strains 1. Analysis of inheritance of the response to cadmium and of the presence of cadmium in fetal and placental tissues. Teratology, 1974, 10(3):243–262.

144. Wolkowski–Tyl, R. M., Strain and tissue differences in cadmium–binding protein in cadmium–treated mice. In Developmental Toxicology of Energy–related Pollutants D.O.E. Symposium Series, 1978, 47:568–585.

145. Woollam, D. H. M. and J. W. Millen, Influence of uterine position on the response of the mouse embryo to the teratogenic effects of hypervitaminosis A. Nature, 1961, 190:184–185.

RECEIVED February 2, 1981.

The Contribution of Epidemiology

MARY WAGNER PALSHAW

Stauffer Chemical Company, Westport, CT 06880

Throughout this week, we will be hearing about the various toxicological and biochemical methods employed in assessing the toxicity of pesticides. This afternoon, I will discuss a complimentary science, epidemiology, and provide some insight into epidemiologic methods for investigating and characterizing health effects in humans which may be associated with exposure to pesticides in the workplace.

Definition and Concerns of Epidemiology

Epidemiology may be defined as the study of factors which contribute to the occurrence, distribution and course of disease in a population group.

Epidemiology is considered the detective branch of medicine because its purpose is to investigate and identify specific agents or factors that may cause disease and also identify people who are at high risk for developing a disease. It, therefore, provides the basis for public health programs designed to prevent and control disease. Prevention may be effected by reducing or eliminating exposure to a specific factor once its importance in producing disease has been demonstrated.

Among the public health programs aided by knowledge resulting from epidemiologic investigations are those directed at the prevention and control of conditions such as cancer, cardiovascular disease and stroke. Epidemiologic methods are also essential to the evaluation of the efficacy of new prevention and therapeutic measures and any possible harmful side effects they may have.

Epidemiology focuses on groups of people rather than on a specific individual. The epidemiologist attempts to determine whether there has been an increase of a disease over the years, whether one geographical area has a higher frequency of the disease than another, and whether the characteristics of persons with a particular disease or condition distinquish them from those without it.

0097–6156/81/0160–0157$05.00/0
© 1981 American Chemical Society

The investigation of a disease begins with a description of its occurrence in a population. The basic information required is the time (day, month, season or year) of onset of the disease, place (country, city, urban or rural residence) and various personal characteristics such as age, sex, race, ethnic group, educational background, socioeconomic status, occupation, biological characteristics such as biochemical levels and cellular constituents of blood, and personal living habits such as tobacco usage, alcohol consumption and diet.

Historically, epidemiology originated in relation to the study of the great epidemic diseases such as cholera, bubonic plague, (often referred to as Black Death in the Middle Ages) smallpox, yellow fever and typhus. These disease were associated with high mortality and, until the twentieth century, were the most important threats to life.

Today, with the exception of influenza outbreaks, major epidemics no longer threaten the United States and other highly developed countries, and most of the more important infectious diseases are reasonably under control.

Of course, there are still some surprises. A mysterious outbreak of pneumonia following an American Legion Convention in Philadelphia, in 1976, captured everyone's attention. The investigation was carried out by epidemiologists from the Center for Disease Control (CDC), in Atlanta, working with epidemiologists and other health professionals from the Pennsylvania and Philadelphia Health Departments.

Eventually, the organism responsible for the outbreak was isolated from lung tissues in four patients and it was discovered to be a previously unidentified bacterium. Since then, epidemiologists have investigated the incidence and geographic distribution of the disease, the environmental sources of the organism and the mode of transmission.

In more recent years, chronic diseases have assumed importance as the major health problems of advanced Western Civilization. The new importance of these diseases stems in part from major changes in the environment and the way of life imposed by industrialization and its related migration of people to the cities. It also relates to the increase in the older age groups in the population which has resulted from the removal of infectious diseases as a common cause of early death.

Cancer, high blood pressure, coronary artery disease, diabetes and arthritis are among the lethal or chronic crippling diseases associated with older age.

The investigation of etiologic (causal) factors of the chronic diseases represents the new epidemiologic frontier. In an era of increasing specialization within medicine, epidemiologists now generally specialize in investigation of either infectious or chronic diseases. Within the field of chronic disease epidemiology several subspecialties have developed. These include cancer, genetic, and environmental epidemiology and occupa-

pational epidemiology which employs methods used in both
cancer and environmental epidemiology.

Overview of Occupational Epidemiology

It is well known that the risk of acquiring many diseases
is directly related to occupation. Some examples of disease hazards related to occupation include the development of bone cancers among workers who applied radium paint to watch dials and
hands, the occurrence of lead poisoning in battery workers, bladder cancers in aniline dye workers and lung cancers in miners of
radioactive ores.

Many of the possible health consequences of various occupational exposures are not easily detectable by observation. Epidemiologic techniques can provide a tool for evaluating possible
causal relationships between occupational exposure and development of medical conditions. Their major usefulness is in the ex-
examination of illnesses and deaths occurring after many years
of exposure, and, possibly even after exposure has ended.

A relationship between a particular occupational exposure and
subsequent ill health may be suspected if similar health related
effects have been observed in animal studies or if the condition
has also been found in workers occupationally exposed to a compound structurally related to the chemical in question.

Although occupational epidemiology officially goes back to
1775, when an English physician named Percival Pott observed an
unusually large occurrence of scrotal cancer in chimney sweeps,
most of the methods currently employed in occupational studies
have been developed in the past twenty years.

The essence of the epidemiologic method is that it measures
the risk of illness (morbidity) or death (mortality) in an exposed population and compares it with the same risks in an unexposed population which is identical in all other respects.

Although epidemiologic studies can't by themselves prove a
cause and effect relationship, they can establish an association
between an exposure and ill health. Conversely, a lack of an
association may provide reassurance that the substance does not
adversely affect human health where laboratory or animal studies
have suggested a problem.

As a result of the shift in emphasis from worker safety to
the larger issues of illness, both acute and chronic, which may
be associated with various occupational exposures, industry, particularly the chemical and petrochemical industry, has begun to
employ occupational epidemiologists to conduct studies of workers.

Working in an industrial setting, the epidemiologist is a
member of the occupational health team. At Stauffer, I am a
member of the Occupational Medicine Department located at
Corporate Headquarters.

Interface with other Scientific Disciplines

When conducting a study, we draw heavily upon the knowledge
and expertise of many other scientific disciplines within the
company.

Chemists. For example, when authorized to conduct a study
of workers who have been exposed to a specific pesticide, it is
important to know the composition of the product and the reactions involved, what its unique properties are, how it acts and
what its applications are. For this, we rely on our company
chemists to help explain or supplement the information available
on the product.

Chemical Engineers. We also need to know and understand the
process for commercial manufacture of the product since most of
the studies focus on workers at the plants which manufacture or
formulate the product. For this, we draw upon the expertise of
the chemical engineers.

Industrial Hygienists. Environmental monitoring measurement
data constitute a major component of an occupational epidemiology
study. It is essential to know the amount of pesticide the employees are currently exposed to as well as their historical exposures so that medical findings can be examined in relation to
the work environment.
     Evaluating this relationship is complicated by the fact that
many different and possibly toxic chemicals are being used and
produced at the worksite, there are many different jobs and processes with qualitatively different exposures, and there is frequent movement of workers from job to job.
     The approach taken by the industrial hygienists is to divide
the plant into distinct areas, and to define the job titles held
within each of these areas and the potential exposures to substances for each job title.
     At a plant, there usually are several process areas, a lab-
oratory, maintenance shop and plant office. Within a particular
process area, there may be such job titles as process engineer,
operator and maintenance mechanic.
     The actual exposure level of the product is determined by
taking representative breathing zone samples for a specified time
period and performing analyses of the samples to quantify the
amount of product present.
     A problem faced by the epidemiologist is that exposure infor-
mation is frequently available for only the most recent years of
production and historical exposures must be estimated. This is
accomplished by having supervisory personnel who have been associated with the manufacture of the product during the study
period rank the exposure intensity of all job titles whether or
not the job was directly involved in the operation. This may be

On a severity scale of 0-3 with zero representing no exposure and three representing heavy exposure.

The ranking is determined after consideration of factors such as availability of industrial hygiene information, process changes, production levels, actual operating conditions, engineering and procedural changes to upgrade working conditions, physical proximity of the job titles to points of exposure and utilization of personal protective devices.

Usually, the only workers who can be considered truly non-exposed are clerical personnel in the plant office who generally have little occasion to go out into the plant.

The exposure rating for a particular job title is then linked with each employee's work history to determine the cumulative Exposure Index to the product each individual has had.

The people we are most interested in at a location are those who have had the greatest exposure for a long period of time. If the product is considered a suspect carcinogen, we are particularly interested in the causes of death for those workers employed fifteen or more years because of the long latency period required for many agents to induce cancer.

As I indicated before, we generally study plant workers when we are interested in evaluating health effects related to a pesticide. However, in the future, it is envisioned that we will have the capability to identify all workers who have been exposed to a specific pesticide.

This will include the initial synthesis chemist who assigned a number to a mixture he had just developed, the analytical and field research station personnel, toxicology personnel involved with the acute toxicity, sub-chronic and chronic testing and reproductive and mutagenicity studies, and pilot plant and manufacturing and formulating personnel so that all persons exposed to any level of the pesticide will be included in the study.

Before undertaking a study, we perform a world-wide search of the scientific literature regarding human health effects associated with exposure both to the product and its component substances. We usually also examine the literature on structurally related products.

Frequently, the search will yield few papers related to a specific product so that we lack clues as to what the health effects in humans might be. This is where the toxicologists come in.

Toxicologists. Upon request, the toxicologists will review the toxicology literature and their own studies concerning the product and advise us of any significant findings. From this information, we can then determine the appropriate type of study to design.

If, for example, a chronic inhalation study in rats showed a statistically significant excess of tumors at a given site, we would undertake a mortality study to determine if there were excesses of cancer, particularly at that site, among workers exposed to the product with attention being paid to those workers exposed fifteen years or more.

If evidence of testicular atrophy was observed, we would be interested in designing a clinical study which would assess reproductive function in the exposed workers. If a two year ingestion study in rats reported a statistically significant incidence of neurotoxicity, we would want to investigate whether neurotoxic effects also were produced in an exposed worker population.

Toxicology studies, therefore, can serve to predict disease in humans. Epidemiology studies can then be conducted to assess whether a health effect observed in animals is reproduced in humans.

Physicians. Epidemiologists utilize the data generated from the physical examinations and special tests performed by the plant physicians as part of the Occupational Health Program to conduct their surveillance of the workforce.

In addition, we call upon physicians to use their knowledge of occupational disease in evaluating medical information on study subjects to determine the work relatedness of health effects observed.

When conducting a clinical study such as that of reproductive function, we work with a physician specialist, in this case a urologist, to design the study. This involves developing a medical history questionnaire and physical examination strategy. The physician will perform the examinations and assist with the interpretation of the study findings.

Biostatisticians. The identification of an association between exposure to a substance and subsequent development of a medical disorder requires increasingly complex, sophisticated statistical concepts and methods. We, therefore, work closely with biostatisticians first, to design studies which will detect an increased risk if it is present and, then, in the analysis and interpretation of the study findings.

Epidemiologic Techniques Used to Assess Health Effects Related to Occupational Exposure to Pesticides

Let us examine the techniques epidemiologists use to assess health effects related to occupational exposure to pesticides.

The two principal techniques we use are studies and surveillance of our workers.

Table I

Techniques Used to Assess Health Effects
Related to Occupational Exposure to Pesticides

A. Epidemiology Studies
    1) Mortality
    2) Morbidity
        a) Studies from medical records (health insurance
            claims and Sickness-Absence Records)
        b) Clinical Studies

B. Surveillance
    1) Physical Examination Reports
    2) Illness-related Worker Compensation Claims
    3) Biological Monitoring - Cholinesterase

Mortality Studies. The mortality study is the usual initial
approach towards assessment of health effects in a worker population. The reason for this is that any serious health hazard is
likely to be reflected ultimately in excess mortality from a
specific cause or group of causes. Another is that detailed information on causes of death in the general population is readily
available.

Some pesticides have been found to be carcinogenic to laboratory animals. The mortality study is the tool used to assess the
risk of cancer in humans exposed to these pesticides.

The investigative strategy for a mortality study involves i-
dentification of all workers at a location who have been exposed
to the product since start-up of the production of the product.
Workers who have retired or left employment are traced with the
assistance of the Social Security Administration to determine if
they are living or deceased. Copies of death certificates are
then obtained for all deceased individuals and the causes of
death as listed on the death certificates are used to compute
death rates. These death rates are then compared to rates in the
national population to determine if there are more deaths attributable to a specific cause than one would expect in the national
population.

If an excess number of deaths due to cancer are observed,
a detailed review of medical and occupational records is undertaken for each individual whose death was ascribed to cancer in
an attempt to investigate a possible association between these
deaths and exposure to the product. We take into consideration
the employment history prior to joining the company, the interval
between exposure and death, cumulative exposure to the product
and length of exposure, the toxicology of the other substances
in the workplace and the nature of the illness.

Morbidity Studies. Morbidity studies are carried out on an active work force and focus on causes of illness. The sources of data for a morbidity study can be existing records such as health insurance claims or sickness related absence records, if the cause of illness is recorded.

Clinical Studies. Clinical studies are designed to assess function of a specific body organ or system. They utilize a medical questionnaire, physical examination and laboratory and/or clinical tests tailored specifically to detect impaired function of the system or organ under study. A control of non-exposed workers is used for comparative purposes and the results from both the exposed and non-exposed workers are analyzed statistically to determine if the exposed workers have an increased incidence of the medical condition.

Surveillance. Surveillance of worker populations exposed to pesticides is done in conjunction with a company Occupational Health Program and its primary purpose is early detection and prevention of occupationally related illness. It involves analysis of the annual physical examination findings, illness-related–Worker's Compensation Claims and biomonitoring results.

Physical Examination Findings. The analysis of employee physical examination findings involves a review of all diagnoses made by the examining physician at each company location which handles pesticides. It also includes a determination of the number of individuals who have specific laboratory tests outside the reference range. If, for example we found that a particular location had a large number of employees with elevated liver funcion tests, we would immediately want to know where these individuals worked in the plant so that we can then examine their potential exposures and industrial hygiene sampling data to evaluate whether any exposures may have contributed to liver damage.

Worker's Compensation Claims. The examination of illness-related Worker's Compensation Claims can serve to identify acute medical conditions occurring in the workforce that are definitely work-related.

OSHA Definition of an Occupational Illness

OSHA has defined an occupational illness as any abnormal condition or disorder other than one resulting from an occupational injury, caused by exposure to environmental factors associated with employment. It includes acute and chronic illnesses or disease which may be caused by inhalation, absorption, ingestion or direct contact.

We review all the Worker's Compensation Claims sent in to the Safety Department by our locations and analyze those claims which conform to the OSHA definition of an occupational illness. We first categorize the illnesses according to the OSHA categories for occupational illness.

Table II

OSHA Categories for Occupational Illness
Occupational skin diseases and disorders
Dust diseases of the lungs
Respiratory conditions due to toxic agents
Poisoning (systemic effects of toxic materials)
Disorders due to repeated trauma
Disorders due to physical agents
All other illnesses

Then we compute incidence rates for Occupational Illness, Lost Workday Cases and Lost Workdays and compare these rates with the rates published by the Bureau of Labor Statistics for workers in chemical and allied product manufacturing.

We find that workers associated with pesticides may develop skin rashes and occasionally some respiratory symptoms due to inhalation of these materials.

Biological Monitoring - Cholinesterase. Exposure to organophosphate and carbamate chemicals may result in the inhibition of the acetyl-cholinesterase enzyme which is vital to the maintenance of effective nerve and muscle function. We are fortunate in having a biological monitoring tool, the cholinesterase test, which allows us to detect a potentially significant exposure to organophosphorous compounds before the onset of clinical symptoms.

Although the carbamate compounds may also inhibit this enzyme, any inhibition which occurs tends to be short in duration owing to rapid biological reactivation of the enzyme.
limits our ability to detect a carbamate-related inhibition.

To conduct health surveillance of workers potentially exposed to organophosphate pesticides, Cholinesterase Biomonitoring Programs have been instituted at our research centers and field research stations, toxicology centers and manufacturing and formulating plants. Many of you probably participate in a similar program.

The personnel in the Program have baselines calculated from blood samples taken at pre-employment or following a long period of time in which they have not been exposed to cholinesterase inhibiting compounds. Subsequent values are then compared to these baseline values. A mild to moderate decrease (10-15%) for baseline in either red blood cell or plasma cholinesterase suggests exposure. A decrease greater than or equal to 30% in either plasma or red blood cell cholinesterase in indicative of an excessive exposure.

For an individual identified as having test results 30% or more below baseline, we examine his work history and plant production records to determine what product was being made in this work area on or just before the date blood was drawn. We also review industrial hygiene sampling results which will provide us with the actual level of exposure. These data then can be used to determine what additional protective measures need to be instituted to prevent a recurrence.

Let us examine some of the epidemiologic research which appears in the open literature.

Review of Selected Epidemiologic Studies Related to Pesticide Exposure

Until recently the emphasis has been on studies of morbidity rather than mortality. The primary reason for this relates to an inability to associate excess mortality with a specific pesticide.

Frequently, work histories have been vague and job titles even more vague. Workers would be classified as "A" Operators or "B" Operators with no indication where the person worked. This made it difficult to assess the potential exposure that a worker might have had to the product under study. The end result was that we studied the entire plant population.

With the introduction of more specific job titles, methods of tracking employees as they move about the plant and the advent of better epidemiologic and statistical methodologies to identify causal agents in a multiple exposure environment, we anticipate that more studies will be directed towards assessment of the mortality experience.

There are studies in the literature which have implicated benzene, arsenic and certain compounds of hexavalent chromium as human carcinogens. The inference is that pesticides which incorporate these substances may be potential human carcinogens.

The findings of an occupational study conducted by Mabuchi and colleagues reported in the Archives of Environmental Health, in 1979, do suggest that occupational exposure to arsenical pesticides increases the risk of cancer.[1] This study involved workers at a plant which manufactured and formulated arsenic-base insecticides, rodenticides, and herbicides. Mortality from lung cancer in male workers was significantly higher than expected for workers with presumed high exposure to arsenicals.

Morbidity Studies. Morbidity studies have revealed a wide variety of toxic effects in workers exposed to pesticides.

Neurological Effects. A major incident of occupationally-related illness associated with a pesticide involved Kepone. Kepone is a chlorinated hydrocarbon insecticide used domestically as an ant and roach poison. In 1975, after workers at a plant

which manufactured Kepone were discovered to have a variety of ailments, a clinical study of 133 employees was undertaken.[2]
The workers reported disorders characterized by onset of tremors, chest pain, weight loss, mental changes, skin rash, muscle weakness, loss of coordination and slurred speech. Over half had experienced tremors following exposure to Kepone. These findings suggested that Kepone produced neurological disorders involving the brain, peripheral nerves and muscle and the liver. In addition to the neurological findings, sperm counts were reportedly decreased. Today, following treatment to help their bodies eliminate the Kepone, most of the affected workers have no remaining signs of Kepone poisoning and are able to work again.

Reproductive Effects. Another clinical study, this one designed to assess reproductive function, involved a fumigant, 1, 2-dibromo-3 chloropropane (DBCP). In 1977, Whorton and Milby investigated the testicular function of 145 employees of a plant which formulated DBCP.[3] They used a questionnaire, physical examination which focused on the reproductive system, sperm counts and blood tests to determine the level of the hormones that stimulate and maintain sperm production.
The findings showed that approximately 45% of the workers tested had sperm counts less than 40 million/ml. of semen. (For this study, the authors considered normal sperm counts to be 40 million/ml. or greater.)
There also appeared to be a direct relationship between exposure duration and sperm count. Workers with sperm counts of 1 million/ml. or less had been exposed for at least three years. No workers whose sperm count exceeded 40 million/ml. had been exposed for more than three months.
One year later, the investigators re-examined twenty-one of the employees who were found to have either no sperm or a sperm count of less than 20 million/ml.[4] Of the men who had no sperm in 1977, they found that none showed improvement in 1978. However, the nine men who had sperm counts less than 20 million/ml. did show evidence of improvement. Their data suggest that DBCP induced testicular dysfunction is likely to be reversible among the moderately affected individuals. However, reversibility among the severely affected men was not detected, possibly because insuffient time had elapsed since cessation of exposure.

Studies On Field Workers. The extent to which farm workers are adversely affected by exposure to pesticide residues on the foliage of treated crops and in the soil is difficult to assess because cases are largely undetected and grossly under-reported.
In an attempt to get some idea of the magnitude of the problem, the California Department of Public Health conducted a study of field workers in Tulare, California.[5] They interviewed 1,120 non-migrant farm workers concerning the occurrence in the previous year of symptoms such as nausea, eye and skin irritation, chronic headaches, and sleeplessness.

They interviewed a group of controls at the same time who were
of the same economic, social and ethnic background and lived in
the same area but were not engaged in agricultural field work.
The field workers reported the symptoms approximately fifteen
times that of the controls.

Ongoing Research. In terms of ongoing surveillance, the California Department of Food and Agriculture reviews annually all
Doctor's First Reports of Worker Injury which are required to be
submitted by physicians that treat any occupationally related
illness or injury. In 1977, more than 1.5 million occupational
illness and injury reports were submitted from the California
workforce of more than 12 million workers. Of these 1,531 cases
were classified as "probably" related to pesticide exposure.

The National Institute for Occupational Safety and Health
(NIOSH) is currently assembling a registry of all workers in the
United States who may have been exposed to dioxin in the manufacture of the herbicide, 2,4,5-T. After the work histories and
medical information have been collected on each employee, they
will proceed with a mortality study to evaluate the mortality
patterns of these workers, with particular attention being paid
to cancer cases.

In addition, the Environmental Protection Agency has an Epidemiologic Studies Program which has several studies in progress.
Examples of some of the types of studies include: determination
of the number of pesticide poisonings in field workers, a determination of the body burden of pesticides and physiological responses the effects of organophosphates on cholinesterase values,
worker safety during reentry into recently sprayed orchards, and
a determination of the relationship between pesticides and urinary excretion of pesticide metabolites.

New Research. Regarding new research, the Department of
Labor and the Environmental Protection Agency plan to co-sponsor
a five-year study of the effects of pesticide exposure on the
health of youths under sixteen years of age who are employed in
agricultural operations. The study will be undertaken to determine: a) actual pesticide exposure and physical effects of such
exposure; b) absorption rates of pesticides into the body, and
c) acute and chronic health effects in relation to duration and
level of exposure.

The Contribution of Epidemiology

The principal contributions that epidemiology makes today are
the continued investigation of and elucidation and characterization of causes of disease in humans and the identification of
factors which contribute to their occurrence.

In an occupational setting, where prevention of work-related illness is a primary goal, analysis of employee medical information serves as a surveillance mechanism for early detettion of such illness.

Well designed occupational studies can identify agents in the workplace which may cause cancer or other disorders. They provide the only means for settling the issue of whether a specific product is a human carcinogen. In addition, they provide the essential information on worker populations required by the federal health agencies to supplement toxicology data in standard setting and re-registration of pesticides and by company management in establishing internal workplace standards.

Both government and the private sector are placing heavy emphasis on providing a safe and healthful workplace. Epidemiology will play a key role in the success of these objectives.

Literature Cited

1. Mabuchi, K.; Lilienfeld, A.; Snell, L. Lung cancer among pesticide workers exposed to inorganic arsenicals. Arch. Environ. Health 1979; 34, 312-320.

2. Taylor, J. R.; Selhorst, J. B.; Houff, S. A.; and Martine, A. J. Kepone Intoxication in Man-1. Clinical Observations. Paper presented in part at the 28th Annual Meeting of the American Academy of Neurology, April 26 - May 1, 1976, Toronto, Canada, 21pp.

3. Whorton, D.; Milby, T.; Krauss, R.; Stubbs, H. Testicular function in DBCP exposed pesticide workers. J. Occupa. Med. 1979; 21, 161-166.

4. Whorton, D.; and Milby, T. Recovery of testicular function among DBCP workers. J. Occupa. Med. 1979; 22, 177-179.

5. California Community Studies on Pesticides: Morbidity and mortality of poisonings. Report to Office of Pesticides, Bureau of State Services (EH), USPHS, January 15, 1970.

6. Maddy, K. T., Peoples, S. A.; and Johnson, L. A. The impact of pesticide exposures on community health services in California. Vet. Hum. Toxicol. 1979; Aug. 21, 262-5.

RECEIVED February 2, 1981.

The Application of Fundamentals in Risk Assessment

ALBERT C. KOLBYE, JR.

Associate Bureau Director for Toxicology, Bureau of Foods, Food and Drug Administration, 200 C Street, SW, Washington, DC 20204

"Risk assessment" is a popular term that appeals to scientists and regulators concerned with the vexing problems associated with evaluating and estimating potential hazards to human health. My preference is to talk in terms of evaluating potential hazards to human health and to avoid using the term "risk assessment." This presentation will focus on aspects of evaluating chemical safety in relation to carcinogenesis, but the fundamental considerations are relevant to many other biological end-points of human disease.

Extrapolation Models

Risk assessment has too many different meanings depending upon the viewpoints of the scientists and non-scientists using the phrase. Today, one meaning of risk assessment concerns usage of various mathematical models to extrapolate dose-response relationships of toxicologic data observed by experimental or epidemiological techniques in order to project estimates of expected disease incidence from populations of animals to humans exposed to significantly smaller amounts of the chemical substance under investigation. These mathematical models vary in the premises assumed to apply to the shape of the dose-response curve as exposures are decreased to zero levels. Extrapolation models are frequently applied by statisticians examining biological dose-response data who have developed a significant volume of literature concerning theoretical considerations of such models. These models represent a very simplistic approach towards data that in reality reflect highly complex biological considerations not easily explained to non-toxicologists and regulators, nor readily understood by lay people, such as the ordinary citizen/consumer.
　　Extrapolation models are usually applied when considering potentially carcinogenic chemicals from a regulatory and social policy-making viewpoint. Many of the models adopt very conservative premises that in effect assume the hypothesis that

there is no "safe" exposure to a carcinogenic chemical and that
any exposure will be associated with some definable risk for
cancer induction in the population at large. This philosophy
implies that there are no threshold phenomena associated with
cancer induction.

A recent comprehensive reference to extrapolation models
can be found in the Food Safety Council's Final Report of the
Scientific Committee, published June 1980.

Damage/Repair Balance Versus No-Threshold Premise

The no-threshold hypothesis evolved in relation to various
experimental and epidemiological studies concerning the biological hazards related to penetrating ionizing radiation. It has
been called the radiomimetic hypothesis, i.e., that carcinogenic
chemicals mimic the carcinogenic effects of penetrating radiation. There are some data that indicate that relatively small
increases in exposure to penetrating radiation are associated
with increases in the incidence of various cancers.

It is appropriate to reflect that penetrating radiation by
definition penetrates through tissues without respecting many of
the physiological barriers such as membranes which have very
important and complex functions to regulate the entry and exit
of chemicals in cells and micro-cellular organelles. We should
keep in mind, however, that while penetrating radiation can and
does induce damage to DNA, there are mechanisms existing in the
mammalian body to repair such biological damage. Unrepaired
damage to DNA can occur either by overwhelming physiological
repair or if the normal repair mechanisms cannot operate in
particular instances to repair certain types of damage. We also
know that error-prone repair may occur and contribute to the net
resulting damage of DNA.

When evaluating human exposures to penetrating radiation or
to chemicals, we should consider the balance between biological
damage and biological repair. If repair is complete, no permanent damage will occur. If repair is incomplete, or potentiates
the damage because error-prone repair is invoked, or if normal
repair is overwhelmed by excessive damage, then adverse effects
relevant to the induction of cancer and genetic damage are likely
to occur, provided the somatic or germinal cells involved
survive.

What Is A "Carcinogen"?

We can now reflect on some of the present philosophies we
currently employ to detect and regulate "carcinogenic" chemicals.
If exposing test animals to a chemical is associated with a
statistically significant increase in the incidence of cancers
in the test animals as compared to unexposed controls, our present
practice is to designate the chemical substance a "carcinogen."

Closer examination of this premise as practiced suggests that every influence upon the incidence of cancer exerted by various exposures to cancer substances, if a positive influence, would be designated as being "carcinogenic." Perhaps it is now appropriate to question the validity and sufficiency of that premise. (1) As scientists and regulators, we tend to generalize. Any generality will have its exceptions, and frequently generalizations are either over-extended or attacked because some exceptions exist. In fact, any generality has its limitations. If we continue to designate "carcinogens" on the basis of whether or not a statistically significant increase in the incidence of cancer is induced, let us examine further where that practice could lead us.

There is no question that some chemicals are strongly carcinogenic. After relevant experiments are performed, one can observe dramatic increases in the incidence of certain types of cancer as dose-related responses to the chemical exposure. But we should remember that a statistically significant increase in incidence does not necessarily represent a dramatic or powerful increase in incidence because, as the number of animals under test increases, the actual differences in incidence patterns deemed to be statistically significant will grow smaller. Thus, a highly significant statistical difference in incidence may, in actuality, be almost negligible in terms of public health importance, although obviously such is not always the case. A particularly relevant consideration is whether the "normal" incidence of cancer has just been shifted from one tumor type to another or merely represents an increased survival to older age of animals having an increasing risk for cancer induction with increasing age. One also wonders about the validity of interpreting epidemiological data concerning the apparently increased incidence of human cancer in one country as compared to another when age specific incidence patterns for all diseases competing for mortality have not been completely accounted for and evaluated in relation to each other. (2)

"Carcinogens" Versus Cancer Risk Factors

A more fundamental question is whether or not all influences on the incidence of cancer in animals or humans are necessarily related to direct "carcinogenic" action per se (in the sense of electrophilic activity leading to covalent bonding with DNA). Many scientists and regulators concerned with the prevention of cancer are insufficiently aware of a very extensive body of scientific literature which documents the fact that many factors can influence the incidence patterns of cancer in animals and humans. Many interactions can and do take place that either potentiate or ameliorate the effective potency of other substances, endogenous or exogenous, that have potential carcinogenicity. Other substances can dramatically influence the biological resistance of animals and humans to cancer

induction, thus acting to increase or decrease biological susceptibility to cancer induction. (3, 4) Obviously, a decrease
in biological resistance to cancer induction will increase susceptibility and, therefore, overall risk for cancer. In this
presentation, these substances are designated as "cancer risk
factors."

Carcinogenesis Involves Progressive Events and Is a Multistage Process

A very substantial body of evidence leads to the conclusion
that the induction of cancer is a multistage process involving
a progression of events leading to a formation of a colony of
malignant cells which then is called a malignant neoplasm.
Cancer cells are "malignant" because they do not respect normal
physiological boundaries and do not accept biological control
by the larger society of cells comprising the organism such as
the animal or human body. Thus, they parasitize the body, invade
into other tissues, and may seed new colonies in distant tissues
to form metastatic lesions. They do not usually attain the
malignant state immediately. Apparently passage through additional generations of cells is required in order for the cells
to attain a state of relative autonomy.
In its simplest form, the progressive events involved with
the induction of cancer have been referred to as the "initiation"
and "promotion" stages of carcinogenesis. This two-stage model
was first observed by scientists such as Berenblum, Shubik, and
Van Duuren who were investigating chemically-induced skin
cancer in rodents. They noted that skin cells could be
"initiated," i.e., selectively damaged in such a way that they
attained and retained a potential for malignant conversion. If
these epithelial cells were then subsequently treated with
certain chemicals capable of "promoting" the conversion of
initiated cells to malignant cells (but incapable of inducing
cancer by their own action alone), cancer would develop. While
these phenomena are not perfectly understood at the present
time, much knowledge concerning etiological mechanisms has been
developed in the past several decades, which is presented in
brief as follows:

Genotoxicity and Initiation

Usually, the first event in this multistage progression
takes place when certain types of damage to DNA are caused by
viruses, radiation, or chemical insult. (5) The latter involves the capability of some electrophilic chemicals to react
with DNA and covalently bind to it (such as by alkylation),
and disrupt normal sequencing of base pairs during replication.
Such chemicals have been described as being genotoxic, although

this term is employed also in a broader sense of describing
other mutagenic capabilities of the chemicals in question, which
may involve different mechanisms leading to genotoxic damage.
When certain types of genotoxic damage to DNA has occurred and
other preconditions are fulfilled, the cell may proceed to the
state of being "initiated" in the sense of now having malignant
potential. The preconditions are that the cell survives the
toxic insult to DNA, that the damage to DNA is not sufficiently
repaired to negate the damage, and that the "critical" unrepaired and damaged DNA can be encoded into the replicating
genome to persist unrepaired in future generations of cells
propagated from the one(s) originally incurring the critical
genotoxic damage.

If an initiated cell for some reason does not progress into
subsequent stages of events, and the state of initiation does
not revert back towards normality, the cells will retain their
potential for malignant conversion. There is evidence from
cell and tissue culture studies that varying degrees of rever-
sion towards normality appear to occur, but that it may well
not be complete in the sense that some initiated cells are likely
to replicate indefinitely in the future, retaining their state
of initiation and thus their potential for malignant conversion.

Promotion

Subsequently, a second stage in the progression of events
leading to formation of malignant neoplasms involves hyperplasia,
replication of critically damaged DNA in the active genome,
increased DNA and increasing degrees of abnormalities observed
in cell structure and function leading to autonomous behavior
and the biological characteristics associated with malignant
neoplasms as described by many cytologists and pathologists.
The subsequent stage of carcinogenesis has been referred to as
the "promotion" stage which has been extensively studied
originally in the experimental induction of skin cancer and
later with respect to the induction of other cancers. (6, 7)
It was first noticed that some chemicals, at the doses given to
skin, would not induce cancer by themselves or did so only after
a prolonged latency period. However, other chemicals, if
applied subsequently, "complete" the induction process by
inducing "promotional" phenomena and thus complete the progressive spectrum of events involved with the induction of
cancer. These phenomena have been reproduced experimentally
or have been observed to occur not only in skin cancer, but also
with respect to the induction of malignancies in liver (8, 9,
10), forestomach (11), lung (12), breast (13), kidney (14),
bladder (15, 16), and colon (13, 17, 18, 19, 20). Apparently
such promoting substances may act in part by influencing
enzymes and inducing the synthesis of certain polyamines, which

in turn stimulate hyperplasia and DNA replication, and induce changes with respect to cell cycling, cellular differentiation and maturation. Other changes may occur that are associated with membrane activity and function. An important effect of accelerated cell division is to accelerate the expression of fixed DNA damage in the replicating genome.

The interaction between biological repair of DNA damage and increased fixation of such damage by accelerated DNA replication is critical, because the timing and effectiveness of physiological DNA repair may be disturbed by the increased mitotic activity induced by hyperplastic toxicity. Error-prone repair can also augment the degree to which damaged DNA is propagated into replicating DNA, causing cellular "initiation" which, if "promoted," can progress to neoplastic growth and the abnormal characteristics associated with cancer.

The classic promoters are not carcinogenic per se, or only weakly so, since by themselves they usually do not induce cancer, but when applied to target cells which have already been initiated by a cancer initiator, promoters will facilitate, enhance, and potentiate the effective potency of the initiators to induce malignant transformations expressed as an increase in tumor incidence and the earlier appearance of malignancies.

Direct-Acting Complete Carcinogens

Direct-acting complete carcinogens have the ability by themselves both to initiate and promote tumor induction so that if critical doses are attained, the lesion induced progresses to frank malignancy, i.e., cancer. Other carcinogens may have less promoting capabilities. Many potentially carcinogenic chemicals require metabolic activation before the metabolite has sufficient electrophilic biochemical activity to damage DNA and initiate the series of steps progressing to cancer induction. (5)

Activation

If activation of procarcinogens to electrophilic metabolites is enhanced, the expectable result is an increase in the number of carcinogenically active molecules capable of initiating cancer. Conversely, as deactivation or detoxification increases, one would expect lesser amounts of potentially carcinogenic metabolites available to initiate cancer induction. Many compounds can enhance or inhibit carcinogenesis without being carcinogenic themselves (or one carcinogen can enhance or inhibit another) by inducing or inhibiting microsomal enzymes. These metabolic pathways important to the activation or deactivation of potentially carcinogenic compounds may be distorted or shunted, thus markedly affecting the degree to which a procarcinogen is activated to electrophilic status.

Many polycyclic aromatic hydrocarbons have this effect, as do
the notable examples of polychlorinated biphenyls and phenobarbital.   In many instances, they induce more deactivation
than activation, but in some instances they augment activation.
These phenomena have been extensively described and documented
in the scientific literature.

Tumor Promoters/Potentiators

As discussed earlier, other chemicals are capable of inducing a series of changes in the same target cells that are
susceptible to initiation of cancer, but these chemicals are
only capable of acting as tumor promoters or tumor potentiators.
Examples of these chemicals range from the croton oil derivatives, such as certain phorbols, on through a broad spectrum
of other naturally-occurring or man-made compounds.   Many of
these compounds have been designated already as potential carcinogens by observing statistically significant increases in
the incidence of cancer associated with exposures to these
chemicals.   It may be of interest to suggest that chloroform
and carbon tetrachloride have tumor promoting or potentiating
activity, as do saccharin, DDT, PCBs, certain phenols, phenobarbital and, believe it or not, such substances as ethanol,
bile acids, citrus oil, and others, including hormones such
as estradiol.   (21)

Modifying Factors

In similar fashion, but now on a broader biological scale,
the sensitivity to cancer induction of an organism, be it a
cell, tissue, organ, or animal, can be substantially influenced
by other "modifying factors."   These indirectly act to influence the milieu in which the target cells exist and modify
biological resistance.   If one immuno-suppresses an animal and
oncogenic viruses are endogenously present, an increase in the
risk for cancer induction will likely be the result.   If one
alters the hormonal status, profound changes in the biological
activity of many cells will occur, affecting cancer risk.
Genetic factors also influence risk for cancer.   Excesses or
deficiencies of certain micro-nutrients such as vitamins or
minerals or macro-nutrients, such as protein or fat, also can
have dramatic influences upon the susceptibility of animals
to the induction of cancer.   Such modifiers appear to act by
modifying or intoxicating enzymes, co-factors, substrates, and
membranes important to the maintenance of normal homeostatic
physiological function and biological defenses against cancer
induction.   Some biological defense mechanisms against toxic
insults, including those from carcinogens, involve substances
like Vitamins A, C, and E.   Glutathione and other sulfhydryl
compounds can also deactivate carcinogens.

Hyperplastic Toxicity

The term "hyperplastic toxicity" is used in this presentation to describe toxicity-induced cell proliferation associated
with increased mitotic activity, increased DNA (as judged by
increased density of nuclear chromatin staining or indicia of
increased DNA synthesis) and other changes associated with
enzyme and membrane functions which are not of a malignant
nature per se.
     Hyperplastic toxicity is a very common cellular reaction
that can occur in almost all mammalian tissues in response to
a variety of toxic insults. The cells proliferate, increasing
in number by reproducing faster. The influence of toxicity
and hyperplasia on cancer induction has been the subject of much
interest and some controversy for many years. The controversy
centered on the nature and extent of the precise role of hyperplasia in relation to carcinogenesis. One question was somewhat
over-emphasized in the minds of many cancer researchers: was
the induction of hyperplasia a necessary precondition for
induction of cancer? (For many centuries, physicians observed
that the onset of cancer appeared to be associated with chronic
irritation and inflammation of tissues, such as scrotal cancer
in chimney sweeps, skin cancer in certain occupations where skin
irritation was observed in association with exposure to chemical
substances, and in this century the association of lung cancer
in people with chronic bronchitis induced by inhaling tobacco
smoke or other irritating fumes.) Beginning early in this
century and continuing to this day, many experiments were performed to provide data to clarify the relationships between
hyperplasia and cancer. The answer to this particular question
seems to be the following: Observable hyperplastic toxicity
is not a necessary prerequisite for the induction of each and
every type of cancer, if by induction of cancer you mean
initiation of cancer.
     But all cancers are forms of malignant hyperplasia. Hyper-
plasia, metaplasia, and dysplasia are observed and documented
progressive stages in the development of malignant neoplasms,
i.e., those new growths of malignant tissue cells we call
cancer. Hyperplastic tissue responses to many toxic agents
involve abnormal acceleration of cell replication which in turn
involves a marked increase in mitotic activity, including an
increase of DNA.
     Not all compounds which induce hyperplasia can act as tumor
promoters, thus while the phenomena associated with tumor promotion include hyperplastic changes, hyperplasia per se is not
precisely the same phenomenon as promotion. However, certain
patterns of hyperplastic toxicity appear to be identical and
coincidental with certain biological phenomena observed to
occur when the classical tumor promoters are administered to

the same target cells. If the toxicity induced by toxic
challenges results in the induction of certain polyamines or
other factors that appear to be capable of stimulating cellular
and DNA hyperplasia and changes in membrane and enzyme functions,
then a tumor promoting action is likely to result which will
facilitate and enhance cancer induction provided that the
target cells have already been initiated by the same or different chemicals. (22)

Having seen that hyperplasia in one form is an integral
part of the expression of an initiated cell into a neoplasm, and
that other forms of hyperplastic toxicity are co-equal with
tumor promotion, we can now ask whether or not pre-existing
hyperplastic toxicity enhances the biological susceptibility
of particular tissues to the induction of cancer? The answer
to this question is yes, at least in many instances, since a
substantial body of evidence again illustrates many situations
where pre-existing toxicity and hyperplasia resulted in an
increase in the biological susceptibility of target tissues to
cancer initiation and further promotion. (14, 23, 24) It
would appear that one factor may involve DNA replication with
an increase in the amount and surface area of DNA available as
a target for alkylation and mispairing. Other factors that may
be involved include: increased permeability of membranes to
toxic agents, distortion and impairment of enzymes, and related
co-factors and substrates. These are important for maintenance
of normal physiological cell functions and biological defense
against toxic insults including those from electrophilic carcinogens. The deactivating activities of the endoplasmic
reticulum and microsomal enzymes also protect against carcinogens. If these protective metabolic pathways are functionally
distorted or impaired, biological susceptibility to cancer
induction can be increased.

Dose-Response Considerations

As mentioned earlier, complete carcinogens can both
initiate and promote cancer by themselves, but differences in
dose-response have been observed and documented, which suggest
that the promoting action of complete carcinogens is related
to higher and repeated dosages of the same chemical, whereas
initiation of malignant transformation may occur at lower
doses. This implies then that the risk for cancer induction
by a complete carcinogen will be increased as exposure to that
carcinogen increases, because not only will a greater amount
of cellular initiation be effected, but the promoting action
of the compound will be more effectively expressed by repeated
exposures and thus these driving forces will result in the
faster induction of cancer.

Dosing Test Animals at Substantial Levels

Our current practice is to dose test animals at substantial
levels to determine whether or not a compound induces cancer.
If such dosing results in cell damage such as hyperplastic
toxicity (25), which may invoke tumor promoting or potentiating
activity, the fact that endogenous initiators (such as nitrosamines) exist in the mammalian body implies that a positive
induction of cancer in a population of test animals will not
differentiate between initiation and promotion. (1, 3) Therefore we will not know from this evidence alone whether the
"carcinogen" is an initiator or a promoter, or both. If the
compound when tested at substantial levels can potentiate the
effective carcinogenic potency of endogenous carcinogens by
other mechanisms, or if it acts to decrease biological susceptibility to cancer induction from endogenous carcinogens,
or from other carcinogens that may exist environmentally, again
we will not be able to tell what category of carcinogenic
activity we are dealing with unless we ask the relevant questions
in the first place and develop the data to provide relevant
answers.

Strategy for Preventing Chemically-Induced Cancer in Humans

The second consideration concerns our strategy for preventing cancer in humans by controlling exposures to chemicals
and setting priorities for testing, regulating, and other forms
of public health action. Since certain forms of toxicity
clearly can enhance the induction of cancer, should not one of
our highest priorities in cancer prevention be to prevent all
potentially toxic exposures to chemicals (not only for cancer
prevention but obviously to prevent all forms of toxicity)? By
preventing toxicity per se, likely we will prevent a significant
amount of cancer. Let us remember that many of the human cancers associated with human exposures to most occupational and
certain environmental carcinogens were well within toxic dose-
response ranges and that tissue damage including hyperplastic
toxicity were frequent concomitants.

The most dangerous carcinogens are likely to be those that
have the ability to initiate and/or promote cancer at subtoxic
doses not likely to attract much attention per se until
identified as such. Should we not place emphasis as a first
priority on detecting and controlling the worst carcinogens
first, i.e., those which are effective initiators and promoters
at relatively low subtoxic doses likely to be well within the
range of anticipated human exposure? If we are interested in
practicing that philosophy, then we need to re-examine our
procedures and practices for detecting carcinogens, modify our
testing protocols, and evaluate potential hazards to human
health from a different perspective than we do now.

First and Second Order Cancer Induction

We then should consider classifying the induction of
cancer into two classes: First Order, those having the capability to induce cancer at subtoxic doses, either because they
are effective initiators or complete carcinogens; and Second
Order, those that act to influence the induction of cancer by
inducing various forms of hyperplastic toxicity by activating
endogenous electrophiles, or by decreasing biological resistance
to other independently operating patterns of cancer induction.
(26, 27)
   The validity of this approach can be tested without going
to the extremes of an ED01 experiment such as that performed
with 2AAF at the National Center for Toxicological Research.
That experiment showed that 2AAF carcinogenicity involved two
patterns of induction operating at different dose-response
curves, one of which was operating within the lower doses
given.
   Simply stated, we can determine from 90-day in vivo studies
whether or not a test substance induces hyperplastic toxicity
in various target organs, and, if so, at what doses. Serial
sacrifices and interrupted dosing schedules are needed to
determine the progressive nature of the lesions noted and the
extent of biological repair. Within the range of inducing
hyperplastic toxicity, one group of animals is then carried
for lifetime dosing. Well below the range of inducing hyperplastic toxicity, other groups are carried through their
lifetimes with serial sacrifices being conducted to note progressive lesions. Another cohort of animals are subjected to
intermittent or interrupted dosing at subtoxic levels. Additional investigations could be conducted such as using phenobarbital and polychlorinated biphenyls to induce microsomal
enzymes and to promote hyperplastic toxicity in organs such as
liver in order to determine the effects of added stress. If
conducted properly, we should have enough in vivo data to
determine whether we are dealing with a first or second order
carcinogen if cancer in fact is induced. Other sources of
biological data may be helpful in this regard, especially
in vitro mutagenicity and transformation data.

Evaluating Cancer Risk Factors

As discussed above, there is a need to go beyond generalizations that a substance does or does not "induce" cancer.
We need to determine whether or not particular factors can
potentially influence cancer induction, and, if so, how and
under what circumstances? (26, 28) We need to evaluate
initiators and "complete carcinogens" in one category as First
Order carcinogens. Second Order compounds should not be called
"carcinogens", even though under some circumstances and at

some doses they may be powerful driving forces that can substantially influence cancer induction. Second order substances
may activate certain first order compounds to electrophilic
initiators of DNA damage important both to carcinogenesis and
mutagenesis. Other second order compounds may promote or
potentiate cancer induction by inducing hyperplastic toxicity
or invoking other mechanisms. Those compounds that interfere
with normal physiological status by disturbing protective
enzymes, substrates, vitamins, nutrients, hormones, immune
mechanisms, etc., are likely to increase risk for cancer induction caused by independently operating first order substances
either manufactured endogenously in the body (such as nitrosamines) or entering into bodily contact from environmental
sources. The role of "toxicity" per se should not be under-
estimated as a potentially powerful force that can substantially
influence biological susceptibility to cancer induction by
decreasing biological resistance.

Such cancer risk factors should be evaluated using available
techniques (from the fields of toxicology, pharmacology, and
nutrition) to study dose-response phenomena associated with
their modes of action, metabolism (both normal and abnormal),
and excretion. Newer techniques for studying interactions
without having to resort to super-scale life-time bioassays
are becoming available every year. In vitro techniques can
supplement short-term in vivo studies, but we should not over-
emphasize in vitro approaches, nor under-estimate the value of
in vivo biological data since it is extremely important to
estimate the toxicity of compounds in the context of the
biological defense-mechanisms available to the mammalian body.
(29, 30) In this regard, epidemiological studies can provide
much needed information concerning biological susceptibility
and resistance factors, particularly in relation to human
metabolism, nutrition, and genetic factors which are extremely
important.

Last, but by no means least, the mode and amount of exposures to various substances should be kept in full consideration
at all times. We have tended to over-emphasize micro-exposures
to substances we suspect of having unusual patterns of biological
activity, at the expense of having ignored the macro-exposures
to the universe of natural compounds present in the foods we
eat. Just because they may be nutrients and Nature-produced
does not mean that they cannot and do not exert powerful influences upon cancer induction. They can and do, both to
protect against or to enhance cancer induction, but that is
another topic for a different time.

"Literature Cited"

1. Kolbye, A.C., Jr., Decision-making Issues Relevant to Cancer Inducing Substances. In "Regulatory Aspects of Carcinogenesis and Food Additives: The Delaney Clause" (Coulston, F.), Academic Press, Inc., New York, N.Y., 1979, 93-102.
2. Kolbye, A.C., Jr. Cancer in Humans: Exposures and Responses in a Real World. Oncology, 1976, 33, 90-100.
3. Kolbye, A.C., Jr., A U.S. Viewpoint: Legislative and Scientific Aspects of Cancer Prevention, Preventive Medicine, 1980, 9, 267-274.
4. Sivak, A., Mechanisms of Tumor Promotion and Cocarcinogenesis: A Summary from One Point of View. In T.J. Slaga, A. Sivak, R.K. Boutwell, eds. "Carcinogenesis - A Comprehensive Survey - Volume 2", Raven Press, New York, N.Y., 1978, 553-564.
5. Miller, E.C., Some Current Perspectives on Chemical Carcinogenesis in Humans and Experimental Animals: Presidential Address, Cancer Research, 1978, 38; 1479-1496.
6. Boutwell, R.K., Biochemical Mechanism of Tumor Promotion. In T.J. Slaga, A. Sivak, R.K. Boutwell, eds., "Carcinogenesis-A Comprehensive Survey - Volume 2", Raven Press, New York, N.Y., 1978, 49-58.
7. Boutwell, R.K., The Function and Mechanism of Promoters of Carcinogenesis, CRC Crit. Rev. Toxicol., 1974, 419-443.
8. Farber, E., Solt, D., A New Liver Model for the Study of Promotion. In T.J. Slaga, A. Sivak, R.K. Boutwell, eds. "Carcinogenesis - A Comprehensive Survey-Volume 2", Raven Press, New York, N.Y., 1978, 443-448.
9. Pitot, H.C., Barsness, L., Kitagawa, T., Stages in the Process of Hepatocarcinogenesis in Rat Liver. In T.J. Slaga, A. Sivak, R.K. Boutwell eds. "Carcinogenesis - A Comprehensive Survey - Volume 2", Raven Press, New York, N.Y., 1978, 433-442.
10. Kitagawa, T., Sugano, H., Timetable for Hepatocarcinogenesis in Rat. In Nakahara, W., Hirayama, T., Nishioka, K. and Sugano, H., eds. "Analytic and Experimental Epidemiology of Cancer", University Park Press, Baltimore, London, Tokyo, 1972, 91-105.
11. Goerttler, K., Loehrke, H., Schweizer, J., Hesse, B., Systemic Two-Stage Carcinogenesis in the Epithelium of the Forestomach of Mice Using 7, 12-dimethylbenz(a)anthracene as initiator and the Phorbolester 12-0-tetradecanoyl-phorbol-13-acetate as promoter. Cancer Research, 1979, April; 39(4), 1293-1297.
12. Witschi, H., Lock, S., Butylated Hydroxytoluene: A Possible Promoter of Adenoma Formation in Mouse Lung. In T.J. Slaga, A. Sivak, R.K. Boutwell, eds. "Carcinogenesis - A Comprehensive Survey - Volume 2", Raven Press, New York, N.Y., 1978, 465-474.

13. Wynder, E.L., Hoffman, D., McCoy, G.D., Cohen, L.A., Reddy, B.S., Tumor Promotion and Cocarcinogenesis as Related to Man and His Environment. In T.J. Slaga, A. Sivak, R.K. Boutwell, eds., "Carcinogenesis - A Comprehensive Survey - Volume 2", Raven Press, New York, N.Y., 1978, 59-78.

14. Pound, A.W., Influence of Carbon Tetrachloride on Induction of Tumours of the Liver and Kidneys in Mice by Nitrosamines. Br. J. Cancer, 1978, 37, 67-75.

15. Hicks, R.M., Chonaniec, J., Wakefield, J.St.J., Experimental Induction of Bladder Tumors by a Two Stage System. In T.S. Slaga, A. Sivak, R.K. Boutwell, Eds. "Carcinogenesis - A Comprehensive Survey - Volume 2", Raven Press, New York, N.Y., 1978, 475-489.

16. Hashimoto, Y., In Vitro and In Vivo Effects of Nitrosamine and Urea on Rat Urinary Bladder Epithelial Cells. In T.J. Slaga, A. Sivak, R.K. Boutwell, eds. "Carcinogenesis - A Comprehensive Survey - Volume 2", Raven Press, New York, N.Y., 1978, 443-448.

17. Narisawa, T., Magadia, N., Weisburger, J., Wynder, E., Promoting Effect of Bile Acids on Colon Carcinogenesis after Intrarectal Instillation of N-Methyl-N'-Nitro-N-Nitroso-guanidine in Rats, J. Natl. Cancer Ins., 1974, 53, 1093-1097.

18. Pollard, M., Luckert, P.H., Promotional Effect of Sodium Barbiturate on Intestinal Tumors Induced in Rats by Dimethyl-hydrazine, J. Natl. Cancer Inst., 1979, 63: 1089-1092.

19. Reddy, B.S., Narasawa, T., Weisburger, J.H., Wynder, E.L., Promoting Effect of Sodium Deoxycholate on Colon Adenocar-cinomas in Germfree Rats. J. Natl. Cancer Inst., 1976, 56, 441-442.

20. Reddy, B.S., Weisburger, J.H., Wynder, E.L., Colon Cancer: Bile Salts as Tumor Promoters In T.J. Slaga, A. Sivak, R.K. Boutwell, eds. "Carcinogenesis - A Comprehensive Survey - Volume 2", Raven Press, New York, N.Y., 1978, 453-464.

21. Boyland, L. Some Implications of Tumour Promotion in Carcino-genesis, I.R.C.S. Med. Sci., 1980, 8: 1-4.

22. Reddy, J.K., Rao, M.S., Enhancement by WY-14, 643, A Hepatic Peroxisome Proliferation of Diethylnitrosamine - Initiated Hepatic Tumorigenesis in Rat, Br. J. Cancer, 1978, 38, 537-543.

23. Anthony, P.P., Precancerous Changes in the Human Liver, J. Toxicol. Environ. Health, 1979, 5,(2/3), 301-313.

24. Craddock, V.M., Cell Proliferation and Experimental Liver Cancer, In H.H. Cameron, D.A. Linsel, and G.P. Varwicks eds. Liver Cell Cancer, Elsevier Scientific Publishing Company, Amsterdam-New York-Oxford, 1976, 153-201.

25. Reitz, R.H., Quast, J.F., Watanabe, P.G. Gehring, P.J., Chemical Carcinogens: Estimating the Risk, Science. 1979, 205; 1205-1207.

26. Hecker, E., Cocarcinogens or Limited Cancer Inducing Agents. Important New Aspects of the Etiology of Human Tumors and the Molecular Mechanisms of the Origin of Cancer. Naturwissen-Schaften, 1978, 65, 640-648.
27. Berenblum, I., A Re-evaluation of the Concept of Cocarcino-genesis, Progr. Exp. Tumor Res, 1969, 11, 21-20.
28. Berenblum, I., Carcinogenicity Testing for Control of Environmental Tumor Development in Man, ISR. J. Med. Sci., 1979, 15(6), 973-479.
29. Kolbye, A.C., Jr., Regulatory Considerations Concerning Muta-genesis, J. Soc. Cosmet. Chem., 1978, 29, 727-732.
30. Kolbye, A.C., Jr., "Impact of Short-Term Screening Tests on Regulatory Action, Applied Methods in Oncology #4", Elsevier Biomedical Scientific Publishing Company, Amsterdam, 1980. (In Press).

RECEIVED February 12, 1981.

Toxicology: A Summary

LEON GOLBERG[1]

Chemical Industry Institute of Toxicology, P.O. Box 12137,
Research Triangle Park, NC 27709

There has never been a more exciting time in Toxicology.
Never before have we had such a wealth of new ideas and concepts
pumped in by the basic sciences, such a multiplicity of new
methods and approaches, such sophisticated and sensitive analytical procedures. Our problem is to assimilate and apply all
these opportunities, which bid fair to revolutionize the classical approaches to safety evaluation. Hence an even greater
source of concern may be expressed thus: will we be afforded
a breathing-space to develop the new tests to a satisfactory
point, before they become a part of government regulation?
    In my introductory address, I referred to a different kind
of regulation, namely the basic biological regulatory and defensive mechanisms that exist within each cell and between cells,
making possible the integrated harmonious functioning of the
whole organism that we call "homeostasis". Superimposed on
this basic concept of the capacity of the organism to adapt
to change is the clear evidence of limits to the capacity of
these defensive mechanisms to cope with endogenous changes (for
instance, caused by disease processes) or exogenous environmental
changes. Equally, the body's defenses can be overwhelmed by
the action of toxicants - both physical and chemical agents.
    Five factors help to determine the impact of a toxic agent
on any population of experimental animals or people. These
are as follows: the potential of the compound to bring about
specific toxic effects; its potency under defined experimental
conditions; the degree and circumstances of exposure of the
population; the range of individual susceptibilities within
that population; and the synergistic or antagonistic interactions
occasioned by simultaneous exposure to a multiplicity of toxic
agents.
    Selection of a compound for toxicity testing often raises
many different questions that need to be addressed in advance
of biological experimentation. These issues include the decision
on specification of the test material, the nature and concentrations of trace impurities, as well as the changes that occur
on storage or admixture with animal diets.

[1] Current address: 2109 Nancy Nanam Drive, Raleigh, NC 27607.

0097–6156/81/0160–0187$05.00/0

In the hierarchy of toxicity testing, the use of structure-activity relationships is becoming an increasingly important predictive tool. The reliability of this tool is, of course, dependent on the accuracy of the data bases which it incorporates. Efforts are under way to make these bases more reliable. Further steps in safety evaluation may follow a "decision tree" approach; any scheme adopted should involve a variety of screening procedures, including tests for genetic toxicity, as well as early studies of metabolism and pharmacokinetics.

Finally, the importance of human studies was emphasized - not only studies in human volunteers, or epidemiological research, but also the use of breath analysis, human lymphocytes, hemoglobin, etc. as indices of exposure and of effects of such exposure. The suggestion of a human liver bank was also discussed.

Consideration was next given (by Dr. B. Schwetz) to the widening concepts of toxicology, specifically in relation to changes in time, space, species, toxins, concentrations and parameters of concern. Time at which toxic effects might be manifested had extended beyond the immediate future to subsequent generations as yet unborn, and toxicological test procedures had been developed to cope with these concerns. The localization and distribution of toxins to remote recesses of the environment and ecosystems now involved a huge range of species and, most immediately, pets, wild animals, beasts of burden and food sources. Among the compounds being studied, there was increasing emphasis on environmental toxicants like PCB's, PBB's, TCDD and other dioxins and dibenzofurans. In parallel with developments in analytical chemistry, toxicology was now concerning itself with amounts as low as a few molecules. Societal concern now embraced the quality of life and, often, a misplaced insistence on zero risk.

The trends in toxicology reflected these developments. More reproductive studies of a more sophisticated kind were following effects on sperm and ova, as well as on the complete reproductive process. Studies of the developing embryo and fetus reflected a more critical and reasonable attitude to thresholds and dose-response aspects of teratogenicity. Greater involvement in behavioral studies had led to combined approaches that permitted analysis of the postnatal consequences of in utero exposure. Finally, the contribution of immunotoxicology was increasing; not infrequently, compounds stimulated the immune system at low levels of exposure and inhibited it at high levels.

Analysis of the biochemical aspects of organ specificity in toxic action (by Dr. J. S. Dutcher) began with a survey of the factors that influence patterns of organ-specific toxicity. Notable among these are the mechanisms of detoxication and metabolic activation, leading to covalent interaction with cellular macromolecules and consequent toxicity.

A superb example of the shifting target of toxic action

had been found in the furanoterpene, 4-ipomeanol. Organ-specific
toxicity reflected tissue levels of alkylation consequent upon
metabolic activation. In the rat, bronchiolar cell necrosis
resulted from activation of the compound in Clara cells. In
the mouse, beside the lung, renal tubular necrosis occurred.
The liver was the target organ at all doses in the Japanese
quail, whereas in the Syrian golden hamster both lung and liver
necrosis was observed. The explanation for these targets hinged
upon local sites of metabolic activation. Possible stability
of the resulting metabolites formed, and subsequent transport
from the site of activation, had been ruled out as a mechanism
for toxicity in other organs. Studies involving microsomal
enzyme induction or inhibition had revealed changes in severity
or site of toxic action. Equally, the protective effect of
glutathione had been established in various organs and species.
      As a starting-point in his discussion of genotoxicity,
Dr. G. M. Williams analyzed animal carcinogenicity from the
standpoint of the definition of a carcinogen, based upon an
operational description, and the classification of carcinogens.
Two broad categories of carcinogens are recognized: those that
are genotoxic and elicit DNA damage, and epigenetic carcinogens
that involve no DNA damage. This dichotomy excludes reversible
binding to receptor sites and action of intercalating agents,
which are mutagenic but not carcinogenic.
      The correlation between evidence of genotoxicity and finding
of carcinogenicity is greatest for DNA repair, resulting from
covalent damage to DNA and amplification of DNA repair synthesis
in response to the lesion in DNA. Good correlation exists for
mutagenicity observed in bacterial and mammalian cell systems.
A correlation has not been established for sister chromatid
exchange and neoplastic transformation of cells in vitro. Thus
the ultimate purpose of in vitro tests is to limit further testing,
particularly to eliminate long-term studies and to provide an
understanding of the mechanism of action of the test compound.
      After reviewing the various available tests, and emphasizing
the shortcomings of the S9 fraction (which entails a selective
loss of detoxication potential), the make-up of a battery of
appropriate short-term tests was discussed. A battery comprising
the following five tests had shown a high degree of sensitivity:
a bacterial test; DNA repair with primary cultures of hepatocytes;
mammalian cell mutagenesis using various cell lines, including
replicating liver epithelial cells; sister chromatid exchange,
the most objective evidence of chromosomal damage; and cell
transformation in vitro. If a test material is positive in
the first two of these tests, there is a definite presumption
of carcinogenicity.
      A brief account was presented of abbreviated in vivo bio-
assays for carcinogenicity. These included skin painting, with
or without a promoting agent (TPA); the production of pulmonary
tumors in strain A mice; the development of breast cancer in

female Sprague-Dawley rats by day 55; and altered foci produced
in rodent liver, which correlate well with subsequent development
of hepatocellular carcinoma.

A positive result in any in vitro bioassay, coupled with
a positive result in one of the limited in vivo bioassays probably reflects carcinogenic potential.

Discussing the mode of action of carcinogenic pesticides,
Dr. Williams dwelt on the polychlorinated compounds that elicit
tumors in rodents, mouse liver being the particular target and,
occasionally, the thyroid. It seemed clear that these were
epigenetic carcinogens that did not form covalent adducts with
DNA nor damage DNA. Various tests for unscheduled DNA synthesis,
point mutations and neoplastic transformation were all negative.
Tests for promotional effect, in systems that revealed phenobarbital to be a prototype promoter, served to establish that the
chlorinated pesticides acted in the same way. Possible mechanisms of promoting action were reviewed.

Prediction of carcinogenic potential of pesticides was
taken a step further by Dr. S. Nesnow, who described EPA's "phased
approach" for the application of short-term tests to these compounds. Phase 1 involved detection of point mutations, DNA
damage and chromosomal effects in appropriate microorganisms.
Phase 2 aimed at verification of any positive findings by use
of higher-order test systems (human lung fibroblasts, recessive
lethality in Drosophila, dominant lethality in the mouse and
neoplastic transformation in cell culture systems). The final
stage called for quantitative risk assessment through the use
of rodents to study gene mutations and chromosomal effects,
as well as long-term carcinogenesis bioassay.

Application of these approaches to 38 pesticides was described in detail. Testing was far from complete as yet - as
an example, for only 13/38 compounds were carcinogenicity data
available in evaluated form.

The fundamentals of testing for reproductive and teratogenic
effects of chemical agents were described by Dr. R. Tyl. She
laid emphasis on teratogenic phenomena in man, one-fifth of
which were attributable to germinal mutations. Developmental
criteria and landmarks in rodents were tabulated to illustrate
the important applications of such data in safety testing.

The role of Epidemiology was presented by Ms. M. W. Palshaw,
with a clear analysis of occupational studies in this field.
She stressed that, while an association between exposure and
ill-health may be established by such means, cause and effect
relationships cannot be proved. The interface between epidemiologists and experts in other scientific disciplines (chemists,
chemical engineers, industrial hygienists, toxicologists, physicians and biostatisticians) was discussed, emphasizing the crucial importance of exposure data. Consideration of the OSHA
categories for occupational illness was followed by illustra-
tions drawn from biological monitoring of cholinesterase levels

in red cells and plasma, as well as neurological and reproductive effects. The contribution of Epidemiology to worker safety puts it in the front line of health protective measures.

In considering the evaluation of risks to human health, Dr. A. C. Kolbye reviewed the mathematical approaches to risk asssessment but concluded that such mathematical models ignore biological variables. Damage and repair, as well as cell replication acting as a fixative of DNA damage in the replicating genome, need to be weighed in relation to the no-threshold hypothesis which was developed by analogy with penetrating radiation.

Dose-response considerations should loom large in risk assessment. Where a compound acts as a promoter, its effectiveness at lower doses is likely to be decreased. By using "Maximum Tolerated Doses", cell and tissue damage is brought about that stimulates hyperplastic activity that not infrequently causes the action of endogenous initiators to be promoted to cancer. Accordingly an appropriate strategy for cancer prevention calls for classification into two categories: first order carcinogens or initiators that are effective at subtoxic doses; and second order promoters that bring about hyperplastic toxicity and associated phenomena. Such second-order compounds should not be regarded as carcinogens. They may influence first-order compounds, depending on the biological defense mechanisms present and the mode and amount of exposure. Experimental procedures are available that permit a clear distinction to be drawn between first- and second-order compounds.

RECEIVED February 12, 1981.

BIOCHEMICAL ASPECTS

Biochemical Aspects: An Introduction

GINO J. MARCO

CIBA–GEIGY Corporation, Biochemistry Department, 410 Swing Road,
Greensboro, NC 27409

One of the meeting grounds of biology and chemistry is in
the realm of biochemistry. In this portion of the conference,
we wish to bring the toxicological concerns of pesticides into
the molecular world more familiar to the chemist. In the first
"transition" paper, Dr. Laishes will give a more molecular
insight into the views of toxicological areas heard yesterday.
Concepts already mentioned will be discussed in chemical terms
on the mechanics of the carcinogenic processes as well as
possible repair mechanisms. Dr. Gillette will concentrate on a
most frustrating, albeit fascinating, aspect of biochemical
studies, the metabolite that is there but difficult to prove;
that is, the reactive intermediate. How much of a culprit is
that entity in toxicological expression? I know we will get an
insight on that question. The threshold concept and the
pharmacokinetics involved will be addressed by Dr. Ramsey. My
own prejudices suggest thresholds must exist and Dr. Ramsey will
have some interesting views along those lines.

In addition to studies on mode of action of pesticides, one
of the largest biochemical efforts in the area of pesticide
chemistry is devoted to the metabolism occurring in a variety of
biological systems; but of most importance in this conference,
the events occurring in mammalian systems. Dr. Ivie has an
intriguing overview of the metabolism area covering some of the
past and present observations with some thoughts about the
future effort. The future direction of biochemical strategies
will be discussed by Dr. Wright. Are there areas of biochemical
studies not addressed in a sufficient manner? His comments,
I'm sure, will stimulate discussions and possibly differences
of opinion.

And finally, as there are in all types of scientific effort,
there are problems and pitfalls. Those in biochemical studies
will be viewed by Dr. Waggoner. As is often the case, posing
problems and pitfalls does not guarantee solutions. However,
Dr. Waggoner will provide food for thought and hopefully
discussions in the meeting room as well as outside the sessions.

0097–6156/81/0160–0195$05.00/0
© 1981 American Chemical Society

With the overlap in our scientific disciplines, a certain
amount of repetition is inevitable; but there will be more
chemical emphasis in this session.

RECEIVED February 9, 1981.

13

Experimental Approaches Towards the Biochemical Analysis of Chemical Carcinogenesis

BRIAN A. LAISHES

McArdle Laboratory for Cancer Research, University of Wisconsin Medical School, Madison, WI 53706

The purpose of this presentation is to highlight developments in chemical carcinogenesis research that are directed towards understanding the development of malignancy at the molecular level. Because of the breadth of this topic, a high degree of selection has been necessary in order to adhere to space limitations, and, unfortunately, many important studies could not be included. An effort has been made to present selected studies that give a historical perspective to certain research developments and to include studies that exemplify efforts to delineate the truly complex biology of cancer development.

Early Epidemiologic Data

Early studies that represent the beginnings of our knowledge of chemical carcinogenesis were reviewed briefly by E.C. Miller (1). The first of these studies, in 1761, was by the physician John Hill of London, England, who reported on the development of nasal cancer as a consequence of excessive use of tobacco snuff (2). Percival Pott, a surgeon in London, reported on the unusually high incidence of cancer of the skin of the scrotum of young men who had worked as chimney sweeps in their childhood (3). The first preventive measures against chemically induced cancer in humans arose 3 years later through the Danish chimney sweepers' guild urging its members to take daily baths (4). During the following century, further observations of higher incidences of specific cancers of the skin and urinary bladder were reported in individuals with particular prior chemical exposures (1).
It was not until the 1930's that the epidemiologic data were mirrored by definitive laboratory data on the carcinogenicity of pure chemicals for experimental animals.

0097–6156/81/0160–0197$05.00/0

198 THE PESTICIDE CHEMIST AND MODERN TOXICOLOGY

Laboratory Models and Pure Chemical Carcinogens

Cancer of the skin became the first experimental model
of chemically induced cancer in 1915 when Yamagiwa and Ichikawa,
in Japan, induced skin carcinomas in the ears of rabbits by
repeated topical applications of coal tar for long periods
(see 5). In 1918, Tsutsui induced skin cancer in mice with
tars and, in 1922, Passey induced skin cancer in mice with
ether extracts of tars (5).
The active molecules responsible for the induction of
skin cancer with tars and tar extracts became the objects of
numerous investigations. One important lead was uncovered
by Hieger, who showed that the fluorescence spectra of products
from the carcinogenic tars and of synthetic benz[a]anthracene
derivatives were similar (6). Thus, in 1930, Kennaway and
Hieger demonstrated the carcinogenicity of dibenz[a,h]anthracene,
the first pure, synthetic carcinogen (7). The carcinogenic
hydrocarbon, benzo[a]pyrene, was soon isolated from coal tar
by Cook, Hewett, and Hieger (8).
Cancer of the liver, the first experimental visceral cancer,
became a model of chemically induced cancer in 1933 when Yoshida
induced liver tumors in rats and mice with oral administrations
of o-aminoazotoluene (2',3-dimethyl-4-aminoazobenzene) (9).
o-Aminoazotoluene is a derivative of the azo dye scarlet red
(1-[4-(o-tolylazo)-o-tolylazo]-2-naphthol), which was used
by Fischer in 1906 to induce proliferative lesions in the skin
of rabbits (10). The skin lesions induced by Fischer did not
become frank cancers, however, and regressed when the applications of scarlet red were stopped.
Cancer of the urinary bladder was introduced as an experimental model in 1938 when Hueper, Wiley, and Wolfe induced
cancer in the urinary bladder of dogs by feeding them 2-naphthyl-
amine (11).
Regarding the induction of cancer by pure chemicals, it is
noteworthy that in 1932 Lacassagne induced mammary cancers in
male mice by estrone treatment, thus pioneering an experimental
model for hormone-induced tumors (12).

Initiation-Promotion

The two-step, initiation-promotion concept was first conceived by Rous and co-workers about 40 years ago (13,14), and
this concept continues to play a prominent role in experimental
designs probing the biology of the cancer disease process.
These investigators studied the roles of irritation and the
stimulation of cell division on the induction of tumors in
rabbit ears previously treated with coal tar. Holes were punched
in the rabbits' ears with a cork borer, and it was found that
tumors appeared along the edge of the wound. The discovery
by Berenblum of the cocarcinogenic properties of croton oil

(with benzo[a]pyrene) (15,16) led to the remarkable discovery
by Mottram that benzo[a]pyrene need be applied only once to
induce tumors when it was followed by repetitive applications
of croton oil (17). It was demonstrated that the dose of the
initiating agent determines the eventual tumor yield, that
the promoting agent determines the duration of the latent period
(18), and that the initiation step is irreversible (19). Reversing the order of treatment, by administering croton oil for
many weeks followed by a single dose of benzo[a]pyrene, resulted
in no tumors (20). Boutwell demonstrated that doses of croton
oil that were either too small or too widely separated resulted
in no promotion, thus documenting the reversibility of the
effect of tumor promoters (21).

The generality of the initiation-promotion, two-step system
is remarkable. Armuth and Berenblum have extended the system
to mouse liver and lung, using dimethylnitrosamine as initiator
(22); to rat mammary gland, using 7,12-dimethylbenz[a]anthracene
as initiator (23); and to a system of two-stage transplacental
liver carcinogenesis in C57BL/6 mice (24,25). Transplacental,
initiation-promotion experiments were reported by Goerttler and
Loehrke, who treated mice prenatally by injecting the initiating agents 7,12-dimethylbenz[a]anthracene or ethyl carbamate
into the pregnant mother (26); the offspring, when treated
between the ages of 12 and 26 weeks with topical applications
of the tumor promoter 12-0-tetradecanoyl-phorbol-13-acetate
(TPA), exhibited tumors in skin and in other organs.

Tumor promotion activity has been demonstrated for a variety
of agents in various organs: butylated hydroxytoluene (BHT)
in mouse lung (27), bile acids in colon (28), saccharin and
cyclamate in rat urinary bladder (29), TPA in an in vivo-in
vitro rat trachea model (30,31), and phenobarbital (32-36)
and polychlorinated biphenyls (34,37,38) in rat liver.

The basic principles of the well-known two-step initiation-
promotion system are outlined in Figure 1 (39). The qualitative
differences in the responses of the target tissue to initiating
or promoting agents are remarkable in that initiating agents,
which are often complete carcinogens at higher doses, can be
administered in low doses that do not produce tumors (21,39).
Similarly, multiple doses of promoting agents that are not
complete carcinogens induce essentially no tumors, whereas
high incidences of tumors arise when these same doses of initiating and promoting agents are administered in sequence. Reversing
the order of exposure abolishes the synergism. Finally, there
is little or no recovery from the effects of initiating agents
(19,40), whereas tissues can recover from the effects of promoting agents (21,41).

Recent studies with the mouse skin tumorigenesis model have
revealed fascinating quantitative differences in the response
of the target tissue to initiating or promoting agents. As
reported by Boutwell, the tumor response induced by repetitive

application of 7,12-dimethylbenz[a]anthracene (DMBA) alone
requires about 10 times as much DMBA as that required when
DMBA (initiation) is followed by repetitive applications of
the tumor promoter TPA (42).

Cocarcinogenesis

Cocarcinogenesis (15,43) was discovered before the two-
stage concept of initiation-promotion (13,14,16,17,18). Cocarcinogenesis has been the subject of recent reviews (40,44,45)
and also of some confusion. Boutwell presented a clear definition: Cocarcinogenesis denotes the situation in which a second
factor (cocarcinogen), when introduced together with the carcinogen, increases the response to the carcinogen. The term cocarcinogenesis has no implication of denoting a specific step in
tumor development (21). The distinction between carcinogenesis
and tumor promotion is not always clear. On the other hand,
some investigators have clearly demonstrated that some cocarcinogens are not tumor promoters and, conversely, that some tumor
promoters are not cocarcinogens (46). With regard to human
lung cancer development, for example, it seems likely that
cigarette smokers are being exposed to cocarcinogens together
with carcinogens in cigarette smoke (e.g., 46). Since the
mechanism of cocarcinogenesis can differ from the mechanism
of tumor promotion, it is appropriate to distinguish between
the two.
     Adherence to a definition of cocarcinogenesis is advisable
for the purpose of clarity, but it is important to realize
that agents that are cocarcinogens can be utilized to exert
influences before, during, or after the application of the
primary chemical carcinogen. Generally, experiments designed
to investigate cocarcinogenesis involve the administration
of the test factor concurrently with the primary carcinogen.
Promoting agents must, by classical definition, be applied
following the completion of "initiating action".

Basic Dose-Response Relation in Carcinogenesis

In 1941, the relation between dosage of carcinogen and
tumor production was quantitatively demonstrated (47) in that
higher doses of carcinogen produced greater numbers of tumors.
Druckrey and Küpfmüller, in 1948, demonstrated that the yield
of tumors within the life-span of the animal depended on the
total dose of the carcinogen administered and not on the size of
daily doses into which the total dose was divided (48). These
investigators showed that increasing the daily dosages shortened
the time of cancer development. When similar results were obtained in quantitative experiments with many other chemical carcinogens (49), it became evident that the carcinogenic effects
of most individual doses persist over the entire life-span of
the rats, "summing up" to the final appearance of tumors.

The decrease in the effective dose required for cancer development with longer administration times (i.e., smaller daily doses) was investigated by Schmähl and Mecke, using the carcinogen 4-dimethylaminostilbene (50). The data were striking in that when the daily dose was 0.6 mg, the median effective total dose was 110 ± 29 mg per animal, whereas one-third of this daily dose (0.2 mg) required a total dose of only 61 ± 9.4 mg for the same tumor incidence. When administered over a longer period, a smaller total dose was required for cancer induction.

Using structurally dissimilar carcinogens, 4-nitroquinoline N-oxide and 3-methylcholanthrene, Nakahara and Fukuoka were able to demonstrate, in mouse skin, that the carcinogenic process started by either compound could be brought to completion by the other even when a long period intervened before treatment with the second carcinogen was begun (51). Thus, they proposed that the mechanism of action of the two structurally diverse carcinogens must be qualitatively similar.

Antagonistic Effects of Carcinogens

Of many experiments investigating the antagonistic effects of chemical carcinogens, the observations of Richardson and Cunningham proved to be of particular significance (Figure 2) (52). These investigators demonstrated that 3-methylcholanthrene was an effective inhibitor of azo dye hepatocarcinogenesis when administered simultaneously with the dye. These observations were extended to other systems, including the inhibition of 2-acetylaminofluorene-induced carcinogenesis (53), but the mechanism of inhibition awaited the discovery, in the Millers' laboratory, of enzyme induction in mammals (54,55,56).

Covalent Binding of Carcinogen Molecules

As the number and variety of identified chemical carcinogens increased, it became apparent that many of these chemicals were structurally dissimilar, and that some carcinogens produced tumors at distant sites regardless of their port of entry. Metabolic activation of carcinogens became probable when, in 1947, the Millers discovered that a metabolite of N,N-dimethyl-4-aminoazobenzene was covalently bound to hepatic proteins of rats fed this dye (57). Similar observations in other laboratories soon followed (see 1).

The incorporation of ^{14}C from a ^{14}C-labeled nitrogen mustard into purine fractions from the RNA and DNA of some mouse tissues was reported in 1957 by Wheeler and Skipper (58). Alkylation of liver RNA by the carcinogens dimethylnitrosamine and ethionine was reported by Farber and Magee in 1960 (59), with subsequent studies by others (reviewed in 1) demonstrating the incorporation

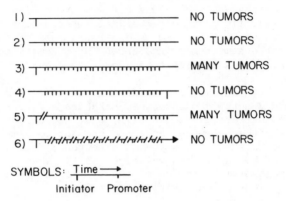

Figure 1.   A generalized schematic of the two-step initiation–promotion system of cancer development.

In Line 1, a single small dose of a chemical carcinogen results in no tumors produced within a specified period of time: these animals are initiated. In Line 2, multiple doses of tumor promoter cause no tumors. In Line 3, multiple doses of tumor promoter applied to the initiated animal result in many tumors. Reversing the order of treatments to promotion–initiation results in no tumors (see Line 4). In Line 5, a long time period may intervene between initiation and promotion with no reduction in tumor incidence. The nonadditivity of doses of tumor promoter is demonstrated with intervening periods between doses of tumor promoter (see Line 6) (39).

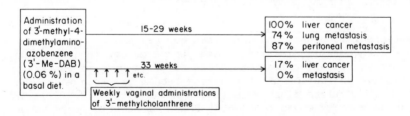

Figure 2.   An outline of the data reported by Richardson and Cunningham (52). The inhibition of 3'-methyl-DAB-induced hepatocarcinogenesis by concomitant administration of 3-methylcholanthrene.

of ^{14}C from ^{14}C-labeled ethionine, 2-acetylaminofluorene, dimethylnitrosamine, and polycyclic hydrocarbons into the DNA
and RNA of target tissues.

Enzyme Induction in Mammals

In 1954, Brown, Miller, and Miller demonstrated that certain
peroxides and hydrocarbons, including 3-methylcholanthrene,
increased the hepatic N-demethylation activity (demethylase
enzyme system) of both rats and mice for N-methyl aminoazo
carcinogens (54). In-depth studies by Conney in the Millers'
laboratory revealed that both demethylase activity and reductase
activity (reduction of the azo linkage) were increased several-
fold by the intraperitoneal injection of small amounts of 3-
methylcholanthrene 24 hours prior to assay (55); the demethylase
enzyme system was localized in the microsomes of rat and mouse
liver (60). The demonstration of lower levels of free and
bound dyes in the liver and of noncarcinogenic 3'-methyl-4-
aminoazobenzene (3'-Me-AB) (Figure 3) in the blood of rats
fed 3'-methyl-4-dimethylaminoazobenzene (3'-Me-DAB) with a
protective hydrocarbon suggested that the hydrocarbon facilitated
maintenance of high levels of deactivating liver enzymes that
metabolized the dye to less active or inactive derivatives
(Figure 4) (61).

Proximate and Ultimate Metabolites of Procarcinogens

About four decades ago, the potent insecticide 2-acetyl-
aminofluorene (AAF) was furnished by the Division of Insecticide
Investigations of the Bureau of Entomology and Plant Quarantine
to the Bureau of Agricultural Chemistry and Engineering of
the U. S. Department of Agriculture, where the carcinogenic
properties of AAF were discovered in 1941 by Wilson, DeEds,
and Cox (62). In 1960, Cramer and the Millers demonstrated
the metabolism of AAF to N-hydroxy-AAF, referred to as the
proximate carcinogenic form of the procarcinogen, AAF, since
N-hydroxy-AAF is a stronger carcinogen than the parent compound
and also active in a wider range of tissues and species (63,64,
reviewed in 1). A major ultimate carcinogenic metabolite of
N-hydroxy-AAF in rat liver appears to be N-sulfonoxy-AAF (65).
The sulfuric acid ester is a very strong electrophile, and
at least three other enzymatic pathways for the conversion
of N-hydroxy-AAF to electrophilic reactants have been observed
in rat liver (reviewed in 1).

Adduct Formation by AAF Metabolites

Both acetylated and nonacetylated adducts have been isolated
from hepatic macromolecules of rats treated with AAF or N-hydroxy-
AAF. Methionine adducts represent about 10% of protein-bound

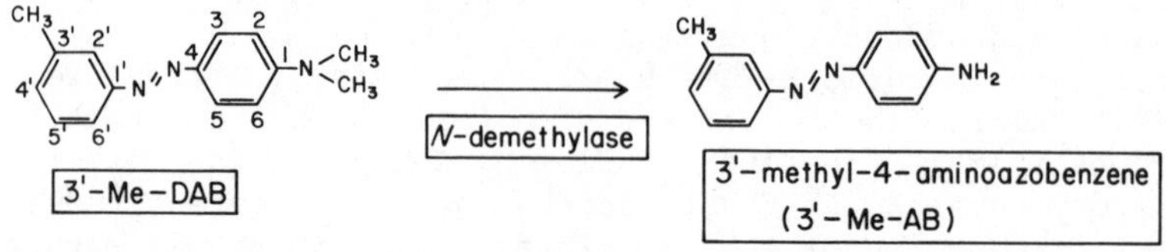

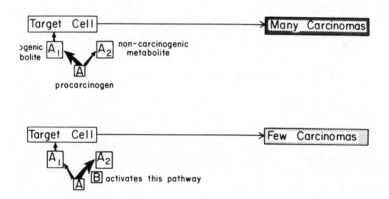

Figure 3. The action of the N-demethylation enzyme on the hepatocarcinogen
3'-methyl-DAB to produce the noncarcinogenic dye 3'-methyl-AB. The N-de-
methylase is the first mammalian enzyme for which induction was demonstrated.

Figure 4. Schematic of a mechanism, based on enzyme induction, by which chemical carcinogen B antagonizes the carcinogenic action of chemical carcinogen A as
observed in the effect described by Richardson and Cunningham (see Figure 2) (52).

The procarcinogen A must be metabolized to the active carcinogenic metabolite A_1.
B induces enzymes that metabolize A to the inactive noncarcinogenic metabolite A_2.
The term "target cell" is used only for convenience and actually may represent one or
more cells in a target tissue. Following initial interaction of the target tissue with a
chemical carcinogen, target cell then may represent one or more cells that are actually
different from the original designated target cell.

fluorene derivatives, and guanine adducts comprise the major
share of nucleic acid adducts formed in rat liver in vivo.
As reviewed by Kriek (66), substitution at the C-8 of
guanine yields N-(guan-8-yl)-AAF, which is the major adduct in
RNA, is a minor adduct in DNA, is alkali- and acid-labile,
locally distorts the DNA helix, and distorts regions digested by
single strand-specific nucleases. The C-8 adduct is a repair-
able lesion in rat liver DNA in vivo. On the other hand, substitution at the 2-amino group of guanine yields 3-(guan-N^2-yl)-
AAF (N-2 adduct), which is not present in RNA, is a minor form
in DNA, is stable to alkali and acid, does not distort DNA
helix, and is not released by single-strand specific nucleases.
The N-2 adduct is essentially a nonrepairable lesion in rat
liver DNA in vivo (66). About 70% of the DNA-bound fluorene
residues do not contain the N-acetyl group, and only about
10-15% of the deacetylated material could be identified as
N-(deoxyguanosin-8-yl)-AF. One possibility is that part of
the latter adduct undergoes hydrolytic cleavage of the imidazole
ring (66).

The fact that DNA in mammalian cells is highly organized
in chromatin may affect DNA repair processes and result in
a dynamic picture of chromatin rearrangement occurring during
excision repair of carcinogen-DNA adducts (67).

New insights into the kinetics of binding and repair of carcinogen-DNA adducts are being accumulated through new techniques
of detecting small quantities of adducts by radioimmunoassay
procedures (68,69). Thus, for example, AAF-DNA adducts can
now be detected by such methods with DNA extracted from livers
of rats given even low doses of dietary AAF (Poirier, True,
and Laishes, unpublished observations).

Molecular Alterations in the Genetic Programs of Target Cells

Target Cell DNA. The molecular alterations in the genetic
programs of target cells that occur as a consequence of interactions with carcinogenic stimuli and that are essential for
the expression of malignant phenotypes are unknown. The strong
relation that exists between the mutagenic and carcinogenic
effects of chemicals (70), together with the covalent binding
of electrophilic metabolites of chemical carcinogens to various
atoms in the DNA molecule (1), has stimulated interest in the
hypothesis that structural DNA modifications are essential
for carcinogenesis. Somatic mutations arising from errors
in replicating carcinogen-altered DNA have been proposed as
manifestations of the "fixation" of DNA damage and as necessary
steps in the production of transformed cells (71).

Gene Expression in Target Cells. The arrangement of genes
in eucaryotic viruses, such as DNA viruses that occur in the
cell nucleus (72), or RNA retroviruses that have a DNA nuclear
phase (73), has demonstrated that many genes are composed of
separated DNA sequences (74,75,76). Thus, tandem fusions of
neighboring coding sequences are accomplished by the post-tran-
scriptional removal of polynucleotides (intervening sequences
or introns [74]) from polycistronic nuclear transcripts. Studies
from many laboratories have thus demonstrated that, for eucaryotic
genes for globin, ovalbumin, immunoglobulin, SV40, and polyoma,
the coding sequences, on DNA that are ultimately translated
into amino acid sequences are not continuous but are interrupted
by DNA sequences that are "silent" and spliced out of the RNA
transcripts (e.g., 74,75,76).

It is, therefore, entirely feasible that chemical carcinogens may act at the level of RNA processing by affecting the
splicing enzymes or specific splicing signals in the RNA base
sequences. The control of gene expression at the RNA level
adds considerable breadth to the possible molecular events that
may yet be discovered as essential components of the carcinogenic process.

Cell Cycle and Cell Differentiation. One of the oldest
problems in experimental biology is that of understanding cell
differentiation. Differentiation has been defined as "those
mechanisms that make information not readily available in a
particular mother cell readily available in its daughter cells"
(77). In some instances differentiation may be initiated only
during a particular phase of the cell cycle. Thus, the maturation
of cells in slowly renewing (e.g., liver) or fast-renewing (e.g.,
skin) epithelial tissues occurs as cells achieve a terminally
differentiated state and cease proliferating (77). Whether
carcinogens act by disrupting lineages of normal differentiation
in epithelial target tissues is an attractive hypothesis but
still a matter for speculation. It is conceivable that the
gene regulatory processes in eucaryotic cells that control cell
differentiation may, in some way, be linked to the processes
that control cell replication.

Cancer Development as a Biological Process

The development of cancer is characteristically a long,
complicated disease process regardless of whether observations
are made with the human population or with laboratory animals.
In addition, in any given epithelial tissue, for example,
primary cancer is most often observed to arise first in only
a few, if not in single, locations. One of the first pieces
of information required to analyze the molecular basis of
carcinogenesis is to determine which cells in the target tissue
are essential components of the carcinogenic process.

Cell Lineages During Carcinogenesis. One of the simplest
approaches to elucidating the biology of cancer development
is to delineate cell lineages that occur during carcinogenesis.
It is, for example, conceivable that epithelial cells comprising
a carcinoma develop by a direct one-step cellular lineage (Figure
5), a two-step cellular lineage (Figure 6), or a multi-step
cellular lineage (Figure 7). Carcinogenic stimuli may induce
many phenotypically altered cell populations. Some of these
may play no role in the development of carcinoma (Figure 5),
may be involved in the development of a cell selection pressure
essential to the development of carcinoma (Figure 6), or may
represent one step in a series of essential sequential alterations
in target cells that are developing the malignant phenotype
(Figure 7). A clearer understanding of the complexities involved
in actually delineating cellular lineages during cancer development may be acquired by a brief review of one well-characterized
laboratory model.

Experimental Induction of Liver Cancer. An extensive
descriptive histopathology has been documented for the induction
of liver cancer in laboratory rats by controlled exposure to
chemical carcinogens (e.g., 78,79). Observations in many laboratories have indicated that new cell populations are generated
prior to the appearance of frank hepatocellular carcinoma;
this finding suggests that certain liver cell lineages may
be essential components of the development of liver cancer
(e.g., 78,79). When rats are exposed continuously to chemical
carcinogens in the diet, their livers exhibit complex mosaics
of lesions superimposed on one another. Approaches are being
developed to reduce the complexity of the biology of experimental
hepatocarcinogenesis by shortening the duration of carcinogen
exposure. Such "pulse doses" of chemical carcinogens provide
means to operationally dissect the carcinogenic process into
a rapid, early phase (during which the carcinogen first interacts
with the target cell(s) and the process is "initiated") and
a later phase of considerably longer duration (during which
the altered target cells develop the capacity to express the
malignant phenotype). Examples of such pulse-dose experimental
regimens include single exposures to a carcinogen after partial
hepatectomy (35,80), brief exposure to a carcinogen followed
by phenobarbital administration (32,35,81), and brief exposure
to a carcinogen followed by the application of a potent cell
selection pressure (82,83). The latter technique produces
demarcated populations of neoplastic cells with an exceptional
degree of synchrony (82,83).
      In order to delineate cell lineages during hepatocarcinogen-
esis, one direct approach is to dissociate the liver tissue
into cellular components during various stages of hepatocarcinogenesis and transfer the "putative premalignant" cells to histo-
compatible (isogeneic) rat strains (Figure 8) (84,85). Physio-

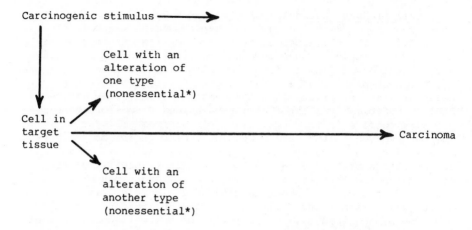

Figure 5.   One-step cellular lineage in which a normal cell is converted to a cancer cell with no intervening essential cellular alterations. (An essential alteration is defined as a change that is necessary for the development of carcinoma.)

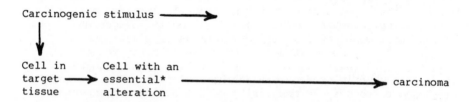

Figure 6.   Two-step cellular lineage in which a normal cell is converted to a cancer cell via an intermediary cell expressing an essential cellular alteration. (See the caption of Figure 5 for the definition of essential.)

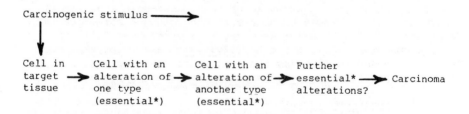

Figure 7.   Multi-step cellular lineage in which a normal cell is converted to a cancer cell via intermediary cells expressing essential cellular alterations in sequence. (See the caption of Figure 5 for the definition of essential.)

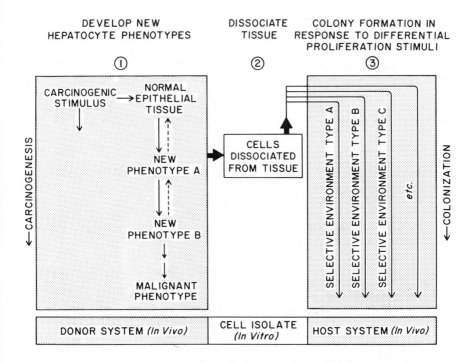

Figure 8. Schematic of one experimental approach designed to search for the cellular components of tissue undergoing chemically induced carcinogenesis that are essential to the carcinogenic process (e.g. Ref 85).

The approach is applicable to the rat liver model of chemically induced hepatocarcinogenesis. In Stage 1, new cellular phenotypes are established in an epithelial tissue treating the donor animal with a chemical carcinogen. The first steps of this hypothetical multistep process are potentially reversible, although it is also possible that new phenotypes may move into normal cell lineages of terminal differentiation that culminate in cell death. In Stage 2, the tissue is dissociated to yield cell suspensions. Stage 2 provides an opportunity to apply cell purification techniques. In Stage 3, the cell suspension can be tested for the presence of progenitor cells with the capacity to form colonies under a number of conditions established in the host animal. Stage 3 may be used to expand numbers of cells, to determine the extent of progenitor cell renewal, and to determine the role of both transplanted cell populations and host animal conditions in the subsequent development of carcinoma.

logic stimuli suspected to play a permissive or inductive role
in promoting the altered cells to carcinoma formation can be
studied in the recipient rats. The in vitro stage of this
three-stage process provides an opportunity to purify cell
populations (86), and the final stage provides a means to determine the colony-forming capacity of the isolated cell population
in the intact animal.

Genotypic Cellular Markers. All of the cellular markers
thus far studied in detail in experimental hepatocarcinogenesis
in the rat have been phenotypic markers subject to physiological
modulations (e.g., 87,88; see 83 for discussion) and are thus
not yet reliable for unequivocal delineation of cellular lineages.
This problem is common to many, if not all models of cancer
induction in epithelial tissues. Stable genotypic markers
will be required. A notable example of the delineation of
cell lineages using unique chromosome markers is that of hemato-
poietic cell lineages derived from stem cells in adult mouse
bone marrow (89,90,91). Myeloid and lymphoid systems are continually replenished by stem cells from the bone marrow of
adult mice, and the stem cells can easily be measured by their
ability to form macroscopic colonies in the spleens of irradiated
mice (89). By means of radiation-induced chromosome aberrations,
it was shown conclusively that each spleen colony was derived
from a single cell and that the stem cells, granulocytes, and
hemoglobin-synthesizing cells were derived from a single precursor (90). More recently, stem cells bearing unique, radiation-
induced chromosome aberrations were used to provide a direct
confirmation of the hypothesis that a common stem cell gives
rise to both myeloid and lymphoid cells (including both B and
T lymphocytes) and to provide evidence for stem cells restricted
in their capacity for differentiation (91). The technical
advantages of experimentation in hematology are remarkable,
since cells can be manipulated with comparative ease either
in the intact animal or in cell culture (e.g., 91). With the
advent of the proteolytic-enzyme perfusion technique of liver
tissue dissociation and the development of orthotopic transplantation of liver cell suspensions in experimental hepatocarcinogenesis (e.g., 84,85,86), a level of technical flexibility
comparable to that of experimental hematology may eventually
be attained. Thus, for cell lineage studies in experimental
hepatocarcinogenesis, it would be desirable to generate unique
chromosome markers.

A second class of exploitable genotypic markers that may
be utilized in experimental hepatocarcinogenesis involves the
use of liver cell membrane alloantigenic determinants, such
as those expressed as products of the major histocompatibility
complex. These determinants can exert marked control of the
development of putative premalignant lesions, as indicated
by the marked reduction in the number of altered liver lesions

following allogeneic orthotopic liver-cell transplantation
between F344 and Wistar-Furth rats (Laishes, Rolfe, and Onnink,
manuscript submitted). In addition, these determinants can
serve as markers for the cells of putative premalignant liver
lesions in the intact animal by transplanting parental-strain,
carcinogen-altered liver cells into F_1 hybrid rat liver (Hunt,
Buckley, and Laishes, unpublished observations). Thus, by
constructing genotypic mosaic livers containing liver cells
of one genotype (parental) surrounded by cells of another
genotype (F_1 hybrid), fluorescent antibodies for alloantigenic
determinants can be used to "tag" cells of either genotype
in vivo. Alloantigenic determinants may also be exploited
to facilitate the purification of specific subclasses of liver
cells from liver cell suspensions prepared from genotypic
mosaic livers at sequential stages of liver cancer development.
     By identifying the cellular components of target tissues
that are essential components of cancer development, a better
understanding of the molecular basis of carcinogenesis will
hopefully be gained.

Acknowledgement

     The author's investigations referenced herein were supported
by Grant Number CA-07175 and Grant Number CA-24818 awarded
by the National Cancer Institute, DHHS.

Literature Cited

1.   Miller, E.C. Cancer Res. 1978, 38, 1479-1496.
2.   Redmond, D.E., Jr. New Engl. J. Med. 1970, 282, 18-23.
3.   Pott, P. Reprinted in Natl. Cancer Inst. Monogr. 1963,
     10, 7-13.
4.   Clemmesen, J. J. Natl. Cancer Inst. 1951, 12, 1-21.
5.   Haddow, A.; Kon, G.A.R. Brit. Med. Bull. 1947, 4: 314-
     326.
6.   Hieger, I. Biochem. J. 1930, 24: 505-511.
7.   Kennaway, E.L.; Hieger, I. Brit. Med. J. 1930, 1, 1044-
     1046.
8.   Cook, J.W.; Hewett, C.L.; Hieger, I. J. Chem. Soc. 1933,
     Parts I, II, and III, 395-405.
9.   Yoshida, T. Trans. Japan Pathol. Soc. 1933, 23, 636-
     638.
10. Fischer, B. Muench. Med. Wochschr. 1906, 53, 2041-2047.
11. Hueper, W.C.; Wiley, F.H.; Wolfe, H.D. J. Ind. Hyg.
     Toxicol. 1938, 20, 46-84.
12. Lacassagne, A. Compt. Rend. 1932, 195, 630-632.
13. Rous, P.; Kidd, J.G. J. Exptl. Med. 1941, 73, 365-390.
14. Friedewald, W.F.; Rous, P. J. Exptl. Med. 1944, 80,
     101-126.
15. Berenblum, I. Cancer Res. 1941, 1, 44-48.

16. Berenblum, I. Cancer Res. 1941, 1, 807-814.
17. Mottram, J.C. J. Pathol. Bacteriol. 1944, 45, 181-187.
18. Berenblum, I.; Shubik, P. Brit. J. Cancer 1947, 1, 383-391.
19. Berenblum, I.; Shubik, P. Brit. J. Cancer 1949, 3, 384-386.
20. Berenblum, I.; Haran, N. Brit. J. Cancer 1955, 9, 268-271.
21. Boutwell, R.K. Progr. Exptl. Tumor Res. 1964, 4, 207-250.
22. Armuth, V.; Berenblum, I. Cancer Res. 1972, 32, 2259-2262.
23. Armuth, V.; Berenblum, I. Cancer Res. 1974, 34, 2704-2707.
24. Armuth, V.; Berenblum, I. Intern. J. Cancer 1977, 20, 292-295.
25. Tomatis, L.; Mohr, U., Eds. "Transplacental Carcinogenesis" IARC Scientific Publication No. 4; International Agency for Research on Cancer: Lyon, 1973.
26. Goerttler, K.; Loehrke, H. Virchows Arch. Pathol. Anat. Histol. 1976, 372, 29-38.
27. Witschi, H.; Williamson, D.; Lock, S. J. Natl. Cancer Inst. 1977, 58, 301-305.
28. Narisawa, T.; Magadia, N.E.; Weisburger, J.H.; Wynder, E.L. J. Natl. Cancer Inst. 1974, 53, 1093-1097.
29. Hicks, R.M.; Wakefield, J.St.J.; Chowaniec, J. Chem.-Biol. Interact. 1975, 11, 225-233.
30. Steele, V.E.; Marchok, A.C.; Nettesheim, P. In: T.J. Slaga, A. Sivak, and R.K. Boutwell, Eds. "Carcinogenesis, Vol. 2, Mechanisms of Tumor Production and Cocarcinogenesis; Raven Press: New York, 1978, pp. 289-300.
31. Marchok, A.C.; Rhoton, J.; Griesemer, R.A.; Nettesheim, P. Cancer Res. 1977, 37, 1811-1821.
32. Peraino, C.; Fry, R.J.M.; Staffeldt, E. Cancer Res. 1971, 31, 1506-1512.
33. Weisburger, J.H.; Madison, R.M.; Ward, J.M.; Viguera, C.; Weisburger, E.K. J. Natl. Cancer Inst. 1978, 54, 1185-1188.
34. Nishizumi, M. Cancer Lett. 1976, 2, 11-16.
35. Pitot, H.C.; Barsness, L.; Goldsworthy, T.; Kitagawa, T. Nature 1978, 271, 456-458.
36. Kitagawa, T.; Sugano, H. Gann 1978, 68, 255-256.
37. Kimura, N.T.; Kanematsu, T.; Baba, T. Z. Krebsforsch. Klin. Onkol. 1976, 87, 257-266.
38. Ito, N.; Tatematsu, M.; Hirose, M.; Nakanishi, K.; Murasaki, G. Gann 1978, 69, 143-144.
39. Boutwell, R.K. In: T.J. Slaga, A. Sivak, R.K. Boutwell, Eds. "Carcinogenesis, Vol. 2, Mechanisms of Tumor Promotion and Cocarcinogenesis"; Raven Press: New York, 1978, pp. 49-58.

40. Van Duuren, B.L. Tumor-Promoring and Co-Carcinogenic Agents in Chemical Carcinogensis. In: C.E. Searle, Ed. "Chemical Carcinogens" ACS Monograph 193; American Chemical Society: Washington, D.C., 1976, pp. 24-51.
41. Hennings, H.; Boutwell, R.K. Life Sci. 1967, 6, 173-181.
42. Boutwell, R.K. In: H.H. Hiatt, J.D. Watson, J.A. Winsten, Eds. "Origins of Human Cancer", Book B; Cold Spring Harbor Laboratory: New York, 1977, pp. 773-783.
43. Sall, R.D.; Shear, M.J.; Leiter, J.; Perrault, A. J. Natl. Cancer Inst. 1940, 1, 45-55.
44. Sivak, A. Biochem. Biophys. Acta 1979, 560, 67-89.
45. Slaga, T.J.; Sivak, A.; Boutwell, R.K., Eds. "Carcinogenesis, Vol. 2, Mechanisms of Tumor Promotion and Cocarcinogenesis"; Raven Press: New York, 1978.
46. Van Duuren, B.L.; Goldschmidt, B.M. J. Natl. Cancer Inst. 1976, 56, 1237-1242.
47. Bryan, W.R.; Shimkin, M.B. J. Natl. Cancer Inst. 1941, 1, 807-833.
48. Druckrey, H.; Küpfmüller, K. Z. Naturforsch. 1948, 36, 254-266.
49. Druckrey, H. In: R. Truhart, Ed. "Potential Carcinogenic Hazards from Drugs. Evaluation of Risks" UICC Monograph Series, Vol. 7; Springer-Verlag: Berlin, 1967, pp. 60-78.
50. Schmähl, D.; Mecke, R., Jr. Z. Krebsforsch. 1956, 61, 230-239.
51. Nakahara, W.; Fukuoka, F. Gann 1960, 51, 125-137.
52. Richardson, H.L.; Cunningham, L. Cancer Res. 1951, 11, 274 (Abst.).
53. Miyaji, T.; Moskowski, L.I.; Senoo, T.; Ogata, M.; Odo, T.; Kawai, K.; Sayama, Y.; Ishida, H.; Matsuo, H. Gann 1953, 44, 281-282.
54. Brown, R.R.; Miller, J.A.; Miller, E.C. J. Biol. Chem. 1954, 209, 211-222.
55. Conney, A.H.; Miller, E.C.; Miller, J.A. Cancer Res. 1956, 16, 450-459.
56. Conney, A.H.; Miller, E.C.; Miller, J.A. J. Biol. Chem. 1957, 228, 753-766.
57. Miller, E.C.; Miller, J.A. Cancer Res. 1947, 7, 468-480.
58. Wheeler, G.P.; Skipper, H.E. Arch. Biochem. Biophys. 1957, 72, 465-475.
59. Farber, E.; Magee, P.N. Biochem. J. 1960, 76, 58P.
60. Conney, A.H.; Brown, R.R.; Miller, J.A.; Miller, E.C. Cancer Res. 1957, 17, 628-633.
61. Miller, E.C.; Miller, J.A.; Brown, R.R.; MacDonald, J.C. Cancer Res. 1958, 18, 469-477.
62. Wilson, R.H.; DeEds, F.; Cox, A.J., Jr. Cancer Res. 1941, 1, 595-608.

63. Cramer, J.W.; Miller, J.A.; Miller, E.C. J. Biol. Chem. 1960, 235, 250-256.
64. Miller, E.C.; Miller, J.A.; Hartmann, H.A. Cancer Res. 1961, 21, 815-824.
65. DeBaun, J.R.; Miller, E.C.; Miller, J.A. Cancer Res. 1970, 30, 577-595.
66. Kriek, E. In: P. Emmelot and E. Kriek, Eds. "Environmental Carcinogenesis"; Elsevier/North-Holland Biomedical Press: Amsterdam, 1979, pp. 143-164.
67. Lieberman, M.W.; Smerdon, M.J.; Tlsty, T.D.; Oleson, F.B. In: P. Emmelot and E. Kriek, Eds. "Environmental Carcinogenesis"; Elsevier/North-Holland Biomedical Press: Amsterdam, 1979, pp. 345-363.
68. Poirier, M.C.; Dubin, M.A.; Yuspa, S.H. Cancer Res. 1979, 39, 1377-1381.
69. Müller, R.; Rajewsky, M.F. Cancer Res. 1980, 40, 887-896.
70. Hollstein, M.; McCann, J.; Angelosanto, F.A. Mutat. Res. 1979, 65, 133-226.
71. Kakunaga, T. Int. J. Cancer 1974, 141, 736-742.
72. Cold Spring Harbor Symp. Quant. Biol. 1977, 48.
73. Krzyzek, R.A.; Collett, M.S.; Lau, A.F.; Perdue, M.L.; Leis, J.P.; Faras, A.J. Proc. Natl. Acad. Sci. U.S.A. 1978, 75, 1284-1288.
74. Gilbert, W. Nature 1978, 271, 501.
75. Crick, F. Science 1979, 204, 264-271.
76. Ohno, S. Differentiation 1980, 17, 1-15.
77. Reinert, J.; Holtzer, H., Eds. "Cell Cycle and Cell Differentiation"; Springer-Verlag: New York, 1975.
78. Farber, E. The Pathology of Experimental Liver Cell Cancer. In: H.M. Cameron, D.A. Linsell, and G.P. Warwick, Eds., "Liver Cell Cancer"; Elsevier/North-Holland Biomedical Press: New York, 1976, pp. 243-277.
79. Farber, E.; Sporn, M.B., CoChairmen. Cancer Res. 1976, 36, 2475-2706.
80. Scherer, E.; Emmelot, P. Eur. J. Cancer 1975, 11, 145-154.
81. Watanabe, K.; Williams, G.M. J. Natl. Cancer Inst. 1978, 61, 1311-1314.
82. Solt, D.B.; Medline, A.; Farber, E. Am. J. Pathol. 1977, 88, 595-618.
83. Ogawa, K.; Solt, D.B.; Farber, E. Cancer Res. 1980, 40, 725-733.
84. Laishes, B.A.; Farber, E. J. Natl. Cancer Inst. 1978, 61, 507-512.
85. Laishes, B.A.; Rolfe, P.B. Cancer Res. 1980, 40, 4133-4143.
86. Laishes, B.A.; Fink, L.; Carr, B.I. Ann. N. Y. Acad. Sci. 1980, 349, 373-382.

87. Müller, E.; Colombo, J.P.; Peheim, E.; Bircher, J.
Experientia 1974, 30, 1128-1129.
88. Kitagawa, T.; Watanabe, R.; Sugano, H. Gann 1980, 71,
536-542.
89. Till, J.E.; McCulloch, E.A. Radiat. Res. 1961, 14, 213-
222.
90. Wu, A.M.; Till, J.E.; Siminovitch, L.; McCulloch, E.A.
J. Cell. Physiol. 1967, 69, 177-184.
91. Abramson, S.; Miller, R.G.; Phillips, R.A. J. Exp. Med.
1977, 145, 1567-1579.

RECEIVED February 2, 1981.

The Elusive Metabolite—The Reactive Intermediate

JAMES R. GILLETTE

Laboratory of Chemical Pharmacology, National Heart, Lung, and Blood Institute, Bethesda, MD 20205

The ideal pesticide kills the target organism without affecting other organisms. But one of the major difficulties in interpreting toxicity data is that we seldom know why a given substance is more toxic in one animal species than in another species. Sometimes species differences in the response to a substance are related to differences in the affinity and number of receptor sites that combine with the substance. But other species differences are caused by the rates at which the substances are absorbed from administration sites, distributed to various body tissues, metabolized and eliminated from the body. Moreover, in recent years, it has become increasingly evident that biological responses to substances may be caused at least in part by metabolites of the substance. In some cases the response caused by the metabolite is similar to that of the parent substance but all too frequently the response caused by the metabolite is entirely different from the parent compound. Furthermore, toxic responses may be caused not only by chemically stable metabolites but also by metabolites that rapidly react irreversibly with various tissue components including proteins, lipids, and nucleic acids. The half-life of these substances can range from milliseconds to several hours. Thus some metabolites may have such short half-lives that they never leave the immediate environment of the enzymes that catalyze their formation, whereas other chemically reactive metabolites have sufficiently long half-lives that they leave the tissues in which they are formed, enter other tissues of the body and are excreted into urine.

The differentiation of the toxic potentials of parent compounds from those of chemically stable metabolites is relatively simple. When a response depends on the reversible binding of the drug or metabolite to receptor sites and appears soon after the administration of drug, the intensity and duration of the response frequently depends on the drug concentration in blood. Studying the relationship between the duration of action of a drug and the concentration in blood, however, will fail when the response is caused in part by a metabo-

lite and the rate constant of elimination of the metabolite
is greater than that of the parent drug. Under these conditions, the half-life of the metabolite during the terminal
phase will appear to be the same as that of the drug and thus
the duration of action may appear to be approximately related to the concentration of the parent drug even when it
is caused solely by the metabolite. A better strategy to
elucidate the toxicological effects of chemically stable and
long-lived chemically reactive metabolites is to isolate and
identify the metabolites, synthesize them, and test them for
their toxic activities. Standard pharmacokinetic concepts
may then be applied to evaluate the relative contributions
of the parent compound and the metabolites in the manifestation of the toxicity.

These strategies will fail, however, when the toxicity is
caused by short-lived chemically reactive metabolites. Such
metabolites are not easily isolated and thus their identity
must be inferred from indirect evidence based on their ultimate
decomposition products. Even if the chemically reactive
metabolites were identified they would not be easily synthesized
or purified. Moreover, their toxic potential is not easily
studied because they would be inactivated during their passage
from the sites of administration to their target organs. Clearly
a different strategy must be employed to determine which chemically reactive metabolites are toxic and which are innocuous.

Several years ago my colleagues and I devised a strategy to
determine whether a given toxicity is caused by a chemically
reactive metabolite. In developing the approach (1,2), we
considered that chemically reactive metabolites conceivably
could cause toxic reactions, such as a cellular necrosis,
through several different mechanisms (Fig.1).

Conceivably, the target of the chemically reactive metabolite
could be an intracellular enzyme or its substrates required for
the function of cells. It could be a phospholipid in cellular
membranes, which control the intracellular compartmentalization
of intracellular components. It could be part of the protein
synthesis machinery required for the normal replacement of intracellular enzymes. It could also be DNA required for cellular
replication. We also envisioned the possibility that the
manifestation of the toxicity might not occur unless several
of these targets were impaired simultaneously.

It occurred to us that in causing alterations of the target
substances, the chemically reactive metabolite might alter the
target substances by combining covalently with them. It was
also plausible, however, that the toxic response might be caused
by mechanisms in which the chemically reactive metabolite is
not covalently bound to the target substance. The chemically
reactive metabolite might react with a lipid or DNA to form reactive endogenous components and thereby cause the toxicity;
for example, the reaction of trichloromethyl free radical with

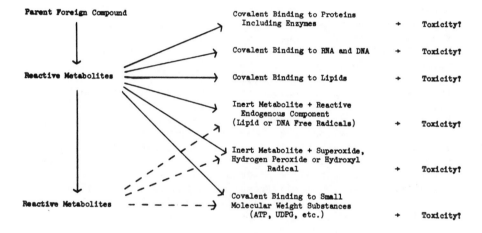

Figure 1.   Postulated mechanisms of toxicity by chemically reactive metabolites

lipid to form chloroform and lipid free radicals has been
suggested as an initial step in the liver necrosis caused by
carbon tetrachloride (3). A chemically reactive metabolite
may react with oxygen to form superoxide, hydrogen peroxide
or hydroxyl free radicals which in turn causes the toxicity
(4).

At the time we were developing our approach, we were
well aware of the many studies by oncologists indicating
relationships between carcinogenesis and the formation of the
chemically reactive metabolites of foreign compounds (5,6,7,8)
These studies profoundly influenced our thoughts. It was
evident that most chemically reactive metabolites do not
react with a single kind of macromolecule, but instead
react with many tissue components including proteins, lipid,
nucleic acids, glycogen and micromolecular substances, such
as ATP, NADPH, NADH and UDPG. It was also evident that the
rates of reaction of a given metabolite with the various
nucleophiles in cells depend on several factors. For example,
the rates of reaction with thiol groups differ markedly from
the rates of reaction with amino groups of proteins and
oxygen or nitrogen groups of the nucleic acids. Hence a
multiplicity of different reaction products may occur. On
the other hand, rather stable chemically reactive metabolites
may combine reversibly with certain sites of some proteins
before the complex rearranges to covalently bound material.
In this case, low concentrations of the reactive metabolite
may combine with relatively few kinds of macromolecules.
Indeed, the inhibition of pseudo cholinesterases by phosphorus
insecticides is an example of this kind of mechanism. For
these reasons, a chemically reactive metabolite may combine
preferentially with certain cellular proteins because they
contain an unusually large number of nucleophilic groups on
the surface of the protein or because the protein has a high
affinity for the reactive metabolite.

Because different chemically reactive metabolites react
with various tissue nucleophiles at relatively different
rates, it seemed likely to us that measuring the total covalent
binding of reactive metabolites to proteins would not provide
a reliable estimate of the relative toxicity of the chemically
reactive metabolites. Indeed it seemed entirely possible that
a chemically reactive metabolite could react extensively with
protein and still be nontoxic. Moreover, it also seemed possible that a toxicant might be converted to a chemically reactive metabolite which combined with protein even though the
toxicity is caused directly by the parent substance.

It occurred to us, however, that we might be able to determine whether a toxicity was caused by chemically reactive
metabolites by studying the effects of various inducers and
inhibitors of the metabolism of the toxicant. According to
our view, the covalent binding of the reactive metabolite to
protein would be approximately proportional to the area under

the cellular concentration curve of the chemically reactive
metabolite and therefore, an indirect measure of the amount
of reactive metabolite in contact with the target component
in cells. Thus, any treatment that results in a change in
the area under the curve of the chemically reactive metabolite
would cause parallel changes in both the covalent binding of
the metabolite to protein and the severity of the toxicity
when the toxicity is caused by the chemically reactive metabolite or a metabolite derived from it. Moreover, the correlation should occur even when the chemically reactive metabolite does not cause the toxicity by covalent binding to any
intracellular component. When the toxicity results from covalent binding of the metabolite the approach may be expressed
mathematically. The amount of metabolite that combines with
a target substance may be expressed as the dose times the
fraction of the dose that is converted to a chemically reactive
metabolite (Ratio A) times the fraction of the chemically
reactive metabolite that becomes covalently bound to the
target substance (Ratio B). Similarly the amount of metabolite
that ultimately becomes covalently bound to protein in the
target tissue may also be expressed as the fraction of the
dose of the toxicant that becomes covalently bound to protein
and this fraction may be expressed as Ratio A times the
fraction of the reactive metabolite that becomes covalently
bound to protein (Ratio B). Thus,

$$\text{Target - Metabolite} = \text{Dose A B}$$
$$\text{Protein - Metabolite} = \text{Dose A B}'$$

It follows, therefore that any treatment that changes Ratio A
or both Ratio B and Ratio B' without substantially changing
the relative rates of the reactions of the metabolites with
protein and the target substance will result in parallel
changes in the severity of the toxicity and the covalent binding to protein even when the target substance is not a protein.
Thus, determining the effects of various treatments that are
known to alter the metabolism of the toxicant or the inactivation of the metabolites is useful in determining whether a
toxicity is mediated by a chemically reactive metabolite.

The concept may also be expanded to include situations in
which the chemically reactive metabolite that reacts with
proteins also is converted to another metabolite that causes
toxicity. In this situation any treatment that causes a change
in the fraction of the dose that is converted to another
metabolite will cause parallel changes in the covalent binding
of the metabolite to protein and the severity of the toxicity.
But a treatment that preferentially alters the conversion
of the chemically reactive metabolite to the toxic metabolite
would cause inversely related changes in the magnitude of the
covalent binding and the severity of the toxicity.

In addition to various treatments that alter enzyme
activities, changes in Ratios A, B and B' may also occur by
changes in the size of the dose of the toxicant. At low

doses the rates of conversion of the parent compound to its
various metabolites including the chemically reactive metabolite and the rates of disposition of the chemically reactive
metabolite will be first order. Under these conditions the
values of the Ratios A, B and B' will be independent of the
dose. But as the dose is increased, the maximum concentration
of the parent compound reached in the organ of elimination
may be sufficient to saturate one or more of the enzymes that
catalyze its metabolism, and thereby Ratio A may be changed.
If the enzyme that has the lowest K_m catalyzes the formation
of the chemically reactive metabolite Ratio A will be decreased.
But if the enzyme having the lowest K_m catalyzes the formation
of an innocuous metabolite Ratio A will be increased. Moreover, increases in the dose of the parent compound will lead
to increases in the amount of chemically reactive metabolite
formed, and in turn may lead to the depletion of intracellular
nucleophiles, such as glutathione; thus increases in the dose
may lead to an increase in Ratios B and B'.

Ratio A in the target tissue may also be changed by an
alteration in the activity of an enzyme in a tissue other
than the initial target tissue and in some cases by changing
the route of administration. An understanding of the kinetics
of these effects is especially important for they can account
in part for a shift of the toxicity from one organ to another
caused by various treatments (9). Under these conditions,
Ratio A would be the amount of substance converted to the
chemically reactive metabolite in the target tissue divided
by the sum of amounts of the substance metabolized and otherwise eliminated in the target tissue and the other tissues
of the body.

With these considerations in mind, we devised the following sequence of in vivo and in vitro experiments by which we
determine whether a given toxicity is caused by a chemically
reactive metabolite.

1) Determine whether the substance causes toxicities in
various species and strains of animals. Obviously one cannot
study the mechanism of a toxicity in an animal species when
the toxicity does not occur in that species. Surprisingly,
however, many investigators spend considerable effort in
elucidating the pharmacokinetics and the pattern of the metabolism of a substance, in a given species before they demonstrate
that the substance is toxic in that species. By first carrying out toxicity studies in different animal species the investigator can choose the species with which further studies
may be carried out.

2) Determine the dose-response relationships of the
substance and the toxicity in the different animal species.
This is a natural consequence of step 1, since the investigator should administer several different doses of the
substance in evaluating species differences in the incidence
and severity of the toxicity.

3) Develop analytical methods for the assay of the substance and its major metabolites formed in the animal.

4) When the toxicity is manifested by a single dose of
the substance, study whether pretreatments that are known to
alter the rate or pattern of metabolism of foreign compounds
will alter the incidence or severity of the toxicity.

5) Compare the effects of the pretreatments on the total
body clearance and the pattern of metabolites of the toxicant.
Frequently, such studies elucidate whether the toxicity is
caused by the parent compound or by a metabolite.

6) Determine whether substances radiolabeled at metabolically
stable positions of the substance become covalently bound to
compounds in the target tissues. Subtoxic as well as toxic
doses should be studied in order to determine whether the
covalent binding follows first order kinetics or whether
there are threshold doses below which covalent binding is unimportant.

7) Determine whether the effects of the pretreatments that
alter the pattern and rates of metabolism of the toxicant cause
parallel changes in the amount of radiolabel covalently bound
to components in the target and other tissues. Parallel changes
indicate that the toxicity and the covalent binding are caused
by a common intermediate and may be caused by the same intermediate. Inversely related changes suggest that the two
phenomena are caused by two intermediates that are formed
from a common intermediate.

8) Determine whether the various treatments alter the in vitro
activity of enzymes that catalyze the formation of the chemically
reactive intermediate not only in the target organ but also
in other tissues (liver for example) that metabolize the
foreign compound. The K_m as well as the V_{max} of the enzymes
should be calculated in order to estimate the intrinsic
clearances of the enzymes (i.e. V_m/K_m) in the different tissues
and to assess the possibility that the concentration of the
substance might reach levels in the body that would result in
nonlinear kinetics. Such studies in combination with studies
in vivo frequently are useful in assessing whether the reactive
metabolite is sufficiently stable to escape the organ of
formation and be carried to other target organs.

9) Identify the decomposition products formed from the chemically reactive metabolites. Studies on the effects of various
nucleophiles such as glutathione, various amino acids and purine
and pyrimidine bases frequently provide clues to the types
of adducts formed from the chemically reactive metabolites.

10) Obtain supportive evidence that the toxicity is
mediated by a chemically reactive metabolite. For example,
when a chemically reactive metabolite reacts with glutathione
to form a conjugate, the concentration of glutathione in the
target tissue is frequently decreased. The severity of the
toxicity thus is frequently increased by the administration
of substances (such as diethyl maleate) that also react with

glutathione and decreased by the administration of alternative nucleophiles or precursors of glutathione. Such studies thus lead to the discovery of antidotes to acute toxicities caused by chemically reactive metabolites.

With this approach we and others have discovered that several commonly used drugs can cause tissue damage through the formation of metabolites (Table 1). In addition the studies on the effects of inducers, inhibitors and potential nucleophiles on the covalent binding of chemically reactive metabolites formed in vitro have helped us to understand the characteristics and properties of the chemically reactive metabolites and the enzymes that catalyze their formation and inactivation.

Our studies on the metabolism of phenacetin and acetaminophen illustrate how we have used this coordinated approach in studying toxicities caused by chemically reactive metabolites.

It is well known that large overdoses of acetaminophen cause fatal hepatic necrosis not only in man (10) but also in several laboratory animal species such as rats (11,12), mice (12) and hamsters (13,14). There is a marked species difference in the sensitivity of the various species to the drug (Table 2). In hamsters, necrosis occurs in most of the animals even at doses as low as 150 mg/kg, whereas in some strains of rats necrosis occurs in less than 10% of the animals even at doses as high as 1.5 g/kg (13).

Acetaminophen administered to animals is excreted mainly as its glucuronide and its sulfate conjugate (15), but a small amount of the drug is excreted as its mercapturic acid and cysteine derivatives in all animals studied including man (16) (Fig.2).

Studies on the covalent binding of the radiolabel to liver protein after the administration of various doses of radiolabeled acetaminophen to mice revealed that only negligible amounts of the drug were covalently bound at doses below 100 mg/kg (17). At higher doses, however, considerable radiolabel was covalently bound to liver protein. Moreover, the covalent binding appeared to be negligible until the liver was depleted of glutathione. Since acetaminophen is chemically inert, these findings thus indicated that it was converted to a chemically reactive metabolite in mice. They further suggested that at low doses of the drug, virtually all of the metabolite is converted to a glutathione conjugate that is ultimately excreted as a mercapturic acid. At high doses of the drug, the glutathione in liver is decreased to such an extent that the reactive metabolite can no longer be completely inactivated by glutathione and thus a portion of it becomes covalently bound to liver proteins. In accord with this view, the proportion of the dose of acetaminophen that is excreted as the mercapturic acid is about 10% when low doses of the drugs are administered to mice and it decreases as the dose is increased (18).

TABLE 1   Examples of the formation of chemically
           reactive metabolites

Compound	Tissue binding*	Pathway Intermediate(s)
Bromobenzene (54)	H,L,K	Bromobenzene-3,4-epoxide (?)
Phenacetin (55)	H	Acetaminophen
Acetaminophen (56,20, 21,17)	H	N-Acetimido-quinone (?)
Furosemide (57)	H	Furosemide epoxide (?)
Ipomeanol (58)	L,H,K	(?)
Various furans (59)	H,L,K	(?)
Isoniazid (60)	H	Acetyl hydrazine
Iproniazid (61)	H	Isopropyl hydrazine
Carbon tetrachloride (3,62,63,64)	H,K	Trichloromethyl free radical
Chloroform (65)	H	Phosgene
Chloramphenicol (66)	H,BM	R-oxalyl chloride
Nitrofurantoin (67)	L	Reduction product
Benzene (68,69)	L,BM	(?)

*H = Liver, L = Lung, K = Kidney, BM = Bone marrow

TABLE 2   Liver necrosis caused by acetaminophen

Dose (mg/kg)	Mice	Incidence (%) Hamsters	Rats
150	0	0	-
200	-	20	-
300	22	89	-
375	46	-	-
425	-	100	-
500	76	-	0
750	99	-	-
1000	-	-	2
1500	-	-	6

Data taken from Mitchell et al. (56) and Potter et al. (70).

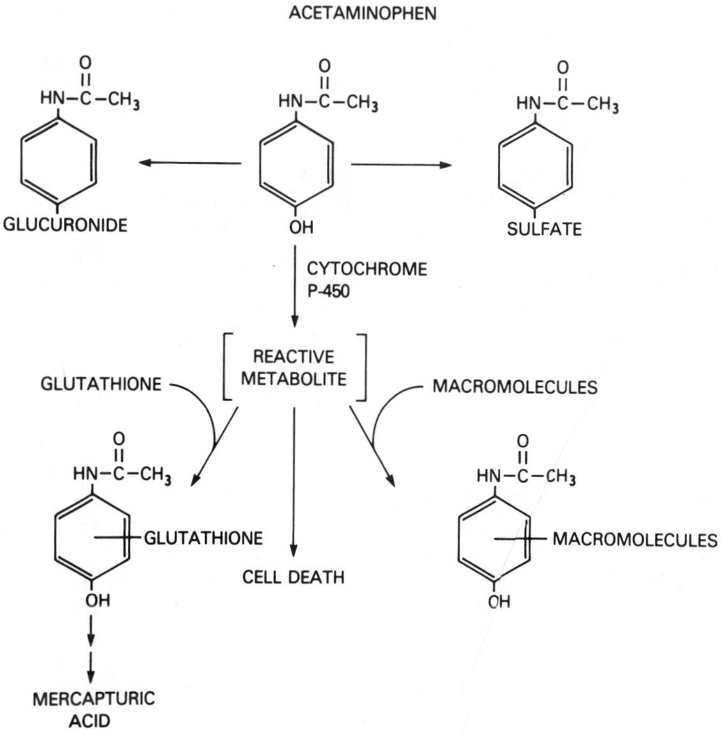

Figure 2.   Principle pathways of acetaminophen metabolism

As the dose of acetaminophen was increased, the incidence
and severity of the liver necrosis in mice was increased
(12). However, an increase in toxicity would be expected to
occur regardless of the mechanism of toxicity. Thus, the
apparent correlation between the increase in covalent binding
and the incidence of toxicity based solely on changes in the
dose (12,17) is only trivial and does not indicate whether
the toxicity is caused by the parent compound, the chemically
reactive metabolite or some other metabolite.

In order to determine whether the toxicity is caused by a
chemically reactive metabolite the animals must be treated
with substances that would alter either Ratio A or Ratio B.
Since the reactive metabolite preferentially combines with
glutathione (17) the depletion of liver glutathione by other
substances that react with glutathione should increase the
covalent binding of the reactive metabolite by increasing
Ratio B, whereas the administration of cysteine should decrease it. In accord with this view diethyl maleate, which
decreases the concentration of glutathione in liver but
does not cause liver necrosis (19), not only increases the
covalent binding of the reactive metabolite of acetaminophen,
but also increases the incidence and severity of the liver
necrosis in mice (17). On the other hand, treatment of mice
with cysteine decreases the covalent binding of the reactive
metabolite and decreases the incidence and severity of the
liver necrosis (17).

Pretreatment of mice with phenobarbital increases the
activity of the enzyme that catalyzes the formation of the
reactive metabolite and thus accelerates the depletion of
hepatic glutathione (17), but apparently does not affect the
enzymes that catalyze the formation of the sulfate or the
glucuronide conjugates because it does not alter the biological half-life of the drug in mice (12). Thus, pretreatment of mice with phenobarbital increases the proportion
of the dose of acetaminophen that becomes covalently bound
to liver protein by increasing Ratio A and increases the
incidence and severity of the liver necrosis (12,20).

Studies with liver microsomes indicated that the formation of the chemically reactive metabolite, as measured
by covalent binding of radiolabeled acetaminophen to microsomal protein, is catalyzed by a cytochrome P-450 enzyme in
liver microsomes (21). They also showed species differences
in the kinetics for the formation of the reactive metabolite.
With liver microsomes from mice the apparent maximal velocity
for the reaction (V_{max}) was about 0.18 nmoles bound/mg protein/
min and the apparent K_m was about 0.36 mM acetaminophen, whereas with liver microsomes from rats, the apparent V_{max} was
0.07 nmoles/mg protein/min and the apparent K_m was 14.8 mM
acetaminophen. Since the intrinsic clearance of a substrate
by an enzyme in the body should be $V_{max}/(K_m + S)$, these findings are in accord with the view that the rate of formation

of the reactive metabolite would be slower in rats than in
mice not only because the V_{max} is lower in rats, but also
because the K_m is higher.

The addition of glutathione to the incubation mixtures
in the presence and absence of the soluble fraction of liver
inhibited the covalent binding of the reactive metabolite to
protein, but resulted in the formation of an acetaminophen-
glutathione conjugate (21,22,23,24). The finding that the covalent binding was blocked at a lower glutathione concentration in the presence of the soluble fraction than in its
absence led to the conclusion that the formation of the
glutathione conjugate was catalyzed by one or more of the
glutathione transferases in liver even though the conjugate
can be formed nonenzymatically. Strangely, the sum of the
covalent binding and the glutathione conjugate also increased
(20,25) as did the rate of disappearance of acetaminophen
(25) as the glutathione concentration was increased. It,
therefore, seems possible that a part is reduced back to
acetaminophen and that glutathione prevents this reduction
by the formation of the conjugate. In accord with this
view, ascorbic acid inhibits the covalent binding of the
acetaminophen metabolite to protein (26) and glutathione
decreases rather than increases acetaminophen dependent
NADPH oxidation by liver microsomes (27).

Thus, the chemically reactive metabolite appeared to be
a short-lived substance that reacts with glutathione and is
easily reducible by ascorbic acid. At first, we suggested
that the metabolite that caused the liver necrosis might be
N-hydroxyacetaminophen (20,22,23,24). This hypothesis was
based primarily on the finding of Calder *et al.* that N-
acetylimidoquinone (N-acetyl-p-benzoquinoneimine) was
an electrophilic compound which could be formed from N-hydroxyphenacetin under acidic conditions (28). Thus, it
seemed possible that liver microsomes might convert acetaminophen to N-hydroxyacetaminophen which in turn undergoes
spontaneous dehydration to the N-acetylimidoquinone. In
support of this hypothesis it was shown that the acetaminophen
analogs p-chloroacetanilide (29,30) and phenacetin (31) were
N-hydroxylated and the treatments of animals which altered
the microsomal N-hydroxylation of these analogs caused
similar changes in the rate of formation of the electrophilic
metabolite of acetaminophen by liver microsomes.

N-Hydroxyacetaminophen was recently synthesized and its
chemical properties examined (32,33). In aqueous solutions
the proposed metabolite was unstable and presumably dehydrated
to the electrophile N-acetylimidoquinone with a half-life of
approximately 15 min. When injected into mice the compound
decreased the glutathione concentration in liver and was
hepatotoxic (32).

Recently, McMurtry et al. showed acetaminophen in Fischer
rats becomes covalently bound to kidney as well as to liver
(34). However, the chemically reactive metabolite in kidney
appears to be produced in the kidney rather than the liver
since 3-methylcholanthrene pretreatment increased the covalent
binding in the liver but not the kidney. Thus it seemed
likely that the chemically reactive metabolite of acetaminophen
formed in the liver has too short a half-life to leave the
liver to any significant extent. Since N-hydroxyacetaminophen
has a relatively long half-life in vitro, the possibility
that the hepatotoxicity of acetaminophen might be mediated
mainly through this metabolite became questionable.

Recent studies have shown that hamster liver microsomes
convert N-hydroxyphenacetin but not acetaminophen to N-hydroxyacetaminophen even though considerbly more acetaminophen is
covalently bound to microsomal proteins than is N-hydroxyphenacetin (35). Moreover, the chemically reactive metabolite
of acetaminophen is apparently not formed by way of acetaminophen epoxide because the formation of 3-hydroxyacetaminophen
is not blocked by glutathione, ascorbic acid or epoxide
hydrolase and covalent binding of acetaminophen is not blocked
by superoxide dismutase (36). Thus, the chemically reactive
metabolite of acetaminophen remains unidentified. It is
still possible that the intermediate is N-acetylimidoquinone
(N-acetyl-p-benzoquinoneimine) because it reacts with glutathione to form a glutathione-acetaminophen conjugate, and is
readily reduced to acetaminophen by ascorbic acid. If N-
acetylimidoquinone is the major reactive metabolite, however,
it must be formed by a hitherto unknown mechanism.

These studies thus indicated that the liver necrosis
caused by acetaminophen in mice is mediated by a chemically
reactive metabolite that combines with glutathione conjugate
to form a conjugate, which ultimately is excreted as a mercapturic acid. The studies further illustrated how a change
in the activity of an enzyme that catalyzes the formation of
a minor toxic metabolite can markedly affect the toxicity
without significantly affecting the biological half-life of
the parent drug. The finding that glutathione is markedly
decreased before the covalent binding of the active metabolite of acetaminophen to protein becomes appreciable led
to the concept of a "dose threshold" for the toxicity.
In mice the "dose threshold" is related to the fraction
of the dose that is converted to the reactive metabolite
(Ratio A) and the amount of glutathione initially present
in the liver. But it should be pointed out that the reason
for "dose threhsolds" may differ in other animal species.
As the dose is increased, the fraction of the dose is converted to acetaminophen sulfate decreases, indicating that
this pathway of inactivation becomes saturated either because the concentration of acetaminophen in liver exceeds
the K_m of the sulfotransferase or because the synthesis of

3'-phosphoadenosine-5'-phosphosulfate (PAPS), the cosub-
substrate of the enzyme, becomes rate-limiting. As the dose
is increased further the concentration of acetaminophen may
exceed K_m of glucuronyl transferase in liver of some animal
species. Indeed, the saturation of both of these enzyme
systems may account in part for the finding that the apparent
half-life of acetaminophen (10) and the fraction of the dose
excreted as the mercapturic acid (37,38) increases as the
dose is increased in man.

The finding that cysteine can prevent the liver necrosis
caused by acetaminophen in mice (17) led to the possibility
that thio compounds might be useful as antidotes, provided
that they are administered while the acetaminophen is being
metabolized. Unfortunately, cysteine is a rather ineffective
antidote except when it is administered intraperitoneally
because it is incorporated into protein by all tissues of
the body and thus is subject to a kind of first pass effect
by these tissues. Most of the emphasis, therefore, has been
toward the development of antidotes that serve as precursors
of cysteine (such as methionine and N-acetylcysteine) and
thus of glutathione or as alternative nucleophiles that
combine with the chemically reactive metabolite.

Cysteamine apparently is an effective antidote not only
in mice (39) but also in man (40). At first it was assumed
that this compound exerted its effect by serving as an alternative nucelophile in the inactivation of the chemically
reactive metabolite. It is also possible, however, that
cysteamine may act by inhibiting the formation of the chemically reactive metabolite (41) and by serving as a precursor
of sulfate, required for the formation of PAPS. Unfortunately,
it is difficult to differentiate among these mechanisms.
The evidence cited in support of the concept that cysteamine
inhibits the formation of the reactive metabolite is based
primarily on the finding that cysteamine decreases the excretion of the glutathione conjugate into bile and of the
cysteinyl conjugate and mercapturic acid into urine. Moreover, no evidence was obtained indicating that a cysteamine
conjugate of acetaminophen is excreted into bile or urine.
However, these results are not definitive. Cysteamine
would cause a decrease in the excretion of the glutathione
into bile and cysteine conjugates and the mercapturic acid
into urine even if it were to exert its protective effect
solely by combining with the chemically reactive metabolite.
Moreover, it is questionable whether the cysteine conjugate
of acetaminophen would be rapidly excreted into bile or
urine before it is converted to other substances by enzymes
such as monoamine oxidase. Furthermore, the fact that high
concentrations of cysteamine inhibit the hydroxylation of
acetanilide *in vitro* (41) may or may not be relevant because
it is not known whether the formation of the chemically
reactive metabolite of acetaminophen is catalyzed by the

same enzyme that hydroxylates anilide or whether the concentrations of cysteamine achieved in vivo approach those
used in vitro. It is also questionable that the increase
in sulfate derived from cysteamine would affect Ratio A by
more than a few percent. It seems to me that the mechanism
by which cysteamine exerts its protective effect must remain
open.

N-Acetylcysteine also prevents the liver necrosis caused
by acetaminophen in animals (42,43,44) and man (45,46). But
again, the mechanism is not entirely clear. It is possible
that N-acetylcysteine may combine directly with the chemically
reactive metabolite to form the mercapturic acid. It is also
possible, however, that N-acetylcysteine is deacetylated to
cysteine and then converted to glutathione (47) or oxidized
to sulfate (48). All of these mechanisms would tend to decrease the toxicity of acetaminophen.

Thus our attempts to identify the toxic chemically reactive
metabolite of acetaminophen have been elusive. But imagine the
greater difficulty in elucidating toxic metabolites when the
substance can be converted to several different chemically reactive metabolites or to the same chemically reactive metabolite by different mechanisms.

Phenacetin can be converted to chemically reactive metabolites that combine with glutathione through at least four
different pathways (Fig.3). 1) Phenacetin is converted to
acetaminophen (9) which is subsequently activated to a chemically reactive metabolite that combines with glutathione
(21). In this pathway the phenolic oxygen in the acetaminophen-SG conjugate originates from the ethoxy oxygen of phenacetin (22,24). 2) Phenacetin is converted to an intermediate
we believe to be phenacetin-3,4-epoxide. The intermediate
decomposes to another chemically reactive metabolite that
reacts with glutathione to form an acetaminophen-SG conjugate.
Exactly 50% of the phenolic oxygen in the conjugate formed
by this pathway originates from atmospheric oxygen and the
other 50% originates from phenacetin (22,24). 3) Phenacetin
is converted to N-hydroxyphenacetin (38). In turn the N-
hydroxyphenacetin can be transformed to N-sulfate and NO-
glucuronide conjugates which decompose to a chemically
reactive metabolite that reacts with glutathione to form
an acetaminophen-SG conjugate (50). The phenolic oxygen in
the conjugate formed by this pathway originates from water
(24). 4) Phenacetin may be converted to N-hydroxyphenacetin
as in pathway 3 but then undergoes oxidative dealkylation to a
chemically reactive metabolite that reacts with glutathione
to form an acetaminophen-SG conjugate (32,35). The phenolic
oxygen in the conjugate formed by this pathway presumably
originates from phenacetin. Another pathway for the formation
of a chemically reactive metabolite may be postulated. In
this pathway acetaminophen is converted to 3-hydroxyacetamino-

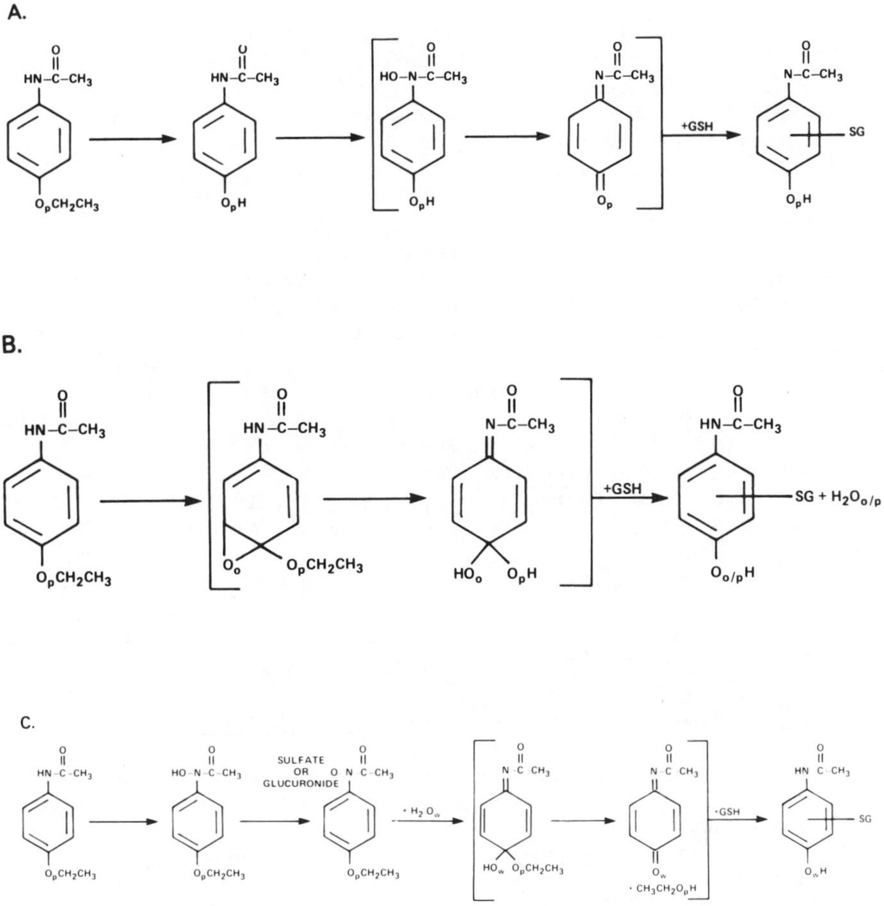

A.

B.

C.

Figure 3. Pathways of phenactin metabolism leading to the formation of glutatione conjugates

phen (<u>36</u>,<u>51</u>), which is a catechol and thus may be oxidized
to a quinone by superoxide (<u>52</u>,<u>53</u>).

Although my Laboratory has used these principles to study
the toxicities caused by large doses of drugs, there is
every reason to believe that these principles will be equally
applicable in studying species differences in the effects
of pesticides. Indeed, it is now believed that compounds
such as piperonyl butoxide and parathion inhibit cytochrome
P-450 enzymes through the formation of chemically reactive
metabolites. The specificity of the effects of these substances presumably occurs either because the chemically
reactive metabolites have an unusually high affinity for
the cytochrome P-450 enzymes or because they are so short-
lived that they never leave the immediate environment of the
active sites of the enzymes. The use of other "suicide
enzyme inhibitors" offers exciting possibilities.

References

1. Gillette, J.R. Biochemical Pharm., 1974, 23, 2785.
2. Gillette, J.R. Biochemical Pharm., 1974, 23, 2927.
3. Recknagel, R.O. Pharmacol. Rev. 1967, 19, 145.
4. Bus, J.S., Aust, S.D. and Gibson, J.E. 1974, Biochem.
    Biophys. Res. Commun. 58, 749.
5. Miller, E.C. and Miller, J.A. 1966, Pharm. Rev.
    18, 805.
6. Miller, J.A. 1970, Cancer Res. 30, 559.
7. Heidelberger, C. and Moldenhauer, M.A. 1956, Cancer
    Res. 16, 442.
8. Magee, P.N. and Barnes, J.M. 1967, Adv. Cancer Res.
    10, 163.
9. Boyd, M.R. 1980 in Environmental Chemicals, Enzyme Function,
    and Human Disease (Ciba Foundn. Symp. #76) Excerpta Medica.
    Amsterdam.
10. Prescott, L.F., Wright, N., Roscoe, P. and Brown, S.S.,
    1971, The Lancet 1, 519.
11. Boyd, E.M. and Berecky, G.M., 1966, Brit. J. Pharmacol.
    26, 606.
12. Gillette, J.R. and Brodie, B.B. 1973, J. Pharmacol.
    Exp. Ther. 187, 185.
13. Davis, D.C., Potter, W.Z., Jollow, D.J. and Mitchell,
    J.R., 1974, Life Sci. 14, 2099.
14. Potter, W.Z., Thorgeirsson, S.S., Jollow, D.J. and
    Mitchell, J.R. 1974, Pharmacology 12, 129.
15. Smith, J.N. and Williams, T.R. 1948, Biochem. J. 42, 538.
16. Jagenburg, R. and Toczko, K. 1964, Biochem. J. 92, 639.
17. Mitchell, J.R., Jollow, D.J., Potter, W.Z., Gillette,
    J.R. and Brodie, B.B. 1973, J. Pharmacol. Exp. Ther.
    187, 211.
18. Jollow, D.J., Thorgeirsson, S.S., Jollow, D.J. and
    Mitchell, J.R. 1974, Pharmacology 12, 251.

19. Boyland, F. and Chasseaud, L.F. 1967, Biochem. J.
     104, 95.
20. Jollow, D.J., Mitchell, J.R., Potter, W.Z., Davis, D.C.,
     Gillette, J.R. and Brodie, B.B., 1973, J. Pharmacol.
     Exp. Ther. 187, 195.
21. Potter, W.Z., Davis, D.C., Mitchell, J.R., Jollow, D.J.,
     Gillette, J.R. and Brodie, B.B. 1973, J. Pharmacol. Exp.
     Ther. 187, 203.
22. Hinson, J.A., Nelson, S.D. and Mitchell, J.R. 1977,
     Mole. Pharmacol. 13, 625.
23. Buckpitt, A.R., Rollins, D.E., Nelson, S.D., Franklin,
     R.B. and Mitchell, J.R. 1977, Anal. Biochem. 83, 168.
24. Hinson, J.A., Pohl, L.R., Monks, T.J. and Gillette, J.R.
     1979, Pharmacologist 21, 219.
25. Rollins, D.E. and Buckpitt, A.R. 1979, Tox. Appl. Pharmacol.
     47, 331.
26. Hinson, J.A., Pohl, L.R., Monks, T.J. and Gillette, J.R.,
     1979, Pharmacologist 21, 219.
27. Hinson, J.A. and Gillette, J.R. 1980, Fed. Proc. 39, 748.
28. Calder, I.C., Creek, M.J. and Williams, P.J. 1974,
     Chem. Biol. Interactions 8, 87.
29. Hinson, J.A., Mitchell, J.R. and Jollow, D.J. 1975,
     Mole. Pharmacol. 11, 462.
30. Hinson, J.A., Mitchell, J.R. and Jollow, D.J. 197
     Biochem. Pharmacol. 25, 599.
31. Hinson, J.A. and Mitchell, J.R. 1976, Drug. Metab.
     Dispos. 4, 435.
32. Healey, K., Calder, I.C., Yong, A.C., Crowe, C.A.,
     Funder, C.C., Ham, K.N. and Tange, J.D. 1978,
     Xenobiotica 8, 403.
33. Gemborys, M.W., Gribble, G.W. and Mudge, G.H.,
     1978, J. Med. Chem. 21, 649.
34. McMurtry, R.L., Snodgrass, W.R. and Mitchell, J.R.
     1978, Toxicol. App. Pharmacol. 46, 87.
35. Hinson, J.A., Pohl, L.R. and Gillette, J.R. 1979
     Life Sci. 24, 233.
36. Hinson, J.A., Pohl, L.R., Monks, T.J. and Gillette,
     J.R. 1979, Pharmacologist 21, 219.
37. Davis, M., Simmons, C.J., Harrison, N.G. and
     Williams, R. 1976, Q.J. Med. 45, 181.
38. Slattery, J.T. and Levy, G., 1979, Pharmacol. Ther.
     25, 184.
39. Mitchell, J.R., Thorgeirsson, S.S., Potter, W.Z.,
     Jollow, D.J. and Keiser, H. 1974, Clin. Pharmacol.
     Ther. 16, 676.
40. Prescott, L.F., Newton, R.W., Swainson, C.P., Wright,
     N., Forrest, A.R.W. and Matthew, H. 1974, Lancet 1, 588
41. Harvey, F.D. and Levitt, T.E. 1970, J. Int. Med. Res.
     4, (Supplement 4, 130).
42. Piperno, E. and Berssenbruege, D.A. 1976, Lancet 2, 738.

236   THE PESTICIDE CHEMIST AND MODERN TOXICOLOGY

43. Gerber, J.G., McDonald, J.S., Harbison, R.D., Villeneve, J.P., Wood, A.J.J. and Nies, A.S. 1974, Lancet 1, 657.
44. Piperno, E., Mosher, A.H., Berssenbruege, P.A., Winkler, J.D. and Smith, R.B. 1978, Pediatrics 62, 880
45. Petterson, R.G. and Rumack, B.J. 1977, J. Am. Med. Assoc. 237, 2406.
46. Prescott, L.F., Stewart, M.J. and Proudfoot, A.T. 1978, Br. Med. J. 1, 856.
47. Orrenius, S., personal communication.
48. Galinsky, R.E. and Levy, G. 1979, Life Sci. 25, 693.
49. Axelrod, J. 1956, Biochem. J. 3, 364.
50. Mulder, G.J., Hinson, J.A. and Gillette, J.R. 1978, Biochem. Pharmacol. Biochem. Pharmacol. 27, 1641.
51. Forte, A.J., McMurtry, R.J. and Nelson, S.D. 1979, Pharmacologist 21, 220
52. Heacock, R.A. 1959, Chem. Rev., 59, 181.
53. Dybing, E. Nelson, S.D., Mitchell, J.R., Sasame, H.A., and Gillette, J.R. 1976, Mole. Pharmacol. 12, 911.
54. Reid, W.D., and Krishna, G. (1973), Expl. Mole. Pathol. 18, 80.
55. Nelson, S.D., Garland, W.A., Mitchell, J.R., Vaishnav, Y., Statham, C.N. and Buckpitt, A.R. (1978), Drug Metab. Disp. 6, 363.
56. Mitchell, J.R., Jollow, D.J., Potter, W.Z., Davis, D.C., Gillette, J.R., and Brodie, B.B. (1973), J. Pharmacol. Exper. Ther. 187, 185.
57. Mitchell, J.R., Nelson, S.D., Potter, W.Z., Sasame, H.A. and Jollow, D.J. (1976), J. Pharmacol. Exper. Ther. 199, 41.
58. Boyd, M.R. (1976) Envir. Health Perspec. 16, 127.
59. McMurtry, R.J. and Mitchell, J.R. (1977), Toxicology Appl. Pharmaco. 42, 285.
60. Mitchell, J.R., Zimmerman, N.J., Snodgrass, W.G. and Nelson, S.D. (1976), Ann. Intern. Med. 84, 181.
61. Nelson, S.D., Mitchell, J.R., Snodgrass, W.G., and Timbrell, J.A. (1978), J. Pharmacol. Exper. Ther. 206, 574.
62. Reynolds, E.S. (1967), J. Pharmacol. Exper. Ther. 155, 177.
63. Uehleke, H. in Biological Reactive Metabolites (D.J. Jollow, J.J. Kocsis, R. Snyder and H. Vainio, eds.), (1977), Plenum Press, New York, p. 431.
64. Ilett, K.F., Reid, W.D., Sipes, I.G. and Krishna, G. (1977), Exper. and Mole. Path. 19: 215, 1973.
65. Pohl, L.R., Bhooshan, B., Whittaker, N.F. and Krishna, G (1977) Biochem. Biophys. Res. Commun. 79, 684.
66. Pohl, L.R., Nelson, S.D. and Krishna, G. (1978), Biochem. Pharm. 27, 491.

67.   Boyd, M.R., Sasame, H.A., Mitchell, J.R. and Catignani, G. (1977), Fed. Proc. 36, 405.
68.   Snyder, R. and Kocsis, J.J. (1975), CRC Critical Rev. 3, 265.
69.   Andrews, L.S., Sasame, H.A. and Gillette, J.R. (1979), Life Sciences 25, 567.
70.   Potter, W.Z., Thorgeirsson, S.S., Jollow, D.J. and Mitchell, J.R. (1974), Pharmacology 12, 129.

RECEIVED March 12, 1981.

15

Pharmacokinetics and Threshold Concepts

JOHN C. RAMSEY and RICHARD H. REITZ

Toxicology Research Laboratory, Health & Environmental Sciences,
Dow Chemical U.S.A., Midland, MI 48640

The fundamental goal of toxicological research is to provide
a rational basis for recommending acceptably safe levels of human
exposure to potentially harmful agents. Chemically induced
cancer is a toxic response that has received primary attention in
recent years. The potential lethality of cancer, its generally
irreversible nature, and relatively long latent (induction)
period are characteristics that have placed carcinogenesis in the
forefront of public concern. Whether an absolute threshold
exists for chemical induction of cancer (a dose or exposure level
below which no carcinogenic event is induced) is at present
debatable.

The concept of one irreversible molecular event giving rise
to the expression of cancer does not allow for the existence of
an absolute threshold. On the other hand, considerations such as
the multistage nature of chemical carcinogenesis, the existence
of DNA repair and immune surveillance mechanisms, and the existence of threshold doses for other pathological responses support
a possible threshold for at least some carcinogenic agents (1).

However, since absolute zero risk is no more realistically
attainable than is the absolute zero of temperature, the problem
remains for toxicologists to estimate the probable finite risk of
carcinogenesis at very low levels of exposure. The magnitude of
risk which may be socially and economically acceptable constitutes
a value judgment that must incorporate many considerations including benefits as well as risk. This judgment can be made with
confidence only when the estimate of risk is based on the application of scientific principles to the best available information.
The purpose of this paper is to investigate the impact of the
pharmacokinetic threshold upon quantitative estimates of carcinogenic risk at low levels of exposure, and to discuss the concept
of a cytotoxic threshold for chemically induced cancer.

0097–6156/81/0160–0239$05.00/0

The Number Zero and Threshold Concepts

Consider the representation of the spectrum of real numbers shown here. There exists an infinite set of positive integers

$$-\infty \underset{\cdots}{} -3 \quad -2 \quad -1 \quad 0 \quad 1 \quad 2 \quad 3 \underset{\cdots}{} \infty$$

as well as infinite subsets of positive fractions, all greater than zero. But only the mathematically unique number zero represents the *total absence* of any quantity. In the complicated continuum of events on a molecular scale that make up biological processes, the number zero has little significance. Furthermore, physical or biological experimentation cannot provide conclusive proof of the total absence of any quantity. Therefore, the concept of an absolute threshold as the attainment of (or departure from) zero biological response is neither theoretically meaningful nor experimentally demonstrable.

On the other hand, we can consider that the spectrum of numbers greater than zero represents a range of relationships between biological quantities. Within this continuum, abrupt changes or transition regions may exist that lead to significantly altered biological relationships. Thus the range of dose levels within which a transition occurs in the relationship between biological quantities and the dose level constitutes a threshold region. This concept of a threshold region has both theoretical and practical significance in toxicology, and within this context we shall investigate the toxicological consequences of both pharmacokinetic and cytotoxic thresholds.

Dose-Response Relationships

The problem of extrapolating an experimentally observable carcinogenic response in laboratory animals to the expected response in humans at low exposure levels is illustrated by the typical dose-response curve shown in Figure 1. The solid line represents the range of observable response, which decreases with decreasing dose level. As the dose level is further decreased, the response diminishes until it virtually vanishes into the normal background incidence of the lesion. The broken lines represent the region into which it is necessary to extrapolate the observed response, and this region is likely to be many orders of magnitude below the observable range. The problem, therefore, is to elucidate the shape of the dose-response curve below the solid line as a *quantitative* function of the dose level.

Pharmacokinetic Principles

Pharmacokinetics is the study of the dynamics of absorption, distribution, biotransformation, and excretion of a chemical from the body. These processes can be described by a set of differential equations which comprise the pharmacokinetic model of the chemical. The types of rate processes incorporated in the model describe the qualitative behavior of the chemical and its metabolites, and quantitation of the rate processes (i.e., numerical values of the pharmacokinetic parameters) provides the means of predicting the concentration of the chemical as a function of time following single or repeated doses. Since most toxic responses (including carcinogenesis) appear to be dependent both on the concentration of the toxic entity at the sensitive site and on the length of time it resides there, the pharmacokinetic characteristics of a chemical are intricately linked to its toxic response.

Most biological processes can be characterized by three types of rate equations, two of which are limiting cases of the third more general rate equation. The first of these is the first order rate equation represented by Equation 1. In this rate process, k represents the first order rate constant and C

$$\text{rate} = k \cdot C \tag{1}$$

represents the concentration (or amount) of the chemical. First order rates can characterize such biological processes as passive diffusion across membranes and glomerular filtration from blood plasma into urine. The unique characteristic of first order processes is that the rate of the process always remains directly proportional to the concentration of chemical. A first order rate can be zero only when the concentration equals zero.

The second type of rate process is known as zero order and is characterized by Equation 2. In this case the rate is equal to the rate constant, and does not change as the concentration

$$\text{rate} = k^{o} \tag{2}$$

of chemical changes. Zero order rates are often encountered as constant rates of input of a chemical, such as an intravenous infusion or continuous uptake of a chemical due to environmental exposure. Saturable biological processes also exhibit zero order properties at sufficiently high concentrations, as pointed out below.

The third, more general, type of rate process follows saturable (Michaelis-Menten) kinetics as shown in Equation 3. This type of process is defined by two pharmacokinetic parameters; V_m is the maximum possible rate of the reaction, and K_m (the Michaelis constant) is the concentration of the chemical when the rate is equal to one-half its maximum value. The unique

$$rate = \frac{V_m \cdot C}{K_m + C} \qquad (3)$$

properties of processes characterized by Michaelis-Menten kinetics lie in the relationship between the concentration C of the chemical and its Michaelis constant K_m. When C is much less than K_m ($C \ll K_m$), the rate will be approximately first order (Equation 1) with an apparent first order rate constant equal to V_m/K_m. However, as the concentration increases, a transition region ensues in which the rate of the process is no longer proportional to the concentration. Finally as the concentration becomes much greater than K_m ($C \gg K_m$), the rate approaches *but does not exceed* the maximum rate V_m. In this concentration range the process will exhibit the zero order properties of Equation 2. Biological processes which utilize a limited resource, such as enzymatically catalyzed biotransformations and active transport, can be characterized by Michaelis-Menten kinetics. In fact, it is likely that virtually all biological rate processes are saturable at sufficiently high concentrations.

Pharmacokinetic Threshold

The behavior of a chemical and its metabolites in the body is described by the parameters of the pharmacokinetic model and is dependent, among other things, on the administered dose level. As a given dose level is repeatedly or continuously administered, the concentration in the body will increase until eventually the rate of input is equal to the rate of output, and the chemical and its metabolite(s) will then maintain a constant (steady state) concentration in the body until exposure ceases. When the dose levels are such that all the biological processes comprising the model are first order (or the concentrations are well below the K_m values for saturable processes) the distribution and concentration of the chemical and its metabolites in the body will maintain values that are directly proportional to the dose level. This direct proportionality (which is a direct result of first order kinetic processes) confers *linear* pharmacokinetic behavior on the chemical within this range of dose levels. However when the dose level is increased until the concentration in the body approaches or exceeds the K_m value for any saturable process, then the direct proportionality maintained at lower dose levels will be lost and the relationship of the concentration of the chemical and its metabolites to the dose level will no longer be a constant value. This deviation from a constant relationship between the administered dose level and the concentrations within the body results in a *nonlinear* pharmacokinetic profile.

The transition from linear to nonlinear kinetics as the dose
level increases constitutes the pharmacokinetic threshold. Since
the biological process represented by a pharmacokinetic model
comprise a continuum of events, the pharmacokinetic threshold
must be considered a gradual transition from linear to nonlinear
kinetics with increasing dose level. Rather than a single precisely defined dose level, the threshold is a range of dose
levels over which this transition occurs. However, the exact
dose range at which deviation from kinetic linearity becomes
apparent is relatively unimportant. The major concern is whether
extrapolations are made from toxicity data obtained at dose
levels either above or below the pharmacokinetic threshold
transition.

Hypothetical Illustration. Since observations of carcinogenic response arise from chronic (i.e., long term) studies, it
will be appropriate to illustrate the existence of a pharmacokinetic threshold based on changes in the steady state level of
a parent chemical and one of its metabolites at successively
increasing dose levels. We will then investigate the cases where
either the parent chemical or its metabolite is the carcinogenic
entity.

The hypothetical model chosen to illustrate the pharmacokinetic threshold is presented in Figure 2. The input of chemical
is considered to occur at an uninterrupted constant rate (k^o)
as might be the case for continuous environmental exposure, and
this input rate is the dose level under investigation. The
parent chemical P can be excreted by a first order process k_p
or metabolized to metabolite M by the saturable process characterized by V_{mp} and K_{mp}. The metabolite can also be excreted by a
first order process k_m or further metabolized by the saturable
process characterized by V_{mm} and K_{mm}.

Biologically plausible values were chosen for the pharmacokinetic parameters of the model, and the steady state concentrations of P (Pss) and of M (Mss) were determined by numerical
integration of the differential equations describing the model of
Figure 2. These steady state concentrations of Pss and Mss were
determined at values of the input rate (i.e., dose level) ranging
from 0.0001 to 300. (See the appendix for a complete description
of the model simulation; the actual units of the dose level k^o
are μmole P/hr).

In order to illustrate the relationship between the steady
state values of P and M to the dose level, each value of Pss and
Mss was divided by the corresponding dose level and the results
are shown in Figure 3.

Linear pharmacokinetics are indicated in this example at the
lower dose levels where the ratio of Pss/dose and Mss/dose
maintain a constant value. For example as the dose level is
increased by a factor of 10, then both Pss and Mss also increase
by a factor of 10. Thus within the dose range where the curves

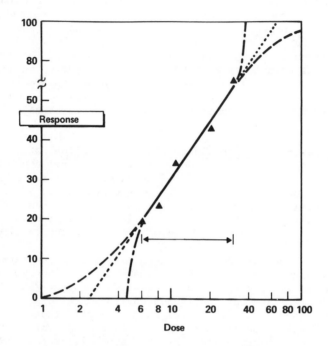

Figure 1. Typical dose–response curve. The solid line represents the experimentally observable range of toxic response. The broken lines at low-dose levels show the region into which it is necessary to extrapolate the observed response.

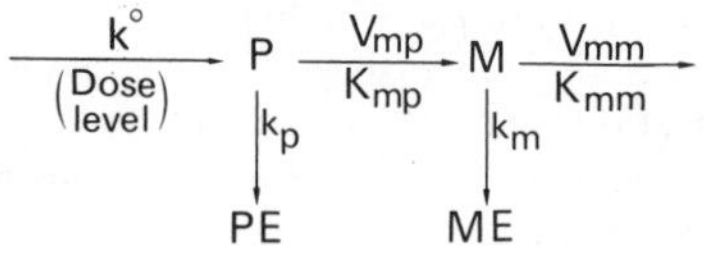

Figure 2. Hypothetical pharmacokinetic model describing the disposition of a parent chemical (P) and its metabolite (M) in the body. The parameters are described in the text.

$$\xrightarrow[\left(\begin{array}{c}\text{Dose}\\\text{level}\end{array}\right)]{k^{\circ}} P \xrightarrow[K_{mp}]{V_{mp}} M \xrightarrow[K_{mm}]{V_{mm}}$$

$$\big\downarrow k_p \qquad\qquad \big\downarrow k_m$$

$$PE \qquad\qquad ME$$

of Figure 3 remain nearly parallel to the abscissa the concentration of the parent chemical and its metabolite in the body remain directly proportional to the dose level, and this constitutes the linear pharmacokinetic range of doses.

However, kinetic nonlinearity becomes evident at dose levels between 0.1 and 1.0. As the concentration of P approaches and then exceeds the value of K_{mp} (the Michaelis constant for the metabolic transformation of P), a dramatic increase in the ratio of Pss/dose is evident. A concurrent increase in the ratio of Mss/dose is also apparent until the rate of formation of M from P becomes virtually saturated. At successively higher dose levels Mss stays almost constant and consequently the ratio of Mss/dose decreases as the dose level increases. Thus at dose levels above the pharmacokinetic threshold the quantitative relationship between the dose level and the steady state concentrations of the parent compound and its metabolite are no longer the same as were maintained at dose levels below the pharmacokinetic threshold region.

Risk Estimation

Numerous models are used for the purpose of extrapolating the carcinogenic response observed at relatively high experimental dose levels to the expected response at much lower dose levels (2, 3). The models differ from each other mainly in the rapidity with which zero response is approached as the dose level approaches zero. Most of the models used for risk estimation make no provision for an absolute threshold for carcinogenic response (i.e., the response equals zero only when the dose equals zero). However a common feature of these models is dependence on the internal concentration of the carcinogenic entity (the effective dose) being directly proportional to the dose level of the parent chemical over the entire range of dose levels. One of the most common models for carcinogenic risk assessment is the one-hit model described by Equation 4 where R_d is the fraction of the population showing a positive response upon exposure to dose level D and β is a sensitivity factor

$$R_d = 1 - e^{-\beta D} \tag{4}$$

relating the dose to the response. Although the one-hit model is the most conservative risk extrapolation model (2, 4), it is representative of the others in its dependence on the effective dose, and can serve as a means of illustrating the impact of the pharmacokinetic threshold upon dose extrapolations.

The values of Pss and Mss obtained from the simulation of the hypothetical pharmacokinetic model were substituted for D in Equation 4, and a value of 0.001 was assigned to the sensitivity factor β. Thus, R_d was calculated for the entire range of dose levels used in the simulation. A portion of the resulting

dose-response curves are plotted as R_d versus dose level in
Figure 4 (in which P is the carcinogenic entity) and 5 (in which
M is the carcinogenic entity). The dose response curve arising
from the parent compound exhibits a nearly sigmoid shape with a
response of 9.26×10^{-1} at a dose level of 300 decreasing to
1.72×10^{-8} at a dose level of 0.0001. The dose response curve
arising from the metabolite shows a maximum response of 1.86×10^{-2}
at the high dose levels, decreasing to 2.09×10^{-7} at a dose
level of 0.0001. The flat portion of the curve at the higher
levels is a consequence of the saturable rate of formation of the
metabolite M, regardless of the increasing concentration of the
parent chemical in the body.

In practice, the carcinogenic response observed at a given
dose level is used to estimate the expected response at a lower
dose level as follows. From the response R_d observed at the
dose level D, the sensitivity factor β is calculated according to
Equation 5 (which is a rearrangement of the logarithmic form of
Equation 4). This value of β is then substituted into

$$\beta = \frac{-\ln (1-R_d)}{D} \tag{5}$$

Equation 4 with the new (lower) value of D to calculate the
response expected at this value of D.

Table I

Calculated response at various dose levels from simulation of
the pharmacokinetic model of Figure 2 (Pss is the toxic entity),
and predicted response at dose level = 0.0001[a].

Dose Level	Calculated Response	Predicted Response at Dose Level=0.0001	Ratio of Predicted Response to Calculated Response at Dose Level = 0.0001
300	9.26×10^{-1}	86.8×10^{-8}	50.5
30	2.08×10^{-1}	77.8×10^{-8}	45.2
3	2.38×10^{-3}	7.94×10^{-8}	4.60
0.1	1.77×10^{-5}	1.77×10^{-8}	1.03
0.01	1.72×10^{-6}	1.72×10^{-8}	1.00
0.0001	1.72×10^{-8}	–	–

[a]Both the calculated (i.e., simulated) response and the predicted (i.e., estimated) response are based on the one-hit
model as described in the text.

The foregoing procedure was used with the simulated response
values (portions of which are shown in Figures 4 and 5) to predict the response at the lowest dose level of 0.0001 used in the

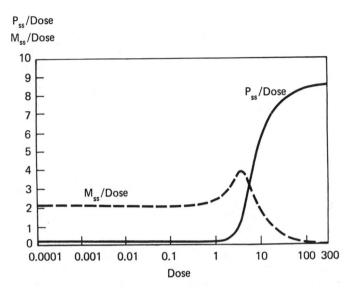

Figure 3.   *The ratio between steady concentration of the parent chemical and dose level (Pss/Dose), and between steady-state concentration of the metabolite and dose level (Mss/dose). Data obtained from simulation of the pharmacokinetic model described in Figure 2.*

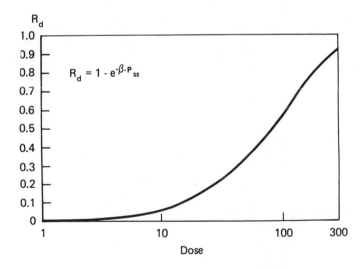

Figure 4.   *A portion of the dose–response curve generated from the pharmacokinetic model of Figure 2 in which the parent chemical (Pss) is the carcinogenic entity and the response is calculated with the one-hit model (see Equation 4)*

simulation. The results for the case in which the parent chemical P is the carcinogenic entity are shown in Table I. The second column in Table I is the calculated response at the indicated dose level, and the third column is the corresponding predicted (estimated) response at a dose level of 0.0001 based on the described procedure. It is apparent that dose levels exceeding the pharmacokinetic threshold are not an appropriate index with which to estimate the response at dose levels below the pharmacokinetic threshold. Conversely, since dose levels below the threshold are directly proportional to the internal concentration of the carcinogenic entity, they can serve as appropriate indices of the response expected at even lower levels.

Similar results for the case in which the metabolite M is the carcinogenic entity are shown in Table II. In this case the predicted response may be either under-estimated or over-estimated when it is based on the observed response at dose levels exceeding the pharmacokinetic threshold. However, as is the case with the parent chemical, dose levels below the pharmacokinetic threshold are proportional to the concentration of the metabolite in the body, and can be used to predict the response at even lower levels.

Table II

Calculated response at various dose levels from simulation of the pharmacokinetic model of Figure 2 (Mss is the toxic entity), and predicted response at dose level = 0.0001^a.

Dose Level	Calculated Response	Predicted Response at Dose Level=0.0001	Ratio of Predicted Response to Calculated Response at Dose Level = 0.0001
300	1.86×10^{-2}	0.062×10^{-7}	0.030
30	1.85×10^{-2}	0.622×10^{-7}	0.30
3	1.16×10^{-2}	3.89×10^{-7}	1.86
0.1	2.13×10^{-4}	2.13×10^{-7}	1.02
0.01	2.10×10^{-5}	2.10×10^{-7}	1.00
0.0001	2.09×10^{-7}	–	–

[a]Both the calculated (i.e., simulated) response and the predicted (i.e., estimated) response are based on the one-hit model as described in the text.

    In both of the foregoing examples, the consistent (linear)
relationship between the concentration of the carcinogenic
entity and the dose level below the pharmacokinetic threshold
yields consistent estimates for the parameter β over this dose
range. However inconsistent estimates of β derived from dose
levels above the pharmacokinetic threshold arise from the nonlinear relationship between the concentration of the entity
inducing the carcinogenic response and the dose level.
    The foregoing simulation illustrates the inadequacy of dose
response data obtained at dose levels above the pharmacokinetic
threshold to predict the response at lower dose levels when the
prediction is based on dose levels alone. It should be emphasized that the magnitude of the error in the predicted response
in this example (pointed out by the ratios in Tables I and II) is
of little quantitative significance. The magnitude of the error
may change by many fold depending on the parameters of the model
employed. It is far more important that the errors in estimated
response are in direct proportion to the extent that the relationship between the steady state level of the carcinogenic entity and
the dose level deviates from the linear relationship maintained
at levels below the pharmacokinetic threshold.
    The pharmacokinetic threshold has significance far beyond
the specialized endeavor of carcinogenic risk estimation. Since
virtually any toxic response is a function of the concentration
x time product of the toxic chemical in the sensitive tissue, the
relationship between steady state concentrations and administered
dose levels is crucial in interpreting and predicting any toxic
response as a function of exposure level. In particular, when
otherwise efficient defense mechanisms or detoxification pathways
are overwhelmed at sufficiently high dose levels dramatic nonlinear increases in toxicity may arise (5, 6).

    Vinyl Chloride. An example of a pharmacokinetic threshold
that relates directly to carcinogenic risk estimation is that of
inhaled vinyl chloride (VC) in rats. VC has been shown to
induce hepatic angiosarcoma in rats at exposure levels ranging
from 50 to 10000 ppm (7), with an essentially flat dose-response
curve at exposure levels from 1000 to 10000 ppm. A reactive
metabolite of VC is likely the carcinogenic entity rather than
the parent compound. Measurements of the amount of VC metabolized
by rats during 6-hour exposure to concentrations of VC ranging
from 1.4 ppm to 4600 ppm were conducted (8). Figure 6 represents
the results of these studies by showing the ratio of VC metabolized
(μg M) to the exposure level plotted versus exposure level. The
data points are mean ± standard deviation, and the solid line was
drawn by inspection. A pharmacokinetic threshold is apparent in
the region of VC exposure levels above approximately 50 ppm.
Therefore, exposure concentrations above this region can not
provide the appropriate index for assessing the potential response
at lower levels.

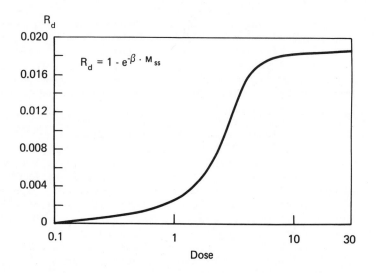

Figure 5.   A portion of the dose–response curve generated from the pharmaco-kinetic model of Figure 2 in which the metabolite (Mss) is the carcinogenic entity and the response is calculated with the one-hit model (Equation 4)

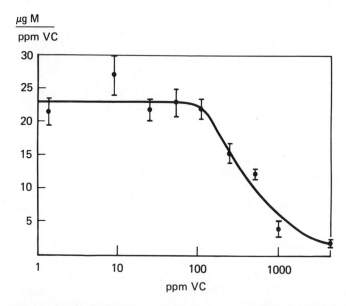

Figure 6.   The ratio between the amount of vinyl chloride metabolized (micro-grams M) and the concentration of inhaled vinyl chloride (ppm VC) in rats (8). Data points are mean ± standard deviation and the solid line was drawn by inspection.

However, Gehring et al. were able to utilize dose response data obtained above the pharmacokinetic threshold by determining the pharmacokinetic parameters describing the saturable metabolism of inhaled VC in rats (8). With knowledge of the V_m and K_m values for the metabolic transformation of VC in rats, it was possible to index the observed response to the internal dose level of the toxic entity (the amount of VC metabolized), rather than to the exposure level.

The observed incidence of angiosarcoma in rats exposed to 50 ppm VC was $1.7x10^{-2}$, and in rats exposed to 10000 ppm VC the incidence was $14.8x10^{-2}$ (7). Using the one-hit model, the tumor incidence predicted at 50 ppm based on the observed incidence at 10000 ppm and the amount of VC metabolized at 10000 ppm is $2.1x10^{-2}$, or only 1.2 times higher than the observed incidence at 50 ppm. However, when the same prediction is based on the exposure level rather than the amount of VC metabolized, the estimated response at 50 ppm is $0.08x10^{-2}$, or almost 21 times lower than the observed incidence. Even though these authors (4) showed that the one-hit model resulted in gross over-prediction of tumor incidence in humans exposed to VC (in this case, the probit model appeared to be most reliable), the above example shows that use of the appropriate index for the concentration of the toxic entity is essential to obtain realistic estimates of the response expected at low levels of exposure.

Cytotoxic Threshold

It is widely believed that at least some types of cancer are related to the induction of mutations in target cells. This theory has as a corollary the hypothesis that since there is only a single copy of each gene in a cell, a single event (i.e., one hit) with a critical genetic component could conceivably produce a mutation which would ultimately lead to cancer. Hence, there can be no absolute threshold under this theory of carcinogenesis (if we ignore the influences of such factors as DNA repair, immunosurveillance and the previously discussed pharmacokinetic threshold). However, since a variety of tumors develop spontaneously in animals even without exposure to exogenous chemicals, any process which enhanced the endogenous stimuli would also be considered as carcinogenic. Since many of these agents appear to exert their activity through interaction with cellular components other than genetic material, there is reason to believe that there may be a threshold in their action.

One example of such a process is chemically-induced recurrent cytotoxicity. In this case individual cells are killed by sufficiently high concentrations of toxic chemicals, thus stimulating replication in the surviving cells to replace the necrotic tissue. Each cellular division has a small but nevertheless finite chance for error in duplicating the genetic material of the cell. Hence the effect of stimulating cellular regeneration in a target organ

throughout a major portion of an animal's lifetime may be to
significantly increase the spontaneous mutation rate in that
tissue. In addition, DNA repair mechanisms seem to be considerably less effective in correcting small amounts of genetic damage
after replication of DNA has occurred (9).

A major characteristic of chemical interaction with non-
genetic components of the cell is that, instead of a single
critical target per cell (the unique molecule of DNA bearing the
genetic information), there are multiple copies of the other
cellular constituents. Loss of a small fraction of these will
not affect the viability of the cell. This multiplicity of non-
genetic components comprises a finite capacity for the cell to
tolerate injury arising from the presence of exogenous chemicals.
Furthermore, if the genetic material is not damaged, there will
be continuing resynthesis of cytoplasmic constituents. In contrast, since DNA acts as its own template, damaged DNA cannot be
replaced by synthesis but must be repaired. Therefore, we can
consider the cytotoxic threshold to be that range of dose levels
below which the rate of destruction of these cytoplasmic components is small relative to their rate of resynthesis. Consequently,
at dose levels below the cytotoxic threshold range there will be
no chemically-induced cell death and hence no stimulus for
increased cell replication rates.

Chloroform. As an example of the cytotoxic threshold, let
us consider chloroform. This chemical induces liver and kidney
tumors in rodents upon prolonged administration of high doses (10).
However, chloroform does not induce mutations in bacterial test
systems (11, 12), nor does it induce significant alteration of
DNA isolated from organs of animals exposed in vivo to chloroform
(13). However, chloroform does induce extensive tissue damage
with subsequent cellular regeneration at the same sites where
tumors later develop (Table 3). When the exposure to chloroform
is reduced to levels which do not produce clinically observable
tissue damage, tumors fail to develop upon chronic exposure (13,
14). Thus it appears that when a chemical influences the carcinogenic process primarily through induction of cytotoxicity rather
than through direct genetic alterations, exposures below the
cytotoxic threshold will not influence the carcinogenic process
appreciably and hence constitute very little risk to man.

Table III

The Presence of Clinically Observable Tissue
Damage and/or Tumors in Various Strains of Mice
Exposed to Chloroform[a]

Tissue Damage	Dose (mg/kg/day)		
	240	60	15
B6C3F1 (Males)	Liver, Kidney	Kidney	None
CD-1 (Males)	Liver, Kidney	Kidney	None
Tumor Development			
B6C3S1 (Males)	Liver, Kidney(?)	–	–
ICI	–	Kidney	None
C57B1	–	None	None
CF/1	–	None	None
CBA	–	None	None

---

[a]Data from references 13 and 14.

Summary

In summary, changes or transition regions in a spectrum of
relationships between biological quantities can have a profound
impact upon the toxic response elicited by exogenous chemicals.
The range of dose levels within which a transition occurs in the
relationship between biological quantities and the dose level of
a chemical constitutes a threshold region.
    Within this context, the range of dose levels of a chemical
causing a transition from a linear to a nonlinear pharmacokinetic
profile comprise the pharmacokinetic threshold dose range for the
chemical. The change in the relationship between the internal
concentration of the toxic entity and the dose level must be
taken into account when extrapolating the observed response at
dose levels above the pharmacokinetic threshold to the expected
response at much lower dose levels.
    The multiplicity of cellular components (other than genetic
material) which may be destroyed or damaged by the presence of an
exogenous chemical before cell death occurrs provides the basis
for a cytotoxic threshold with respect to the range of dose
levels necessary to cause clinically observable cytotoxicity in
the target tissue. Increased cell replication rates arising from
this recurrent cytotoxic injury may lead to the induction of
cancer through the attendant increase in the probability of
unrepaired DNA replication errors. Therefore, to the extent that
an exogenous chemical induces cancer by a cytotoxic mechanism,
dose levels above the cytotoxic threshold range may lead to
enhanced tumor formation, whereas dose levels below this range
should cause no increase in the incidence of tumors.

Literature Cited

1. Gehring, P. J. and Blau, G. E. "Mechanisms of Carcino-
    genesis: Dose Response" J. Environ. Path. Toxicol.
    (1977), 1, 163-179.

2. Van Ryzin, J. "Quantitative Risk Assessment" J. Occup.
    Med. (1980), 22, 321-326.

3. Scientific Committee, Food Safety Council "Proposed
    System for Food Safety Assessment" Fd. Cosmet. Toxicol.
    (1978), 16, (Suppl. 2), 109-136.

4. Gehring, P. J., Watanabe, P. G. and Park, C. N. "Risk of
    Angiosarcoma in Workers Exposed to Vinyl Chloride as Pre-
    dicted from Studies in Rats" Tox. Appl. Pharmacol. (1979),
    49, 15-21.

5. Ramsey, J. C. and Gehring, P. J. "Application of Pharmaco-
    kinetic Principles in Practice" Fed. Proc. (1980), 39,
    60-65.

6. Watanabe, P. J., Young, J. D. and Gehring, P. J. "The
    Importance of Non-Linear (Dose-Dependent) Pharmacokinetics
    in Hazard Assessment" J. Environ. Path. Toxicol. (1977),
    1, 147-159.

7. Maltoni, C. and Lefemine, G. "Carcinogenicity Assays of
    Vinyl Chloride" Ann. N.Y. Acad. Sci. (1975), 246, 195-224.

8. Gehring, P. J., Watanabe, P. G. and Blau, G. E. "Risk
    Assessment of Environmental Carcinogens Utilizing Pharmaco-
    kinetic Parameters" Ann. N.Y. Acad. Sci. (1979), 329,
    137-152.

9. Bermann, J. J., Tong, C. and Williams, G. M. "Enhancement
    of Mutagenesis During Cell Replication of Cultured Liver
    Epithelial Cells" Cancer Lett. (1978), 4, 277-283.

10. National Cancer Institute "Carcinogenesis Bioassay of
    Chloroform" Nat. Tech. Inf. Service (1976), No. PB264018/AS.

11. Uehleke, H., Werner, T., Greim, H. and Kramer, M.
    "Metabolic Activation of Haloalkanes and Tests In Vitro
    for Mutagenicity" Xenobiotica (1977), 7, 393.

12. Simon, V. K., Kauhaner, K. and Tardiff, R. G. "Mutagenic
    Activity of Chemicals Identified in Drinking Water" Second
    International Meeting of the Environmental Mutagen Society
    (1978), Edinburgh, Scottland.

13.   Reitz, R. H., Quast, J. F., Stott, W. T., Watanabe, P. G.
      and Gehring, P. J. In: Water Chlorination: Environmental
      Impact and Health Effects (Vol. III) (1980), Jolley, R. L.,
      Brungs, W. A. and Cumming, R. B. (Ed.), Ann Arbor Press,
      Ann Arbor, Mich. Chap. 85.

14.   Roe, F. J. C., Palmer, A. K. and Worden, A. N. "Safety
      Evaluation of Tooth Paste Containing Chloroform. I. Long
      Term Studies in Mice" J. Environ. Path. Toxicol. (1979),
      2, 799-819.

15.   Belvedere, G., Cantoni, L., Facchinetti, T. and Salmona, H.
      "Kinetic Behavior of Microsomal Styrene Monooxygenase and
      Styrene Epoxide Hydratase in Different Animal Species"
      Experientia (1977), 33, 708-709.

Appendix

The differential equations describing the pharmacokinetic
model in Figure 2 are:

$$\frac{dP}{dt} = k^o - \left\{ \frac{V_{mp} \cdot P}{K_{mp} + P} \right\} - k_p \cdot P$$

$$\frac{dM}{dt} = \left\{ \frac{V_{mp} \cdot P}{K_{mp} + P} \right\} - \left\{ \frac{V_{mm} \cdot M}{K_{mm} + M} \right\} - k_m \cdot M$$

$$\frac{dPE}{dt} = k_p \cdot P$$

$$\frac{dME}{dt} = k_m \cdot M$$

The following values for the pharmacokinetic parameters were
scaled up from in vitro determinations of styrene monooxygenase
and epoxide hydratase activities in mouse liver (15).

$$V_{mp} = 3.414 \ \mu mole/hr$$

$$V_{mm} = 4.947 \ \mu mole/hr$$

$$K_{mp} = 0.5976 \ \mu mole$$

$$K_{mm} = 10.91 \ \mu mole$$

The values of the first order rate constants were arbitrarily chosen as:

$$k_p = 0.114 \ hr^{-1}$$

$$k_m = 0.015 \ hr^{-1}$$

Steady state values of P and M were then determined at values of k^o ranging from 0.0001 to 300 μmole P/hr by numerical integration. Steady state concentrations were considered to be attained when neither the concentration of P or M changed by more than 1 part in 10^5 over 2 consecutive 24-hour time periods.

RECEIVED February 2, 1981.

Metabolic Aspects of Pesticide Toxicology

G. WAYNE IVIE

Veterinary Toxicology and Entomology Research Laboratory, Agricultural Research, Science and Education Administration, U.S. Department of Agriculture, College Station, TX

S. KRIS BANDAL

Agricultural Products, 3M Company, St. Paul, MN 55101

At least 1500 organic and inorganic chemicals are used in a manner such that they can be called pesticides (1). These chemicals are indispensable in the management of a seemingly endless variety of pest organisms, including insects, weeds, fungi, bacteria, pest birds and mammals, and others. Pesticides are intentionally applied to many components of the environment, and they or their degradation products often move quite freely through the environment by mechanisms such as runoff, leaching, and volatilization. The production and use of pesticides on a world scale exceeds 3 billion pounds annually (1), and it can safely be said that residues of various pesticides interact at some level with virtually all components of the environment.

Pesticides by design are meant to be toxic! Although a major goal of the discipline of modern pesticide chemistry is to develop pesticides and consequent use patterns that confine pesticide toxicity to pest organisms, such a goal is seldom attained easily. All living organisms have much in common biochemically, and successful exploitation of the often relatively minor biochemical differences between pest and non-pest species is almost always difficult and is, in fact, sometimes impossible. Thus, it is often necessary to use pesticides that are toxic not only to the pest species but to other organisms as well. Even when we succeed in developing what appear to be highly efficacious yet selective pesticides, we are always concerned that interactions of these chemicals or their transformation products with non-target species, particularly man, may result in some unforeseen toxic consequences.

From the human perspective, the direct toxicological implications of pesticide use to our own species merit the most thorough and serious consideration. Most would agree that the judicious use of pesticides contributes in a positive way to many aspects of human welfare, but we also recognize that these chemicals have genuine potential for adverse human effects. Therefore, if the proposed use patterns of a pesticide create a substantial likelihood that interactions with man may occur, it is prudent to

0097–6156/81/0160–0257$07.00/0

define both the extent of these interactions and their
toxicological significance. Our discussion will center on the
role played by metabolism in the expression of pesticide toxicity
and the evaluation of toxicological significance. We will briefly
discuss the importance of metabolism studies in developing more
efficacious and selective pesticides. We will discuss the
rationale and appropriate methodology used by metabolism
scientists in the design and execution of such studies. Finally,
and most importantly, we will attempt to show how the metabolism
of pesticides may affect their toxicity, and how the data from
pesticide metabolism studies are used in the process of evaluating
toxicological risk to man.

The Nature of Metabolic Reactions

Pesticides are transformed by living organisms through a
great diversity of metabolic reactions. These reactions can be
conveniently grouped into two categories, primary or phase I
reactions, which are those that create or modify functional
groups, and secondary or phase II reactions, which are conjugations. A few examples are shown in Figure 1. Some authors (2)
feel that the terms phase I and phase II are not totally
satisfactory because numerous examples are known of phase II
reactions preceding phase I reactions (e.g., direct conjugations
of chlorinated phenols, Figure 1). Most pesticides, however, do
not lend themselves to phase II reactions without prior phase I
modifications. Although it is generally true that phase I
metabolism of pesticides effects partial or complete
detoxification, at least from an acute toxicity standpoint,
metabolic activations do occur and can be of great toxicological
significance. Phase II or conjugation reactions more often than
not serve to render pesticides or their metabolites more polar for
more efficient excretion (e.g., in urine of mammals) or to
facilitate transport for internal storage in organisms that lack
efficient excretory systems (e.g., plants). It is probably
correct that most living organisms can metabolize pesticides via
both phase I and phase II metabolic pathways.

The schematic shown in Figure 2 is designed to represent the
major metabolic and disposition patterns that different pesticide
types might undergo in higher animal systems. We have somewhat
arbitrarily grouped pesticides into four categories, based on
polarity. A very few pesticides, primarily some organochlorine
insecticides and particularly the insecticide mirex, are highly
lipophilic, are quite metabolically stable, and tend to be stored
in fat with minimal or no metabolism. Direct elimination through
lipid containing animal byproducts (milk or eggs) tends also to
be an appreciable to major disposition mechanism for such highly
lipophilic compounds. Most insecticides are lipophilic, yet are
rapidly metabolized by both phase I and phase II reactions and are
ultimately excreted from the body. Some pesticides, including

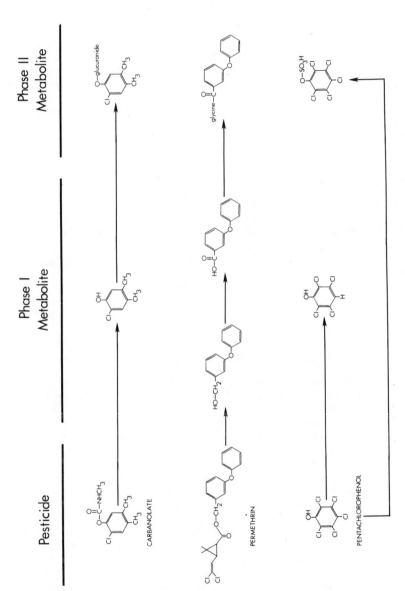

Figure 1. Examples of Phase I and Phase II metabolites of the carbamate insecticide carbanolate, the synthetic pyrethroid insecticide permethrin, and the wood preservative pentachlorophenol

phenolics, amines, etc., are reasonably polar compounds that
generally have functionalities that permit direct conjugation
reactions. Others, such as herbicides formulated as salts, or
compounds that contain moieties that readily ionize at
physiological pH, can be considered hydrophilic and are often
excreted rapidly without any metabolism at all. Phase I and phase
II pesticide metabolites, and possibly even the parent pesticide,
may have the potential for chemical sequestration (i.e., covalent
binding, Figure 2) with tissue components that may ultimately lead
to the expression of chronic toxicity.

Most organisms, regardless of complexity, share a number of
biochemical pathways for metabolizing pesticides. Examples can
readily be found to show that many types of plants and animals
metabolize pesticides by each of the four basic types of metabolic
changes: oxidation, reduction, hydrolysis, and conjugation (3).
Of course, species do differ in the metabolism of pesticides,
these differences are sometimes quite dramatic, and they can be of
great significance in interpreting comparative toxicological
effects. Also, species differences in pesticide metabolism, once
identified, quite often provide impetus to the development of more
selective pest-control agents.

It is not our purpose here to extensively review the
literature on the metabolism of individual pesticides by a variety
of living organisms. Numerous such reviews are available, some
are periodically updated, and we refer the reader to several of
these for an overview of the voluminous literature in this field
(4-14).

Metabolic Basis for Pesticide Selectivity

In the use of pesticides, attempts are always made to direct
their toxic actions toward an individual or group of pest species,
and it is a major goal of the pesticide scientist to develop
efficacious pesticides and use patterns such that little or no
toxicity to other life forms occurs. Such an approach is clearly
desirable from an environmental standpoint, but it often has
definite economic advantages also (e.g., protecting predators and
parasites while controlling a pest insect). In some circumstances, a degree of selectivity is absolutely essential for the
intended use (e.g., herbicides cannot be lethal to the protected
crop). Metabolism studies in the pest species, in the species
being protected, and in associated nontarget organisms, can and
often do provide a wealth of useful information. Such studies may
lead to a more thorough understanding of the mechanisms of
pesticidal action, and this knowledge often leads in turn to the
development of more efficacious, selective, and environmentally
acceptable pest control agents.

While not always so, selective toxicity can quite often be
attributed primarily if not totally to metabolic differences
between species, either in the rate of metabolism or the nature of

products formed.   The   insecticide   malathion   (Figure 3)   is   a
well-known example of such selectivity.   Malathion is highly toxic
to a number of pest insect species, yet it is very low in toxicity
to mammals.   These differences are explained by the fact that
mammals readily metabolize malathion to a nontoxic monocarboxylic
acid,   whereas   this   reaction   occurs   at   a   much   slower   rate   in
susceptible   insects   (15).   The   selectivity   of   the   herbicide
linuron (Figure 4) is also due to differences in metabolic rates
between species, at least in some cases.   Carrot, a tolerant crop,
rapidly   metabolizes   linuron   by   N-demethylation   and
N-demethoxylation to nonherbicidal products, whereas in ragweed,
which is susceptible to linuron, metabolic detoxification occurs
much more slowly (16).   Malathion and linuron are examples of
pesticides   in   which   the   rate   of   metabolic   detoxification
determines selectivity.   Another kind of metabolic influence on
selective toxicity could involve the nature of the metabolic
process that occurs.   While no dramatic examples with pesticides
come readily to mind, the carcinogen 2-acetylaminofluorene (AAF,
Figure 5) serves as an illustration.   This compound is metabolized
in mammals by N-hydroxylation and subsequent conjugation to yield
carcinogenic metabolites, or by 7-hydroxylation to inactive
metabolites.   In the guinea pig and lemming, N-hydroxylation does
not occur to any appreciable extent, and AAF is not carcinogenic
to these species (Table I).

Table I.
Mammalian Carcinogenicity of 2-Acetylaminofluorene (AAF)
as Related to Species Differences in Its Metabolism

Species	% of Dose		Carcinogenicity
	N–OH	7–OH	
Guinea pig	0	72	—
Lemming	trace	42	—
Rat	1–15	19–27	+
Rabbit	13–30	15–29	+
Hamster	15–20	35–39	+
Dog	5	1	+
Man	4–14	25–30	?

From Smith (Ref. 17).

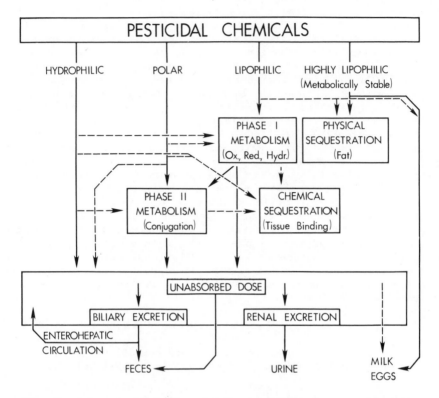

Figure 2.   Schematic of the major metabolic and disposition patterns of pesticides in higher animal systems. Pathways indicated by dashed lines are generally minor ones from a quantitative standpoint.

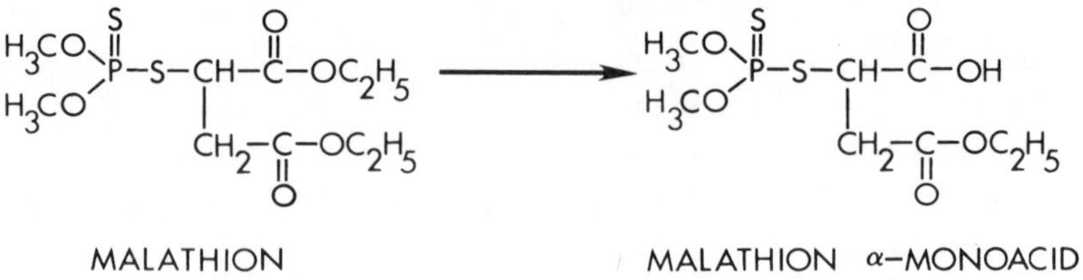

MALATHION   MALATHION α—MONOACID

Figure 3.   Metabolic detoxification of the insecticide malathion

LINURON
Cl
NH–C–N
CH3
OCH3
O
H

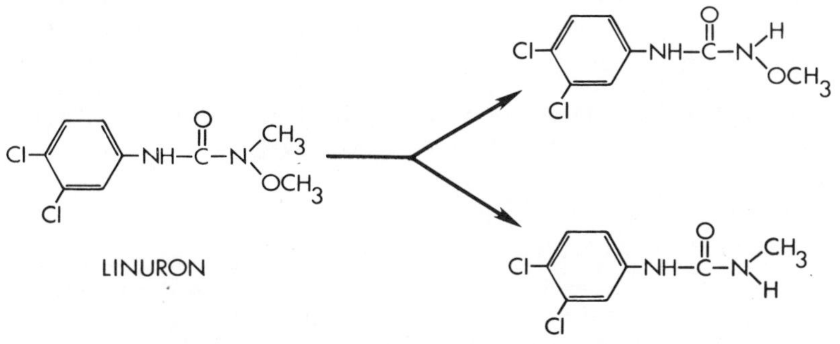

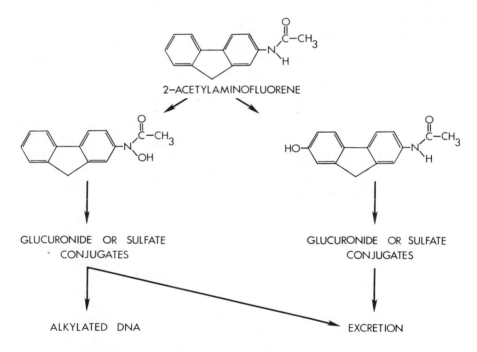

Figure 4. Metabolic detoxification of the herbicide linuron

2–ACETYLAMINOFLUORENE
C–CH3
OH
HO
GLUCURONIDE OR SULFATE CONJUGATES
GLUCURONIDE OR SULFATE CONJUGATES
ALKYLATED DNA
EXCRETION

Figure 5. Mammalian metabolism of AAF to carcinogenic and noncarcinogenic metabolites

With the rat, rabbit, hamster, and dog, however, AAF is metabolized to appreciable amounts of the N-hydroxy metabolite, and AAF is carcinogenic to these animals. Man likewise metabolizes AAF by N-hydroxylation, and while the carcinogenicity of AAF to man is not clearly established, the implications are obvious (17). More detailed treatments of the metabolic basis for pesticide selectivity are available (4, 15, 18).

Metabolism Studies and Safety Evaluation

Evaluating the toxicological significance of pesticides to man is seen to be a highly complex affair when one considers the various ways that pesticides are used, the routes by which man may be exposed to them and, perhaps most importantly, the multitude of chemical transformations that pesticides often undergo before man's exposure to them. Thus, while it is surely appropriate to define the toxicological interactions of a pesticide's active ingredient in experimental animals for extrapolation to man, it is entirely possible that such studies may in some cases have little relevance to real world human exposure. Because of the environmental instability of most organic pesticides, it seems reasonable and in fact likely that the great majority of human exposure to pesticide residues is to products of their decomposition rather than to the parent molecule. Thus, not only must we as metabolism scientists delineate the biochemical pathways of pesticides in experimental animals that are representative of man, we must as well clearly define the nature of their environmental transformations, if the products generated are likely to interact with man. While environmental transformations of pesticides may occur as the result of either biochemical (metabolic) or physico-chemical (e.g., photochemical) reactions, and both have toxicological implications, our purpose here is to consider only metabolic transformations.

For any given pesticide and use pattern, it is easily seen that several types of metabolism studies may be needed to provide a framework for evaluating the toxicological significance of the compound to man. As an example, we can consider a systemic insecticide used as a soil-incorporated granular formulation for insect control on corn. Because corn is consumed by both man and his food animals, several types of metabolism studies are appropriate, including studies of the pesticide itself in one or more laboratory monogastric mammals considered to be human models. Metabolism studies in corn are needed to determine the nature of residues to which man may be exposed through consumption of corn from treated crops. Studies are also needed in food animals that are given corn in the diet (e.g., cattle, swine, and poultry) to assess the extent to which the pesticide or its metabolites may appear in meat, milk, poultry or eggs intended for human consumption. Data from a soil metabolism study might likewise be needed if potentially toxic soil metabolites are assimilated by the treated crop. With other pesticides and use patterns, additional or alternative metabolism studies may be appropriate. It must be emphasized that, although we are attempting here to segregate biochemical or metabolic changes from physicochemical ones, such distinctions are from a toxicological standpoint rather arbitrary and in any case may be difficult to make under field conditions.

We must always recognize that pesticide metabolism studies cannot be considered as an end in themselves; but rather, they are a means toward an end. For the ultimate value of a metabolism study, be it in microorganisms, plants, birds, laboratory mammals, or whatever, is its yield of data valuable toward further assessment of the toxicological significance of the pesticide in question to lower organisms (i.e., its environmental impact) or, more importantly, to assess toxicological significance to man himself.

Methodology, Goals, and Regulatory Considerations

Although the word "metabolism" (Gr. metabole, change) has a rather limited connotation, a "pesticide metabolism study" is usually considered in a broad sense to encompass not only the metabolic alterations of the chemical in question but also the absorption, transport, storage, and excretion or elimination of the parent pesticide and its metabolites by the exposed organism. The schematic in Figure 6 shows that pesticide "metabolism" can be considered as more or less synonomous with the toxokinetic phase of a pesticide/organism interaction. Of course, any metabolic transformation that occurs in the gut prior to absorption of the pesticide would be considered, and is in fact, metabolism.

Because pesticide use patterns often dictate that metabolism studies be conducted in a number of widely divergent life forms, it is clear that no single approach is appropriate for all circumstances. Thus, metabolism studies in microorganisms, plants, mammals, etc., require specialized approaches based on the inherent nature of the organism and the goals of the study itself. Quite often too, the potential use patterns of pesticides may dictate differing methodologies for studies in the same species. For example, metabolism studies in cattle with a pesticide used on feed grains or forage clearly need be done only with oral administration, but if a product is to be used for ectoparasite control on cattle as a dermal spray, the dermal route of exposure would also be appropriate.

Given a suitable experimental design, what then is our goal as metabolism scientists in conducting such a study? It is, simply put, to define accurately and to the fullest extent possible the kinetic and metabolic behavior of the pesticide under study in an appropriate organism under the conditions chosen. We want to know how and at what rate the pesticide is absorbed into the living system, to what products it is metabolized, and to where and to what extent these products are transported, stored, and excreted. Our most important and usually most difficult task is, of course, to definitively characterize the chemical nature of as many of the metabolites as possible, given the limitations of our analytical and spectrometric techniques and of our own scientific capabilities. If the study is designed to define the metabolism of a pesticide in laboratory monogastric mammals (e.g., the rat) for extrapolation to man, then all aspects of the pesticide's kinetics and metabolism are crucially important. Other studies may have aspects of various importance. For example, the characterization of low levels of residues in the seed of food crops (e.g., rice) is more significant than comparable identification of possibly much higher residues in other, but inedible, portions of the plant. For the same reason, residues retained by edible tissues or secreted into the milk or eggs of treated food animals, such as cattle or poultry, are of more toxicological significance than residues in urine or feces.

One of the burdens the metabolism scientist must bear is that the products of pesticide metabolism that are often of the greatest potential toxicological significance (e.g., those in the edible parts of many plants or in milk, eggs, or edible tissues of food animals) are often present only in exceedingly low concentrations. Such properties of a pesticide are, of course, highly desirable ones that more often than not represent accomplished goals of pesticide development. However, the characterization of such residues usually demands the full capabilities of both the scientist and his instrumentation, and in some cases these residues cannot be identified with the technology currently available.

Of increasing importance to the design and execution of pesticide metabolism studies is the impact of the regulatory requirements of pesticide-regulating agencies. In the United States, such regulations are issued by the U.S. Environmental Protection Agency, and they must be carefully considered before initiating most pesticide metabolism studies, particularly those that have direct implications for human health. In its most recent issuance of proposed guidelines for registering pesticides in the United States (19), the Agency states several major purposes for mammalian metabolism studies. These include: 1) to identify and quantify significant metabolites, 2) to determine possible bioaccumulation or bioretention of the test pesticide or its metabolites, 3) to determine absorption as a function of dose, 4) to characterize routes and rates of pesticide excretion, 5) to relate absorption to the duration of exposure, and 6) to evaluate the binding of the test pesticide or its metabolites in potential target organs. The proposed rules contain rather general requirements for dosage levels, dosage routes, and other aspects of such studies, including sample analysis (19).

Toxicity Assessment

If the metabolism scientist has done his job well, i.e., has thoroughly defined the metabolic fate of a particular pesticide in the appropriate living systems, how does the toxicologist use this information to evaluate the toxicological significance of the metabolites generated? Generally, the first step is the assessment of the inherent toxicity of the products to laboratory monogastric mammals (e.g., rat, rabbit, dog) that can be considered representative of man. Chemical synthesis of the appropriate metabolites is often required to provide sufficient quantities for definitive toxicology studies. Comparative acute toxicity tests of the parent pesticide and its metabolites give a rapid and usually reliable estimate of the acute hazards of the products in question. Far more difficult to accurately assess are the chronic or subchronic toxicological hazards that may be posed by pesticide metabolites. Full-scale feeding studies to evaluate carcinogenicity and other chronic effects are almost always done with only the parent pesticide and not with its metabolites. This limitation is partly due to the tremendous time and money expenditures required by such studies, but some would argue that requirements for separate studies of the chronic toxicity of pesticide metabolites would be difficult to justify in any case under most circumstances (20).

Minor Versus Major Metabolites. It is usually neither appropriate nor possible to evaluate the toxicological behavior of every metabolite of a given pesticide that may possibly interact with man. Some of the products may be of such chemical constitution that they can be judged, on the basis of preexisting data, to represent little or no hazard of any sort. Others may present difficult synthetic problems that preclude detailed toxicologic tests. Limitations of time and money are significant factors. Perhaps because of such limitations, it has become fashionable to consider metabolites as being either "major" or "minor" based usually upon the relative amounts formed in a given system. A natural consequence of this distinction has been that "major" metabolites somehow are often construed to be of greater potential toxicological significance, at least in the regulatory sense, than those in the "minor" category (19). However, it seems clear that such a semiquantitative classification has essentially no toxicological significance because closely related chemicals often have toxicological potentials that differ by orders of magnitude; in fact, some of the most toxicologically significant metabolites are likely to be those that are formed in small amounts and are highly reactive. Thus, the selection of "major" over "minor" metabolites for animal studies to predict toxicological significance in man appears to be without biologic foundation (20, 21, 22).

Toxicological Significance of Pesticide Metabolites

Detoxification and Activation Reactions. From an acute toxicity standpoint, the metabolism of pesticides by most organisms usually results in their conversion to products of lesser biological activity. There are several reasons why such would be expected, not the least of which is the fact that the detoxification systems of living organisms have evolved for just such a purpose. Certainly, too, structure-activity relationships are usually so critical that toxicity, especially in the acute sense, is often greatly reduced or totally eliminated as the result of essentially any chemical transformation. Numerous examples of metabolic reactions leading to more-or-less complete pesticide detoxification could be cited, but the o-deethylation of chlorfenvinphos and the ester hydrolysis of carbaryl, both insecticides, are shown as somewhat representative examples (Figure 7).

While most metabolic reactions result in total or nearly total detoxifications, some do not, and it is such transformations that most concern those who attempt to evaluate the toxicological significance of pesticide metabolites. Classical examples of metabolic activation are the oxidative desulfuration of phosphoro-thionates and the N-hydroxymethylation of schradan (Figure 8). While parathion and schradan per se are essentially nontoxic, the indicated metabolic reactions convert them to potent anticholinesterases, and thus metabolism is obligatory to their toxicity. Other pesticide metabolites often have degrees of acute toxicity that are only moderately above or below those of the parent compounds. Examples of moderate activation include the sulfoxidation of methiocarb and the 5-hydroxylation of propoxur to yield metabolites that are 8- to 10-fold more active as anticholinesterase agents (23, Figure 9). An example of metabolic transformations that lead to moderate detoxification is the N-hydroxymethylation of N-methylcarbamates such as mexacarbate to products that are somewhat less anticholinergic (23, Figure 10). It should be emphasized that even if the products of pesticide metabolism retain partial or full inherent toxicity, the structural alterations that result from metabolism may facilitate rapid elimination from the body or further metabolism to nontoxic products which, of course leads to greatly reduced toxicological potential. As an example, aromatic hydroxylation of a given pesticide may not always diminish inherent toxicity, (e.g., propoxur) but the presence of the hydroxyl group in the molecule would be expected to lead to rapid conjugation and excretion by mammals.

Pesticide Conjugates. Although the primary metabolism of pesticides does not necessarily result in a diminution of acute toxicity, secondary or conjugative reactions almost always do. Pesticide conjugates are usually highly polar (e.g., glucosides,

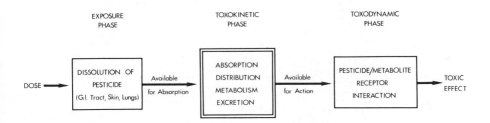

Figure 6. *Aspects of pesticide–organism interactions*

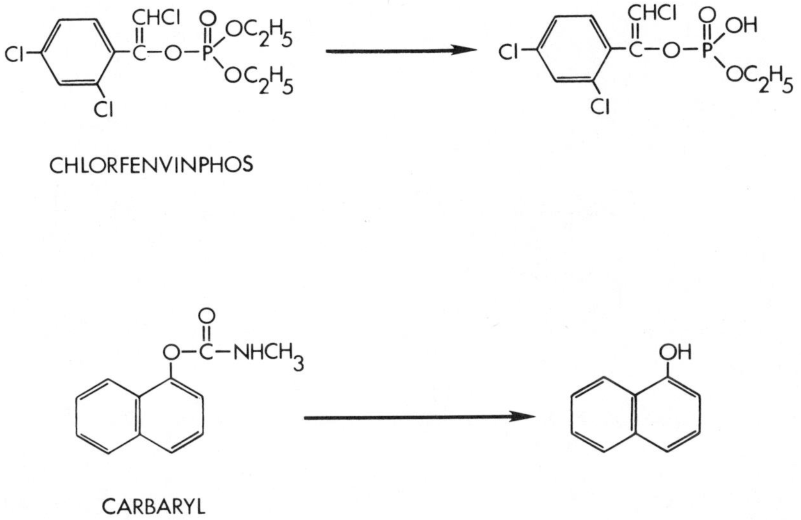

CHLORFENVINPHOS

CARBARYL

Figure 7. *Examples of metabolic detoxification of the insecticides chlorfenvinphos and carbaryl*

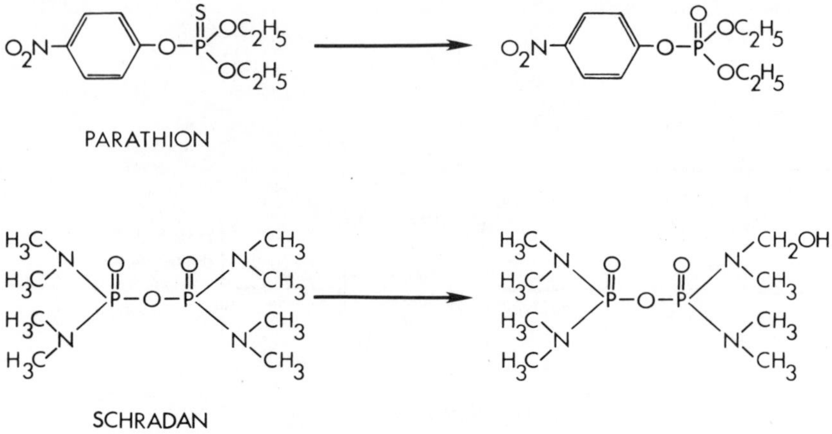

PARATHION

SCHRADAN

Figure 8.   Metabolic activation of the insecticides parathion and schradan

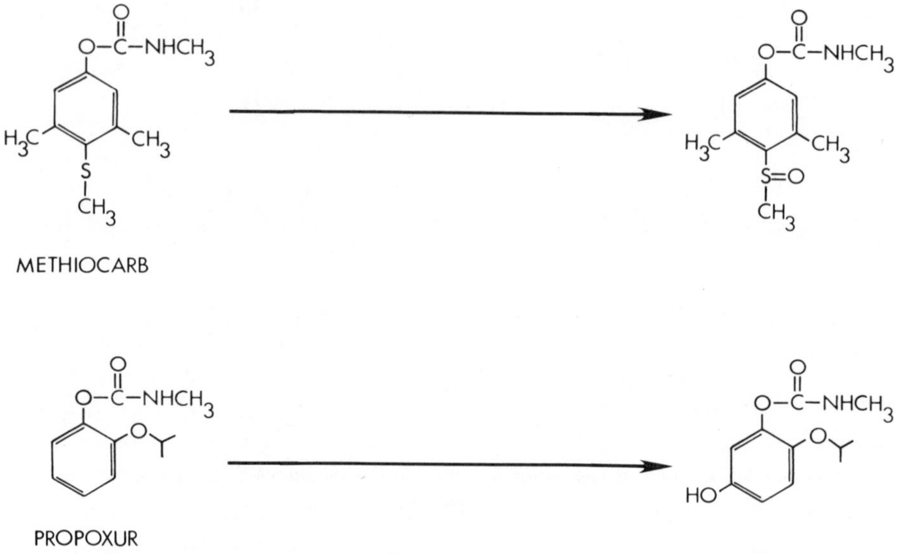

METHIOCARB

PROPOXUR

*Figure 9.   Metabolic transformations leading to moderate activation of the insecti-
cides methiocarb and propoxur*

glucuronides, sulfates, mercapturic acids, etc.), they are readily excreted by mammals, and they are usually devoid of significant acute biological effects. There is, of course, always the possibility that an otherwise innocuous conjugate may be metabolically cleaved to regenerate a toxic metabolite, and studies have shown that this potential does indeed exist. The glucoside conjugate of 1-naphthol, for example, which is a major plant metabolite of the insecticide carbaryl, is extensively hydrolyzed in rats (Figure 11). However, the liberated 1-naphthol is rapidly reconjugated with glucuronic acid or sulfate and is excreted in the urine (24). Although 1-naphthol is not significantly toxic, such reconjugation reactions would presumably offer a significant degree of protection in cases where the exocon (aglycone) is toxicologically significant.

It is true that pesticide conjugates almost always represent a reduced toxicological hazard from an acute standpoint, but such may not be the case regarding their chronic toxicological effects. It is well known that glucuronide or sulfate conjugates are formed as reactive intermediates of some carcinogens (vide infra); thus the consideration of similar pesticide conjugates as innocuous compounds may be totally inappropriate in some circumstances.

Bound Residues. Most pesticide metabolism studies are conducted using radiotracer techniques, and frequently a portion of the radioactivity defies all attempts at removal from the matrix under study. Questions regarding the toxicological significance of such residues naturally arise. If bound residues occur in matrices such that human dietary exposure is likely to occur, such as in edible plant or animal tissues, some estimation can be made of their potential toxicological significance, or at least of their bioavailability, by mammalian feeding studies. Fortunately, bound residues of several pesticides have been found not to be appreciably absorbed from the digestive tract of monogastric mammals (25, 26, 27), and it may be that such chemically unidentified residues from most pesticides will have little or no toxicological significance.

Mutagenicity and Carcinogenicity

The ability to accurately gauge the chronic toxicity of pesticides, particularly their effects on hereditary material that can produce mutagenic or carcinogenic responses in mammals, is the most important aspect of pesticide toxicological evaluation. A discussion of the molecular events leading up to the expression of mutagenic or carcinogenic effects, or of the relationships between mutagenicity and carcinogenicity (and also teratogenicity, which is generally considered to be an acute toxicological phenomenon) is beyond the scope of this paper. Rather, the reader is referred to published reviews of these subjects (28, 29) and to several other papers in this volume.

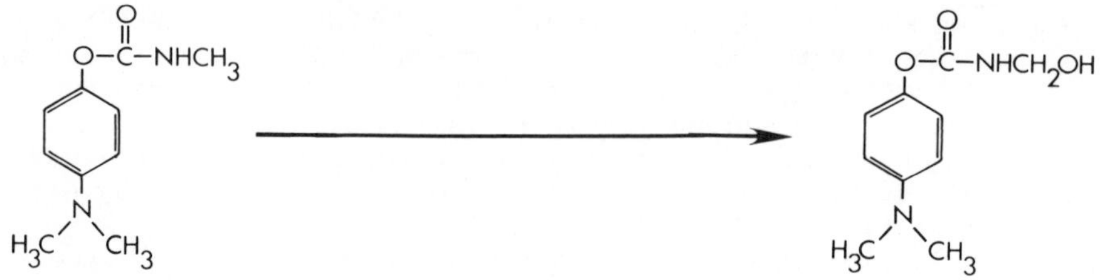

MEXACARBATE

Figure 10.   Metabolic N-hydroxymethylation leading to moderate detoxification of N-methylcarbamate insecticides such as mexacarbate

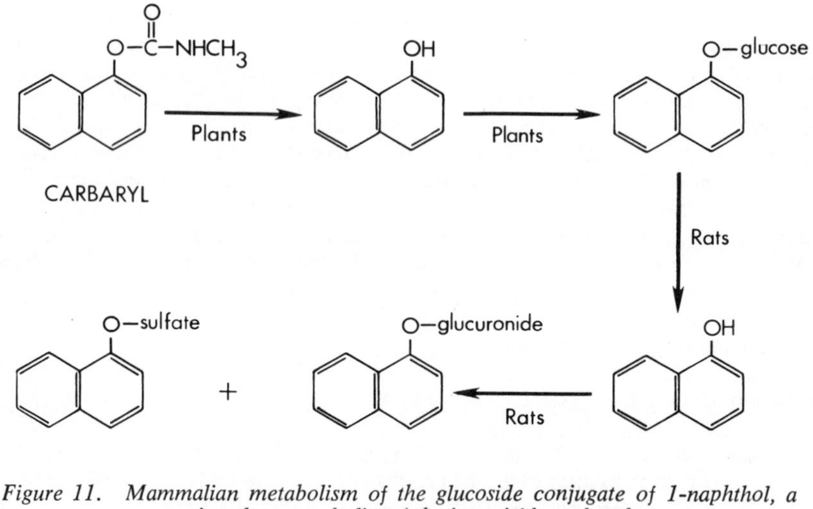

CARBARYL

Figure 11.   Mammalian metabolism of the glucoside conjugate of 1-naphthol, a major plant metabolite of the insecticide carbaryl

Carcinogenicity is certainly the most feared of the potential chronic effects of pesticides, and an assessment of carcinogenic potential is usually required prior to the approval of pesticides for use, at least in the United States and most developed countries. Tests may be required in two or more mammalian species (usually the rat and mouse), and they generally involve chronic exposure of the animals to relatively high doses of the test pesticide over their normal lifespans (18-24 months). As would be expected, there is a large volume of literature on the carcinogenic effects of pesticides in various species, mostly mammals, and some of this has been reviewed (28). However, since long-term carcinogenicity tests are almost invariably done with the parent pesticide and not any of its metabolites, it is usually impossible to make direct correlations of carcinogenicity, when it occurs, with metabolism.

Not only do we seldom if ever obtain direct experimental data on the carcinogenicity of pesticide metabolites, it is likewise impractical to routinely subject most or all pesticide metabolites to a battery of in vivo or in vitro mutagenicity tests. How then can pesticide metabolites be properly evaluated for mutagenic or carcinogenic hazard? In most cases there would appear to be pervading logic that consideration of the chronic toxicological behavior of metabolites separate from that of the parent compound is neither necessary nor appropriate. If a mammalian pesticide metabolite is under consideration that would, by extrapolation, likely be a metabolite in humans also, studies of the metabolite per se, probably in doses far in excess of what would likely be formed from the parent compound in vivo, could give "false positives" by overwhelming otherwise more than adequate protective mechanisms, and conceivably even "false negatives" as a consequence of disposition patterns totally different from the endogenous metabolite (20, 30). In such circumstances, it is doubtful that anything could be learned about the safety of a given pesticide from such metabolite studies that could not more reliably be obtained from proper studies of the parent pesticide itself (20). It would seem that the only appropriate circumstance in which pesticide metabolites might require separate study for mutagenic or carcinogenic effects is in instances where humans are likely to be exposed to metabolites that they would not generate in vivo from the parent compound. Examples might be pesticide metabolites of plant or animal origin that are novel in structure, that appear likely to be of considerable potential toxicological significance, and that could enter the human food chain through contaminated foodstuffs.

Certainly, the most prevalent and perhaps the most logical means of making judgments about the mutagenic or carcinogenic hazards of pesticide metabolites is simply by relating their chemical structures to those of recognized mutagens or carcinogens. This process may be imprecise, but it is probably the best procedure available for determining what pesticide metabolites merit concern or more detailed study. Even if targeted metabolites give positive results in in vivo or in vitro tests for mutagenicity, it must continually be remembered that such findings can only be considered, at most, suggestive evidence of a potential mutagenic or carcinogenic hazard to man. Further, consideration of the mutagenic potency of the metabolites, the probable extent of human exposure to them, and other considerations, may often indicate that a mutagenic or carcinogenic hazard to man, even if it exists, is exceedingly low.

Metabolic Aspects of Pesticide Carcinogenicity and Mutagenicity

In recent years, it has become evident that for many well-studied chemical carcinogens, metabolic activation to a reactive intermediate in the host is required in order for reaction with DNA and other cellular macromolecules to occur (31, 32). Thus, many carcinogens appear to be precarcinogens, which are metabolized in vivo to their reactive forms, or ultimate carcinogens. The ultimate carcinogens identified or postulated so far, although they often have no common structural features per se, contain relatively electron-deficient atoms that can react covalently, without the aid of enzymes, with electron-rich or nucleophilic atoms in cellular components, especially in such macromolecules as the nucleic acids and proteins (32). Thus, carcinogenic polycyclic aromatic hydrocarbons are metabolized to several carcinogenic electrophiles, including epoxides, radical cations, and dihydroxy epoxides (Figure 12). Carcinogenic aromatic amines, amides, and nitro compounds appear to be subjected to N-hydroxylation, then conjugation with glucuronic acid or sulfate to a more reactive species (Figure 13). With nitroso compounds, some of which are potent carcinogens, the ultimate alkylating species is likewise thought to be an electrophilic metabolite, probably a diazonium or carbonium ion (Figure 14).

On the basis of structure-activity relationships among known carcinogens, some generalizations can be made regarding the types of reactive functionalities in pesticides or their metabolites that might convey mutagenic or carcinogenic potential. Because electrophilicity is associated with many ultimate mutagens and carcinogens, any pesticide transformation to an electrophilic species could be of potential significance. However, upon review of the multitude of mechanisms through which various pesticides are, or can be metabolized, one quickly realizes that the generation of potentially reactive species, or of their precursors, is rather commonplace. Aromatic and aliphatic epoxidations, N-hydroxylations, the generation of amines that can form nitrosamines, and other reactions of potential significance are well documented in the pesticide literature, yet there is little indication that most pesticides constitute any significant

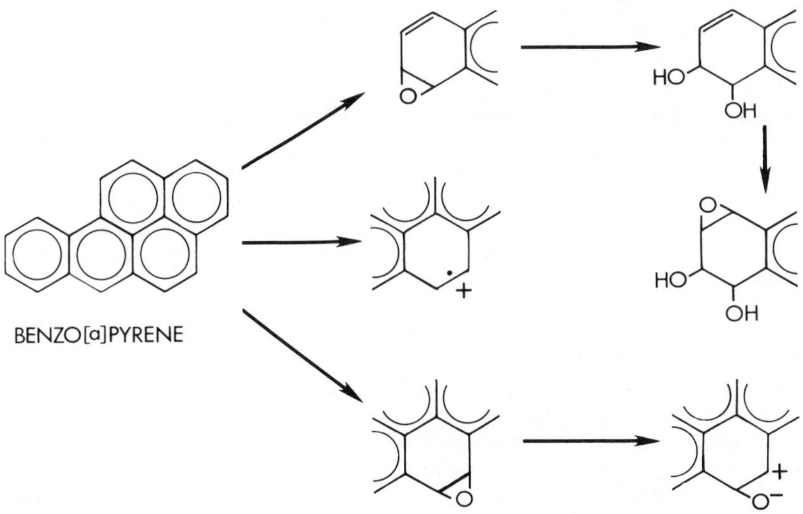

Figure 12.   Examples of metabolic activation of polycyclic aromatic hydrocarbons to reactive electrophiles

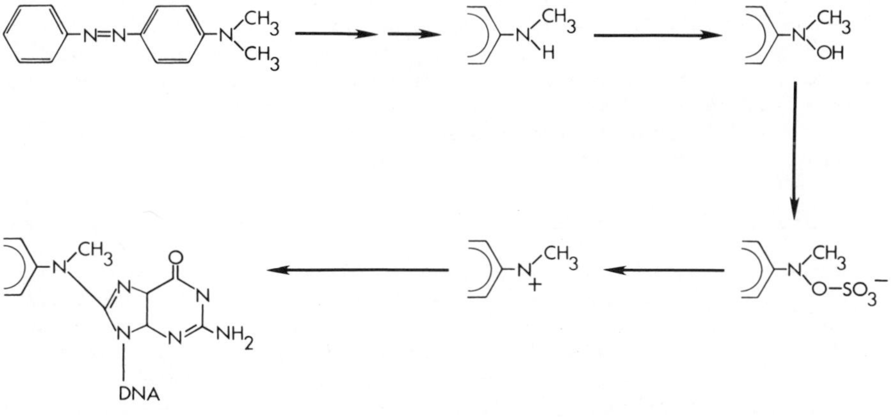

Figure 13.   Metabolic activation of an aromatic amine that ultimately can lead to the formation of a reactive electrophile and alkylation of DNA

276 THE PESTICIDE CHEMIST AND MODERN TOXICOLOGY

mutagenic or carcinogenic hazard. Clearly, the mere generation of
reactive metabolites does not assure that an expression of
toxicity will follow. Subsequent rapid detoxication of reactive
metabolites no doubt occurs in many instances, the reactive
species may form adducts with noncritical macromolecules or other
body constituents, and even if reactive metabolites do alkylate
essential cellular macromolecules, subsequent events, such as DNA
repair mechanisms, may negate any potential toxic effects (33).

In most if not all cases in which pesticides have in fact
been shown to be carcinogenic (28), there has been no clear
definition of the role that metabolism to reactive intermediates
may or may not have played in causing such effects. On the basis
of our current understanding of the mechanisms of chemical
carcinogenicity, metabolism of at least some carcinogenic
pesticides to reactive electrophiles in vivo may occur as an
activation step. Alternatively, it may be that most carcinogenic
pesticides are epigenetic carcinogens rather than genotoxic
carcinogens, i.e., they are cancer promoters rather than
alkylating agents. It is generally accepted that some chemicals
may induce tumor formation without directly initiating neoplastic
changes in any cell. Thus, chemicals that depress immune
responses or alter the hormonal balance in a particular tissue
might provide the appropriate conditions for the preferential
growth of preexisting tumor cells (32). Further, chemicals that
induce or inhibit the action of drug metabolizing enzymes may
promote cancer by enhancing the activation or inhibiting the
detoxification of other chemical carcinogens. It is therefore
possible that metabolism to reactive electrophiles may not be
involved at all in the expression of carcinogenic action of many
or most carcinogenic pesticides. One or more of such promotion
mechanisms might explain the carcinogenicity of the insecticide
mirex, which is reported to be a hepatocarcinogen in mice (34),
even though there is strong evidence that laboratory rodents are
unable to metabolize this insecticide (35, 36).

Pesticide Metabolites and the Regulatory Process

All pesticides can be considered to present at least a
potential toxicological hazard to man, and certainly the primary
goal in the regulation of these chemicals is to minimize such
risks as much as possible. Because risk to man is clearly a
function of exposure, risks are generally minimized by the
regulation of exposure. This is done through the setting of
tolerances. Tolerances represent maximum limits (expressed
usually in parts per million) of a pesticide, its metabolites, or
both, that may legally appear in human foodstuffs, animal feeds,
etc., as a result of pesticide use. The determination of
whether a tolerance will be granted and at what level it will be
set can be a complicated process, but several factors are usually
involved. These include the inherent toxicity of the pesticide

and/or its metabolites, the nature of their toxic action, the proposed use of the contaminated commodity (e.g., as animal feed or human food), the likely extent of human exposure from all sources, the need for the proposed pesticide application, the sensitivity of the available analytical methods, and other considerations. In all cases, tolerances are set at sufficiently low levels to, presumably, assure the absence of significant risk to man.

Depending upon their toxicological significance, pesticide metabolites may or may not be included as components of a pesticide tolerance. Some metabolites may be judged to be of sufficiently low toxicological importance that their inclusion within tolerance limits is deemed unnecessary. Others may retain significant toxicological properties and therefore will most likely be included under tolerance. Examples of both types of metabolites can be seen in the organophosphate insecticide sulprofos, which is metabolized in both plant and animal systems by P=S to P=O conversion, sulfur oxidation, and ester hydrolysis (37, 38, Figure 15). The sulfoxide and sulfone analogs of the intact phosphate esters retain anticholinesterase activity and are considered from the regulatory standpoint to be toxicologically equivalent to sulprofos. The phenols, on the other hand, are of minimal toxicological importance and are not included under the sulprofos tolerance for any commodity.

It is quite possible that the toxicological characteristics of individual pesticide metabolites, particularly demonstrated mutagenic or carcinogenic behavior, could form the basis for denying registration for pesticide use or for revocation of existing registrations. To our knowledge, however, no such actions have yet been taken by a regulatory agency on the basis of demonstrated toxicity of a pesticide metabolite, unless the same toxicological effect is seen with the parent compound.

Extrapolation to Man: The Problem of Species Variations

The primary purpose of evaluating the metabolic and toxicological behavior of pesticides is to assess the risk to man that may result from their use and subsequently to take appropriate regulatory steps to minimize such risks. Obvious ethical and other considerations prevent direct studies of pesticides in humans except in most unusual circumstances, thus extrapolations to man must usually be made on the basis of data obtained with monogastric laboratory mammals. Unfortunately, laboratory research animals are generally chosen more for convenience than for rational, scientific reasons. The handling and housing requirements, incidence of disease, supply and, perhaps most important, cost, are among the factors considered in choosing a species for research (39). For pesticide metabolism studies, the rat and/or mouse is usually the species of choice. We quite willingly assume, perhaps because no obvious alternatives

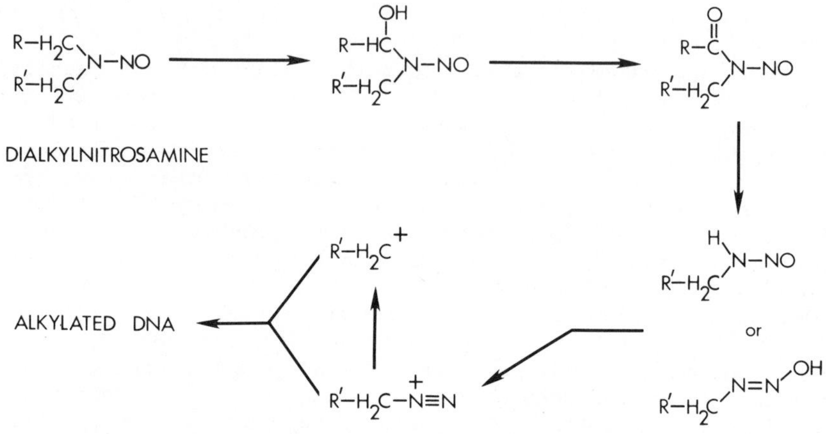

Figure 14.   Metabolic activation of a dialkylnitrosamine leading to the generation of reactive electrophiles and ultimately to the alkylation of DNA

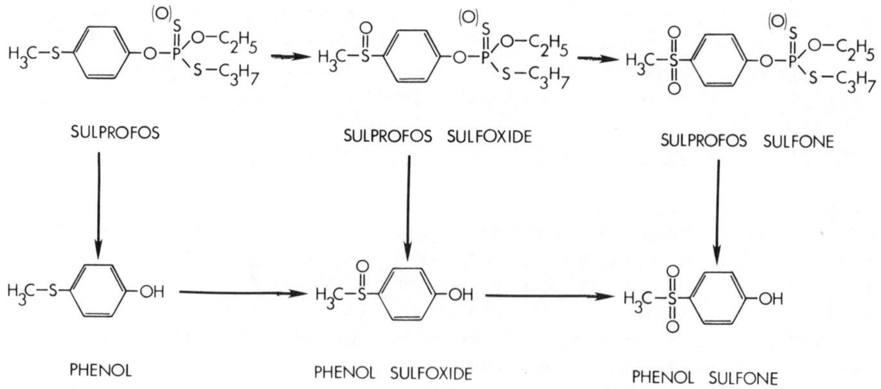

Figure 15.   Structures of the insecticide sulprofos and its major plant and animal metabolites

exist, that results from metabolism studies with these animals are in fact predictive of what will happen in man, or at least that any differences will not be toxicologically "significant." Yet there are clear indications that, in metabolism as well as other toxicological phenomena, considerable species differences do indeed exist (21, 40). Laboratory rodents, in fact, appear to be poor metabolic predictors for man! In a comparison of the metabolic pathways for 21 drugs and other compounds in the rat and man (41), the rat provided a "good" metabolic model for man with only 4 compounds and was a "poor" or "invalid" model (metabolic pathways quite different) with 15 of the compounds studied (Table II). However, the rhesus monkey or marmoset provided "good" metabolic models for man with 16 of the 21 compounds. It is reasonable to assume that similar results would be seen with various pesticides, and thus many of the metabolism studies currently used as a basis for extrapolating toxicological results with pesticides to man may be of limited predictive value. The potential toxicological consequences of this are, of course, unknown.

Table II.
Comparison of Laboratory Rodents and Sub-human
Primates as Metabolic Models for Man

COMPOUND	METABOLIC SIMILARITY TO MAN	
	RAT	MONKEY
Amphetamine		
Chlorphentermine		
4-Hydroxy-3,5-diiodobenzoic acid		
Indolylacetic acid		
Norephedrine*	Invalid	Good
Phenmetrazine*		
Phenylacetic acid		
Sulphamethomidine		
1-Naphthylacetic acid		
Sulphadimethoxine		
Sulphadimethoxypyridine	Poor	Good
Halofenate		
Methotrexate		
Sulphasomidine	Fair	Good
Hydratropic acid		
Diphenylacetic acid	Good	Good
Indomethacin	Poor	Fair
Morphine	Fair	Fair
Oxisuran		
2-Acetamidofluorene	Good	Fair
Phencyclidine	Poor	Poor

From Smith and Caldwell (Ref. 41). *Marmoset, all others rhesus monkey

Pesticide Metabolism: Prospects and Problems

Pesticide metabolism studies are, without question, very important components in the evaluation of the toxicological significance of pesticides to man. The rate, extent, mechanisms, and products of metabolism are inevitably linked to the expression of toxic action, and a clear definition of pesticide biotransformation is often a necessary prerequisite to understanding mechanisms of toxicity and to the formulation of approaches for assessment and management of potentially undesired toxic effects.

What does the future hold? Can pesticide metabolism studies and the data they generate be more effectively used in the safety evaluation process? Can these studies be made more predictive and thus more toxicologically relevant to man? It is, of course, difficult if not impossible to foresee the future accurately. We will, however, make a few observations on these and other matters.

Only a few years ago, a pesticide metabolism study was considered successful if only the major metabolites were characterized, and this was often done solely by chromatographic means -- without spectral confirmation of structure. Today it is not uncommon to see reports in which most if not all of the detected metabolites of a pesticide in a given system are fully and unequivocally characterized by spectral means. Several factors have contributed to such advancements, including the fact that many of us now have available in our research laboratories a full complement of up-to-date, often state-of-the-art analytical, chromatographic, and spectrometric instrumentation. Advances in our capabilities to characterize organic compounds, particularly advances in microspectrometric techniques such as GLC-mass spectroscopy, FT-NMR, and FT-IR make possible the identification of many metabolites at the microgram level. The versatility, accessibility, and overall importance of radiotracer techniques to the metabolism scientist have never been greater. Stable isotopes (e.g., ^{2}H, ^{13}C, ^{15}N) are beginning to find more use in pesticide metabolism studies, and with mass spectroscopy or NMR, stable isotopes can be very useful tools for both metabolite characterization and mechanistic studies (42). In the metabolism study of the future, there will continue to be, and rightly so, great emphasis placed on definitive characterization of all metabolites possible. Hopefully, we will see in the future continuing advances in our capabilities to more fully characterize pesticide conjugates and "bound" residues, because these products often comprise the bulk of the total residue and their toxicological significance, particularly chronic effects, is far from clear.

Species variations that may seriously affect the validity of laboratory animal metabolism studies as predictive models for man are a problem without apparent solution. For proper evaluation of the toxicological significance of pesticides to man, metabolism studies in humans are clearly needed, yet just as clearly are totally inappropriate. It seems, however, that subhuman primates should be far more effectively used than at present as more acceptable and accurate human models of pesticide metabolism. Because of the evolutionary position of these animals with respect to man, and because they represent a very limited resource, subhuman primates are a scientific treasure that merits respect and the wisest use. For these reasons, conventional toxicological investigations of pesticides using primates may be totally inappropriate in most or even all circumstances. However, such restrictions should not apply to in vivo pesticide metabolism studies with primates because these studies need not be destructive of life. With the judicious use of radioisotopes and proper dosage levels, single or small groups of primates could provide invaluable metabolic data over a period of many years for a large number of pesticidal chemicals. Such an approach would require adjustments in regulatory philosophy and could possibly present some problems, such as low level induction of pesticide metabolizing enzyme systems. However, it seems to us that the potential disadvantages are, realistically, quite minor in contrast to the advantages--the likely yield of data of much better predictive value to man.

The discipline of pesticide metabolism chemistry, as well as other disciplines involved in the toxicological evaluation of pesticides, will no doubt be required to become even more responsive to pesticide regulatory agencies in the future. Historically, regulatory requirements have become more and more specific as time has progressed, and there is little reason to believe that such a pattern will not continue in the future. Few would question the wisdom or propriety of regulatory agencies in requiring detailed metabolism studies to support pesticide registrations. However, there is concern on the part of some metabolism scientists that further moves toward specific methodology requirements in metabolism studies may well be counterproductive in that imaginative and innovative research in this field may be discouraged.

Literature Cited

1. Ridgway, R. L.; Tinney, J. C.; MacGregor, J. T.; Starler, N. J. Env. Health. Persp., 1978, 27, 103.

2. Jenner, P.; Testa, B. Xenobiotica, 1978, 8, 1.

3. Williams, R. T. "Detoxication Mechanisms"; John Wiley and Sons: New York, N.Y., 1959.

4. Baldwin, B. C., In "Drug Metabolism -- From Microbe to Man", Parke, D. V.; Smith, R. L., Eds., Taylor and Francis Ltd.: London, Eng., 1977; p. 191.

5. Hathway, D. E., Ed. "Foreign Compound Metabolism in Mammals, Vol. 1"; The Chemical Society, Burlington House: London, Eng., 1970.

6. Hathway, D. E., Ed. "Foreign Compound Metabolism in Mammals, Vol. 2"; The Chemical Society, Burlington House: London, Eng., 1972.

7. Hathway, D. E., Ed. "Foreign Compound Metabolism in Mammals, Vol. 3"; The Chemical Society, Burlington House: London, Eng., 1975.

8. Hathway, D. E., Ed. "Foreign Compound Metabolism in Mammals, Vol. 4"; The Chemical Society, Burlington House: London, Eng., 1977.

9. Kearney, P. C.; Kaufman, D. D., Eds. "Herbicides: Chemistry, Degradation, and Mode of Action"; Marcel Dekker: New York, N.Y., 1975.

10. Menzie, C. M. "Metabolism of Pesticides"; Bureau of Sport Fisheries and Wildlife, Special Scientific Report -- Wildlife No. 127: Washington, D.C., 1969.

11. Menzie, C. M. "Metabolism of Pesticides: An Update"; U.S. Department of the Interior, Fish and Wildlife Service, Special Scientific Report -- Wildlife No. 184: Washington, D.C., 1974.

12. Sieber, S. M.; Adamson, R. H., In "Drug Metabolism -- From Microbe to Man", Parke, D. V., Smith, R. L., Eds., Taylor and Francis Ltd.: London, Eng., 1977; p. 233.

13. Smith, J. N., In "Drug Metabolism -- From Microbe to Man", Parke, D. V.; Smith, R. L., Eds., Taylor and Francis Ltd.: London, Eng., 1977; p. 219.

14. Wit, J. G., In "Drug Metabolism -- From Microbe to Man", Parke, D. V.; Smith, R. L., Eds., Taylor and Francis Ltd.: London, Eng., 1977; p. 247.

15. O'Brien, R. D. "Insecticides: Action of Metabolism"; Academic Press: New York, N.Y., 1967.

16. Kuratle, H.; Rahn, E. M.; Woodmansee, C. W. Weed Sci., 1969, 17, 216.

17. Smith, C. C., In "Proc. 1966 Conf. on Nonhuman Primate Toxicology," Miller, C. O., Ed., U.S. Govt. Printing Off.: Washington, D.C., 1966; p. 57.

18. Hollingworth, R. M., In "Insecticide Biochemistry and Physiology", Wilkinson, C. F., Ed., Plenum Press: New York, N.Y., 1976; p. 431.

19. Anonymous. Federal Register, 1978, 43, 37336.

20. Weiner, M.; Newberne, J. W. Toxicol. Appl. Pharmacol., 1977, 41, 231.

21. Gillette, J. R., In "Drug Metabolism -- From Microbe to Man", Parke, D. V.; Smith, R. L., Eds., Taylor and Francis Ltd.: London, Eng., 1977; p. 147.

22. Weiner, M., In "Drug Metabolism -- From Microbe to Man", Parke, D. V.; Smith, R. L., Eds., Taylor and Francis Ltd.: London, Eng., 1977; p. 431.

23. Nakatsugawa, T.; Morelli, M. A., In "Insecticide Biochemistry and Physiology", Wilkinson, C. F., Ed., Plenum Press: New York, N.Y., 1976; p. 61.

24. Dorough, H. W., In "Bound and Conjugated Pesticide Residues", Kaufman, D.; Still, G.; Paulson, G.; Bandal, S., Eds., American Chemical Society Symposium Series 29: Washington, D.C., 1976; p. 11.

25. Dorough, H. W. Pharmac. Ther., 1979, 4, 433.

26. Paulson, G. D.; Jacobsen, A. M.; Still, G. G. Pestic. Biochem. Physiol., 1975, 5, 522.

27. Sutherland, M. L., In "Bound and Conjugated Pesticide Residues", Kaufman, D.; Still, G.; Paulson, G.; Bandal, S., Eds., American Chemical Society Symposium Series 29: Washington, D.C., 1976; p. 153.

28. Fishbein, L. In "Insecticide Biochemistry and Physiology", Wilkinson, C. F., Ed., Plenum Press: New York, N.Y., 1976; p. 555.

29. Wilson, J. G., In "Mutagenic Effects of Environmental Contaminants", Sutton, H. E.; Harris, M. I., Eds., Academic Press: New York, N.Y., 1972; p. 185.

30. Gehring, P. J., Watanabe, P. G., Blau, G. E., In "Advances in Modern Toxicology", Mehlman, M., Shapiro, R., Blumenthal, H., Eds., Hemisphere Pub. Corp.: Washington, D.C., 1976.

31. Irving, C. C., In "New Methods in Environmental Chemistry and Toxicology", Coulston, F.; Korte, F.; Goto, M., Eds., International Academic Printing Co.: Tokyo, Japan, 1973; p. 99.

32. Miller, J. A.; Miller, E. C., In "Biology of Radiation Carcinogenesis", Yuhas, J. M.; Tannant, R. W.; Regan, J. D., Eds., Raven Press: New York, N.Y., 1976; p. 147.

33. Caldwell, J. Xenobiotica, 1979, 9, 33.

34. Innes, J. R. M.; Ulland, B. M.; Valerio, M. G.; Petrucelli, L.; Fishbein, L.; Hart, R.; Pallota, A. J.; Bates, R. R.; Falk, H. L.; Gart, J. J.; Klein, M.; Mitchell, I.; Peters, J. J. Natl. Cancer Inst., 1969, 42, 1101.

35. Gibson, J. R.; Ivie, G. W.; Dorough, H. W. J. Agric. Food Chem., 1972, 20, 1246.

36. Ivie, G. W.; Gibson, J. R.; Bryant, H. E.; Begin, J. J.; Barnette, J. R.; Dorough, H. W. J. Agric. Food Chem., 1974, 22, 646.

37. Bull, D. L.; Ivie, G. W. J. Agric. Food Chem., 1976, 24, 143.

38. Ivie, G. W.; Bull, D. L.; Witzel, D. A. J. Agric. Food Chem., 1976, 24, 147.

39. Krieger, R. I.; Miller, J. L.; Gee, S. J.; Clark, C. R.; In "Fate of Pesticides in Large Animals", Ivie, G. W., Dorough, H. W., Eds., Academic Press, New York, N.Y., 1977; p. 77.

40. Williams, R. T., In "Fundamentals of Drug Metabolism and Drug Disposition", LaDu, B. N.; Mandel, H. G.; Way, E. L., Eds., The Williams and Wilkins Co.: Baltimore, Md., 1971, p. 187.

41. Smith, R. L.; Caldwell, J., In "Drug Metabolism -- From Microbe to Man", Parke, D. V.; Smith, R. L., Eds., Taylor and Francis Ltd.: London, Eng., 1977; p. 331.

42. Baillie, T. A., Ed., "Stable Isotopes: Applications in Pharmacology, Toxicology and Clinical Research", University Park Press: Baltimore, Md., 1978.

RECEIVED February 2, 1981.

New Strategies in Biochemical Studies for Pesticide Toxicity

ALAN S. WRIGHT

Shell Research Ltd., Shell Toxicology Laboratory (Tunstall), Sittingbourne Research Centre, Sittingbourne, Kent, ME 9 8AG, U.K.

There are two main branches of experimental toxicology:-
1. Environmental toxicology in which the objective is to provide information needed to prevent or minimise adverse effects of chemicals on the capacity of the environment to sustain its life-forms.
2. Mammalian toxicology in which the main objective is to provide information needed to safeguard the health of present and future human populations.
Both branches of toxicology are concerned with the study of the fate of chemicals and the assessment of the toxicity of chemicals. Nevertheless, a clear distinction exists between these subjects and this difference arises because man is not an experimental species. The environmental toxicologist can often study the species at risk or, alternatively, a very closely related species. However, in order to identify human hazards and evaluate the risks to humans, the mammalian toxicologist must resort to the use of biological models.
A wide variety of biological models are employed in mammalian toxicology. They range in complexity from _in vitro_ models, e.g. microbial mutation, to _in vivo_ models, e.g. carcinogenicity in experimental mammals. However, it is clear that the biological models employed to detect human hazards and estimate human risks must be appropriate for these purposes. The response of the model should be relevant to the human situation and the quantitative data generated in the model should be suitable for estimating the human risk with at least a reasonable degree of precision.

The Aims of Biochemical Studies in Toxicology

Quantitative and apparent qualitative species differences in susceptibility to chemical toxicants, including chemical carcinogens, are common even among mammals. The occurrence of such species differences cautions against the direct extrapolation of experimental toxicity data to humans. It also emphasises the tenet that an understanding of the mechanisms of action of chemical toxicants is a basic requirement, not only in the rational development of new tests for the detection of toxic effects, but also in devising approaches to meet additional important objectives such as the evaluation of the toxicological significance of results obtained using experimental species/systems. The selection of suitable experimental species for the detection of human hazards and assessment of human risks is also aided by this knowledge.

In the main, biochemical research in mammalian toxicology is focussed on these objectives in an attempt to provide the sound theoretical basis needed to replace the empiricism that characterises much of current toxicology. It is, therefore, not surprising that, in formulating appropriate research strategies, heavy emphasis is placed on the acquisition and application of mechanistic knowledge and particularly on considerations of the critical events in intoxicating processes and host-dependent (endogenous) factors that influence or determine these effects.

These guiding principles are common to both the older and newer strategies employed in biochemical approaches in toxicology. However, while there has, perhaps, been little or no fundamental change in philosophy there is no doubt that recent advances in the understanding of life processes and the natures of chemical toxicants coupled with the development of techniques to exploit this knowledge have resulted in a dramatic increase in the power and scope of biochemical or molecular approaches. The evolving role of metabolism studies in toxicology is illustrative of these changes, particularly changes in emphasis or focus that are dependent upon increases in basic knowledge.

Studies of the metabolism of xenobiotics were the first biochemically-orientated approaches to find wide application in toxicology. The metabolic biotransformations undergone by foreign compounds were initially viewed as detoxification reactions and as such, were generally regarded as important determinants of the quantitative aspects of toxic effects. Early metabolism studies were, therefore, primarily concerned with the determination of the persistence of chemicals in vivo and with establishing the natures and rates of formation of the end products of metabolism in experimental species and, where possible, in humans. In evaluating the significance of differences between the results obtained in humans and experimental species, there was a distinct tendency to attach greater importance to the persistence and overall kinetics of metabolism of a chemical than to differences in biotransformation pathways. However, the study of the mechanisms of

metabolic biotransformation reactions gradually came to the
fore and, due largely to the insight of J. A. Miller, J. R.
Gillette and others (see Commentary by Gillette, (1, 2)), it
is now generally recognised that the operation of such
biotransformation pathways can often lead to the generation
of toxicologically active species from inactive or less active
precursor compounds.

This knowledge had been widely exploited in toxicology,
e.g. in the development of rapid tests for chemical carcinogens,
and has led to the important generalisation that electrophilic
reactivity is responsible for the adverse biological effects
manifested by many genotoxic (and cytotoxic) agents (3).
Furthermore, these developing insights into the natures of
major classes of chemical toxicants has focussed attention on
nucleophilic centres in biomacromolecules as potential critical
targets. One consequence of these advances is that the emphasis
in metabolism studies is switching, to an increasing extent,
from the study of excretory products to the detection and
assay of covalent products formed in vivo by reactions of
intrinsically reactive parent compounds, chemically-reactive
metabolites or metabolic intermediates with biomacromolecules
as a means of detecting and evaluating toxic
effects (1, 2, 4, 5, 6). Such developments not only illustrate the evolving role of metabolism studies in toxicology
but also point to the major contribution of the study of the
mechanisms of metabolic biotransformation reactions to the
understanding of the nature of intoxication processes; this
is the key to future biochemical strategies in risk assessment.

The Nature of Intoxication Processes

    It seems to be axiomatic that the adverse biological
effects of a toxic chemical must ultimately be dependent upon
an initial interaction between the chemical and one or more
critical targets. Such primary, critical interactions which
may be chemical or physical in nature may be conveniently
described as key intoxicating reactions because they trigger
the sequence of events leading to the development of the
overt biological effect (Figure 1).

    The intoxicating process may consist of one or more
steps. For example, reaction with a chemical may directly
inactivate a vital enzyme. Provided the threshold at which
the enzyme is depleted below its optimal operational
concentration is exceeded, such reaction will lead directly
to the appearance of overt toxic symptoms. In other instances,
an additional step or steps may be required. For example, the
induction of a mutation by direct chemical interaction with a
DNA base, represented by an asterisk in Figure 2a, requires a
subsequent miscoding error either during the repair of the

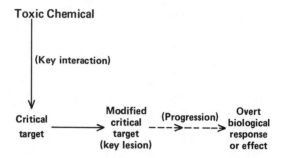

Figure 1.   Stages in the development of toxic effects

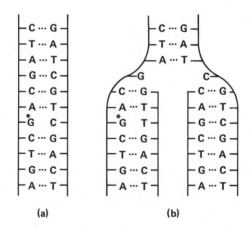

Figure 2.   Schematic of the induction of a mutation by direct reaction of an electrophile with a DNA base (see text for details)

key lesion or during the replication of the chemically-modified
DNA (Figure 2b). In such instances, the key lesion and the
ultimate lesion are fundamentally distinct.

Chemical carcinogenesis, although less well-understood
than mutagenesis, provides an example of a multi-step process
in which it is unlikely that the key lesion(s) and the ultimate
lesion(s) are entirely synonymous. Thus, while many toxic
phenomena may be regarded as single-step processes, i.e. the
toxic effect may be directly ascribed to the key intoxicating
reactions, other toxic processes such as chemical mutagenesis,
chemical carcinogenesis and teratogenesis involve at least
two critically- and temporally-linked stages:-
1. The generation of key (primary, critical) lesions.
2. Progression - the progressive modification of the key lesion
by interaction with other cellular components or exogenous
factors to form the ultimate lesion or overt biological effect.

Each of these stages may comprise a number of discrete steps
and both are strongly influenced and often determined by host-
dependent factors which may vary according to tissue,
individual, strain and species.

Determinants of Toxicity

1. Nature of Key Interactions. The nature of the
initial interaction between a toxic chemical and its critical
cellular target is undoubtedly a key determinant of the
ensuing biological effect and is entirely dependent upon the
physico-chemical properties of the interacting components
under the conditions prevailing in the micro-environment of
the target.

It is well-established that mammalian enzymes catalysing
the metabolism of toxic chemicals are among the most important
endogenous factors that influence the concentration of a toxic
chemical at its target and, consequently, the rate and
magnitude of the key interaction. In the case of the majority
of precursor agents, it is clear that the operation of these
enzymes also determines the chemical structures of the ultimate
toxic agents and thus the nature of the key lesions. Such
enzymes must therefore be regarded as key determinants not
only of the magnitude but also the nature of the adverse
biological effects of such precursor toxicants (Figure 3).

There are, therefore, three broad classes of toxic
agents:-
1. Intrinsically effective or reactive agents (ultimate
toxicants).
2. Precursor agents that are converted into ultimate toxicants
by spontaneous chemical reaction within the target organism.
3. Precursor agents that are converted into ultimate toxicants
by enzyme-mediated reaction within the target organism.

Certain intrinsically effective agents may, of course, be converted into additional intrinsically toxic substances within the target organism.

As discussed above, attention is increasingly being focussed upon electrophilic reactivity as the fundamental cause of the toxicity of many cytotoxic and genotoxic xenobiotic compounds. In many instances, the electrophilic centres responsible for this reactivity are generated during the metabolic biotransformations of the foreign compounds, e.g. the formation of epoxide groups during the oxidative metabolism of alkene or aromatic hydrocarbons.

Provided that energetic and stereochemical requirements are satisfied, such electrophilic centres will undergo nucleophilic substitution reactions with nucleophilic centres in informational or important structural or functional macromolecules such as DNA, RNA and proteins according to the general mechanisms shown in Figure 4. (No attempt has been made to balance ionic charges because there are several possibilities). It is envisaged that such nucleophilic substitution reactions constitute the primary chemical lesions resulting ultimately in cytotoxic or genotoxic effects.

For example, DNA is established as the ultimate target for chemical mutagens. While indirect mechanisms cannot be discounted it is generally held that the heritable changes in DNA stucture induced by treatment with electrophilic mutagens, such as the powerful alkylating agents, are a consequence of primary interactions between electrophilic centres in the mutagens and nucleophilic centres in DNA. These primary structural modifications in DNA structure are realised as mutations by miscoding during DNA replication or DNA repair or by failure to repair these lesions. It is, however, important to to point out that there is substantive evidence that not all of the primary structural modifications introduced into DNA by reaction with electrophiles, even mutagenic electrophiles, are pro-mutagenic (7, 8).

2. Magnitude of Key Interactions. The amount of key lesions formed in a given time is a function of the exposure of the target (target dose). According to the concept developed by Ehrenberg (4) for genotoxic agents, target dose is defined as the concentration-time integral of exposure of DNA to the ultimate genotoxic reactant. Thus, the rate of formation of the key lesions is a function of the physico-chemical properties of the reactants and the concentration of the ultimate toxic form of the chemical in the micro-environment of the target. The amount of key lesions present at any time is a function of their rate of formation, the rate of repair of these lesions or the rate of turnover of the target and the duration of the exposure of the target (Figure 3).

Determinants of Toxicity

A. Nature of key interactions

1. Physico-chemical properties of the ultimate toxicant(s) in the microenvironment of the target.

2. Physico-chemical properties of the target.

3. The nature(s) of the enzyme(s) catalysing the biotransformation of precursor toxicants into ultimate toxicants.

4. "Spontaneous" (non-enzyme-mediated) conversion of precursor toxicants into ultimate toxicants.

B. Magnitude of key interactions

1. Physio-chemical properties of the ultimate toxicant(s) in the microenvironment of the target.

2. Physio-chemical properties of the target.

3. Target dose, defined as the concentration of the ultimate toxicant at the target locus and the duration of such exposure (determined by route of exposure, dose level, exposure time and also by the topology, efficiencies and capacities of activating and deactivating enzymes, membrane permeability, and solution in or adsorption on intracellular and extracellular components).

4. Rate of repair of key lesions or turnover-rate of target.

Figure 3.   Determinants of the nature and magnitude (or amount) of key lesions

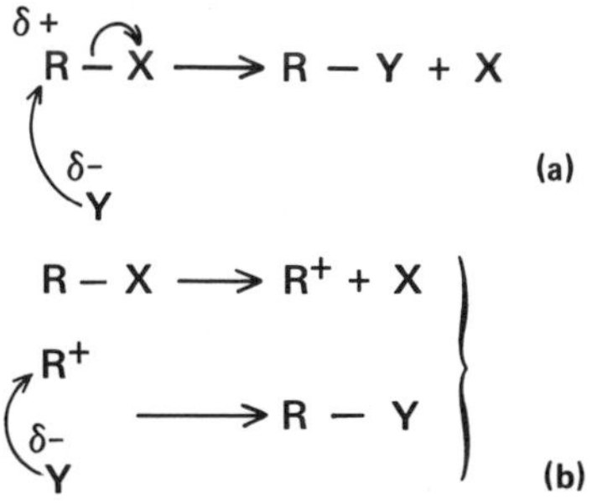

$$\overset{\delta+}{R} \overset{\frown}{-} X \longrightarrow R - Y + X$$

$$\underset{Y}{\overset{\delta-}{\Big\langle}}$$

(a)

$$R - X \longrightarrow R^+ + X$$

$$\underset{Y}{\overset{R^+}{\Big\langle}} \longrightarrow R - Y$$

(b)

Figure 4.   General mechanisms of reactions of electrophiles with nucleophilic centers in biomacromolecules: (a) bimolecular (S_N^2); (b) unimolecular (S_N^1).

Target dose, which is, of course, the relevant dose is
determined not only by the rate, route and duration of exposure
of the organism but also by host-dependent factors such as
the efficiencies and capacities of intoxifying and detoxifying
enzyme systems, membrane permeability and adsorption on or
solution in non-target, cellular and extracellular components.
Thus, the magnitude and, in the case of precursor toxicants,
the nature of the key lesions are determined by the integrated
operation of numerous factors and processes many of which are
host-dependent. Among these endogenous determinants of primary
toxic interactions, the intoxifying and detoxifying enzymes
have, possibly, the greatest significance although species-
differences in the natures of the biological targets should
not be overlooked. In the case of intrinsically effective
toxicants, the efficiencies and capacities of the detoxifying
enzymes are of paramount importance. In the case of precursor
agents, it is the balance between the intoxifying and
detoxifying enzymes that is important.

The estimation of the individual contributions of each
of these endogenous factors in determining the nature and
amount of key lesions is indeed a complex task. However,
provided that the critical target has been identified, the
net effect of all of these host-dependent factors can be
measured by determining the nature and measuring the amount
of the key interactions between the chemical and its biological
target.

The target dose (molecular dosimetry) concept which
provides the basis of this new strategy for assessing the
relationships between applied dose and the dose of ultimate
toxicants arriving at cellular targets, may be viewed as a
refinement of previously-developed and widely-used procedures
for assessing the relationships between applied dose and
endogenous concentrations of toxicants. These latter methods
were based on measurements of the concentrations of the parent
compound and/or its metabolites in the tissues and blood and
also on the kinetics of metabolic biotransformation reactions
in vivo and in vitro. Viewed from a slightly different
perspective, the target dose concept may also be considered
to be a development and formalisation of similar approaches,
e.g. the determination of effects on the activities of tissue
enzymes, in which the magnitude of a specific effect is a
direct function of the concentration of a toxicant at its
site of action. Irrespective of its origins, the target dose
concept nevertheless possesses considerable potential as the
basis of a new strategy for risk assessment.

Target Dose and Risk Assessment

Any estimation of the risks posed to humans by exposure

to a toxic chemical must be based on considerations of the
quantitative relationships between exposure and the adverse
biological effect. In most instance, such risk assessment must
necessarily be based on dose-response data generated in an
experimental species coupled with an estimate of human exposure
(Figure 5). Such an approach is highly empirical unless due
account is taken of potential species differences in the
correlation between exposure and response. In order to improve
the quality of risk assessment, it is necessary to take
account of differences between the experimental species and
humans with respect to factors that influence or determine the
nature and the magnitude of the key lesions and also with
respect to factors that determine the progression of the key
lesions into the overt biological response.

    As discussed above, the nature and magnitude of the
target dose is a prime determinant of the nature and amount
of the key lesions. In certain instances, target dose can be
accurately assessed in experimental species. For example, DNA
is the ultimate target of all chemical mutagens and is also
the key (primary, critical) target for most chemical mutagens.
The target dose for such chemicals is DNA-dose and this can
be estimated by determining the nature of the adducts formed
by reaction of the ultimate mutagen with DNA and by measuring
the amounts of these adducts present in the tissues after a
given exposure time. Additional requirements for the calculation
of target dose are: the reaction rate constant(s), the
biological half-lives of the adducts, which may vary according
to dose, and the duration of the exposure.

    Of course, the purpose of determining DNA-dose, i.e. the
time integral of the concentration of the electrophile at DNA
(adapted from Ehrenberg, Hiesche, Osterman-Golkar and Wennberg
(4)), is to permit calculations of the rates and amounts of
specific chemical reactions between the ultimate genotoxic agent
and DNA. If the amounts of specific adducts can be measured by a
direct procedure then, because they take account of both the
rate of formation and rate of repair of the key lesions, such
measurements are arguably more meaningful and useful than the
measurement of target dose.

    Provided that the physico-chemical aspects of the
interactions between the ultimate genotoxic agent and DNA
have been established, then the quantitative relationships
between overall exposure, target dose and biological response,
such as mutation frequency or tumour incidence, can be
determined in an experimental species. If it can also be shown
that the ultimate genotoxic reactant(s) are identical in the
experimental model and in humans, then the estimation of human
target dose would allow the exposure values in the extrapolative
model for risk assessment (Figure 5) to be substituted by
estimates of target dose (Figure 6).

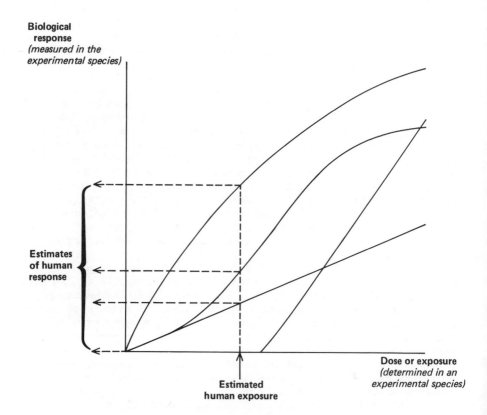

Figure 5.   Current extrapolative model for risk assessment

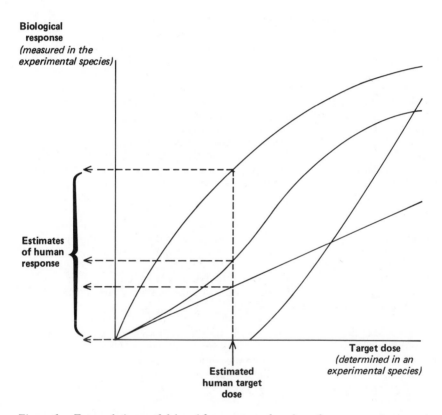

Figure 6.   Extrapolative model for risk assessment based on the concept of target dose

From a theoretical standpoint, this modification represents
a major advance in methods for risk assessment. The substitution
of exposure values by target dose improves the quality of
risk assessment by taking into account all differences between
humans and the experimental species with respect to factors
that influence the concentration of the toxicant at its
critical target.

The Estimation of Target Dose

1. Qualitative Aspects. The estimation of target dose,
particularly human target dose, poses major theoretical and
practical problems. For example, it is essential that the
ultimate genotoxic reactants(s) are identical in the experimental
species and in the human. This correlation between species,
is, of course, a prerequisite for the selection of any model
used to evaluate toxicological risks. However, the identity
of the ultimate toxicant is often assumed rather than
experimentally established. The implicit dangers of such
assumptions are illustrated by the behaviour of polycyclic
aromatic hydrocarbons in the Ames test. Thus, there is evidence
that the incubation of the precursor carcinogens, 7-methyl-
benz(a)anthracene, 7,12-dimethylbenz(a)anthracene and
benzo(a)pyrene with Salmonella or isolated preparations of DNA
in the presence of the post-mitochondrial supernatant (S9)
fraction from the livers of rats, variously pre-treated with
microsomal enzyme inducers, gives rise to DNA adducts that do
not occur in the skin of mice after in vivo exposure to these
compounds or in mouse embryo cells treated in vitro (9, 10, 11).
These findings emphasise that the appropriateness of the
metabolising systems of the model should be experimentally
established rather than assumed.
    Radiotracer techniques can be applied to determine the
natures and quantities of specific DNA adducts in the tissues
of experimental species. Such direct methods cannot, however,
be applied in humans. Differences between DNA adducts formed
with polycyclic aromatic hydrocarbons after in vivo metabolism
or in vitro metabolism catalysed by subcellular fractions
illustrate the potential limitations to the use of human
subcellular fractions in predicting the nature of DNA adducts
formed in vivo. However, there is evidence that intact cells,
e.g. primary cell cultures, may effectively mimic in vivo
metabolism, at least in qualitative terms, and may therefore
provide a useful tool for inter-species comparisons of the
natures of ultimate genotoxic reactants (11, 12, 13).

    2. Quantitative Aspects. Ehrenberg and co-workers have
proposed that the assay of haemoglobin adducts may provide
a suitable although indirect method for the determination of

the natures of the ultimate genotoxic reactants in humans and also for estimating DNA dose in humans exposed to suspect or proven mutagens and carcinogens (14, 15, 16).

As discussed above, the majority of ultimate chemical mutagens and chemical carcinogens and also many cytotoxic agents possess electrophilic reactivity. It is generally believed that reaction between electrophilic centres in such toxicants with nucleophilic centres in informational or important structural or functional macromolecules is the key event in the toxicity of such compounds. Furthermore the determination of covalent adducts with proteins and nucleic acids may provide the basis of a valuable approach not only for the detection of cytotoxic and genotoxic activities but also in discriminating between these activities.

In a homogeneous system, a particular electrophile at low concentration would react randomly with nucleophiles at rates determined by stereochemical factors and by the nucleophilic strength of each nucleophile. The contributions of these factors can be measured and it is, therefore, possible to utilise measurements of the rate of reaction at a particular nucleophilic centre to calculate the rate of reaction with another nucleophile. Although an electrophile absorbed by, or formed within an animal would tend to react at random with cellular nucleophiles, many of these nucleophiles are organised in a highly ordered manner within cells, tissues and organs. This non-random organisation of cellular nucleophiles provides a theoretical objection to the use of a nucleophilic centre in a molecule such as haemoglobin as a dose monitor for say hepatocellular DNA. Such objections may, however, be largely illusory. Thus, the same intracellular enzymes are likely to be the prime determinants of the concentrations of many ultimate toxicants within cells and in the blood. Nevertheless, the use of haemoglobin as a dose-monitor for DNA adducts would be entirely inappropriate if a major difference exists between the human and the experimental model with respect to factors that influence the relationships between haemoglobin dose and DNA dose. The investigation of such a species difference is not amenable to direct experimentation and, in order to generate a satisfactory level of confidence in the use of haemoglobin as a dose monitor for DNA, it is necessary to demonstrate that the proportional relationships between haemoglobin adducts and DNA adducts are approximately constant in experimental species displaying disparate rates of metabolism of the test compound.

In certain instances, the physico-chemical properties of endogenously-formed, ultimate toxicants may prevent their migration from their cellular sites of formation into the blood. An alternative to haemoglobin would be required as a dose-monitor for such agents. Secretory proteins of potential

target tissues, e.g. plasma albumin, may have some utility in
this respect. Furthermore, while the measurement of haemoglobin
adducts may provide an estimate of the total mean exposure of
DNA, it is difficult to perceive how this approach per se can
provide accurate estimates of the DNA dose within specific
tissues in either experimental species or humans, particularly
in the case of compounds requiring metabolic activation.
Information on the in vitro metabolic capabilities of tissue
fractions or primary cell cultures from experimental species
and humans may assist in this respect although it is often
difficult to mimic in vivo metabolism in the in vitro situation.
    At least one additional cautionary note should be added
concerning the use of haemoglobin to estimate the dose received
at critical nucleophilic targets. The N_1 and N_3 atoms of
histidine, the amino nitrogen atom of N-terminal valine residues
and the sulphur atom of thiol groups of cysteine residues of
haemoglobin all undergo nucleophilic substitution reactions
with electrophiles. Reaction at each of these centres has been
proposed as potentially suitable for the estimation of the
amount of specific electrophiles entering or formed in the
body. Studies with the S-oxides of symetryn and cyanatryn,
which incidentally show no propensity to react with DNA, have
revealed major species differences in their reactivity towards
haemoglobin (17, 18). These species differences correlated with
the presence of highly reactive thiol groups in certain
species, e.g. the rat, and their absence in other species,
e.g. the human. These results illustrate a need to be aware
of such species differences which, if undetected, could
confound the use of haemoglobin as a dose monitor.

    3. Analytical Problems. Apart from their indirect
nature, caused by the general inaccessibility of critical
target molecules in humans, methods for the quantitation of
human risks based on the target dose approach also pose major
analytical problems. Radiotracer techniques cannot be used
for the assay of blood protein adducts in humans and currently
available, alternative procedures, e.g. GC-mass spectrometry,
lack the sensitivity needed to measure these adducts in humans
exposed to very small quantities of unlabelled mutagens or
carcinogens. Specific and sensitive 'cold' techniques are
needed for this purpose. Such methods would also be of great
utility in assessing target dose in experimental studies,
e.g. carcinogenicity studies.
    These needs provide scope for considerable innovation
and one procedure that is receiving considerable attention
has its basis in immunochemistry. Thus, specific and sensitive
immunochemical methods are being developed for the detection
and assay of specific protein and DNA adducts formed by reaction
with ultimate genotoxic agents. Immunological methods have

been published for the assay of benzo(a)pyrene-deoxyguanine
adducts, formed by reaction of calf thymus DNA with (+) 7β,
8α-dihydroxy-9α,10α-epoxy-7,8,9,10-tetrahydrobenzo(a)pyrene (19),
for the detection of the C-8 guanine adduct of the carcinogen
2-acetylaminofluorene (20, 21) and for the assay of O^6-ethyl-
deoxyguanosine in DNA treated with the carcinogen N-ethyl-N-
nitrosourea in vivo and in vitro (22) (Figure 7). The further
development of immunochemical procedures for the detection
and assay of specific protein and nucleic acid adducts necessitates joint endeavour by chemists and immunologists.

Limitations in the Target Dose Approach

    Leaving aside the technical problems associated with the
determination of target dose in humans, the incorporation of
the target dose concept into the extrapolative model for
human risk assessment (Figure 6) represents a major advance
over earlier methods illustrated in Figure 5. However, the
improved model retains one of the flaws of the original model:
the risk estimate still relies upon an experimentally-determined
correlation between target dose and biological effect. Thus,
although the target dose approach is designed to take account
of differences between the biological model and the human in
factors that determine the rate of formation of the key
lesions, the risk model takes no account of differences
between the test species (system) and humans in factors that
determine the progression of the key lesions into overt
biological effects. This is, of course, a major defect which
is particularly apparent in the case of genotoxic agents
where there is no doubt that host-dependent factors such as
DNA-repair, rates of cell replication, susceptibility to
promoting agents and, possibly, immune status can variously
inhibit or exacerbate the progression of the key lesions. In
the case of chemical carcinogenesis, there is little doubt
that at least some of these 'modulating' factors exert a
major or, possibly, an overriding influence on the development
of the key lesions into tumours and also that these factors or
processes display marked tissue and/or species dependency.
    In order to improve the quality of the extrapolative
model for the assessment of genetic risks or other toxicological
risks which contain a 'progression component', the response
of the biological model must be corrected to take account of
relevant species differences in factors determining the
progression phase. The identification of these factors and
assessment of their quantitative relevance to processes such
as chemical carcinogenesis are largely tasks for the future.
    There seems little doubt that the target dose approach
has much potential, offering considerable advantages over
previous methods for estimating toxicological risks particularly

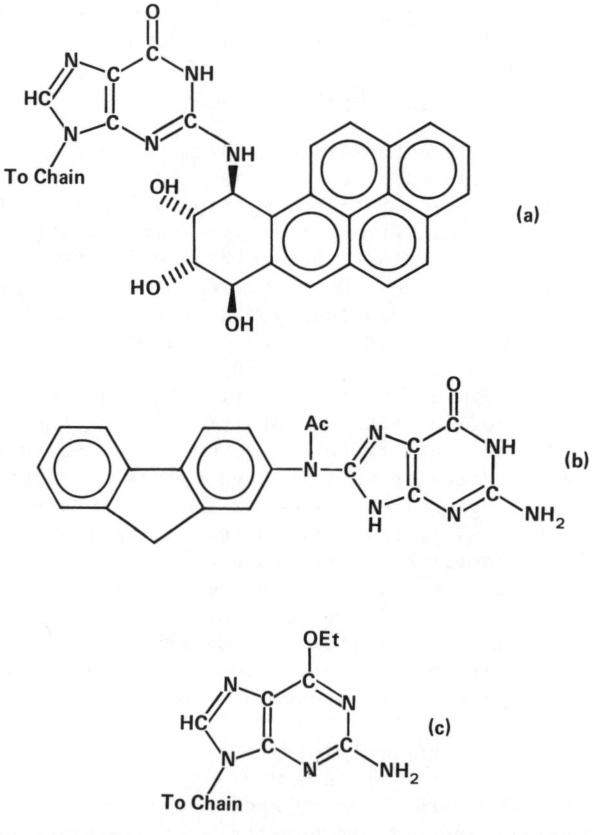

Figure 7. Illustration of the range of DNA adducts currently detected/assayed by immunochemical procedures: (a) presumed procarcinogenic adduct formed from benzo(a)pyrene; (b) principal DNA adduct formed from 2-acetylaminofluorene; (c) presumed procarcinogenic adduct formed from ethylating agents.

genetic risks. Nevertheless, the risk estimate will retain a major empirical component until due account can be taken of the roles of host-dependent factors that limit or enhance the progression of the key toxic lesions into overt biological effects.

Currently, human risk assessment utilising the target dose approach must be based on quantitative dose-response data generated in one or more presumed-sensitive biological models. The risk estimates developed using this approach must be carefully checked against the results of current and future human epidemiological studies.

Applications of the Target Dose Approach in Pesticide Toxicology

The target dose approach for the assessment of genetic risks in man is currently being applied to high volume chemicals such as vinyl chloride and ethylene oxide. The method has not yet been applied in the pesticide field although the approach has been employed in studies to determine the relevance of the results of bacterial mutation tests for the prediction of genetic risks in mammals exposed to the organophosphorus insecticide and anthelmintic dichlorvos (1; Figure 8).

The intrinsic alkylating reactivity of fully esterified alkyl phosphates, phosphonates and their thio-analogues has been recognised for a number of years. In vitro experiments with a range of nucleophiles have shown that dichlorvos possesses weak alkylating reactivity (23, 24). In this context, methylation of bacterial and mammalian DNA has been detected in suspensions of cells exposed to high concentrations of dichlorvos(25, 26). The mechanism of these nucleophilic substitution reactions appeared to be predominantly S_N2 and the N_7 atom of guanine moieties was the principal site of methylation in DNA (Figure 8).

By analogy with the biological effects of powerful alkylating agents, these findings led to the speculation that this compound might be a mutagen and carcinogen. However, although dichlorvos has been shown to induce mutation in bacteria and yeasts (27), there is no evidence that this mixed triester of phosphoric acid produces genotoxic effects in mammals (for reviews see references (27) and (28)). Thus, dichlorvos has been thoroughly evaluated for mutagenicity and carcinogenicity in mammals and the results of these tests have been entirely negative.

It seemed likely that the failure of dichlorvos to induce mutations or tumours in mammals was due to the limiting effect of the known rapid metabolic degradation of this compound on the extent of methylation of DNA in vivo. Certainly in the case of such insecticidal organophosphorus compounds,

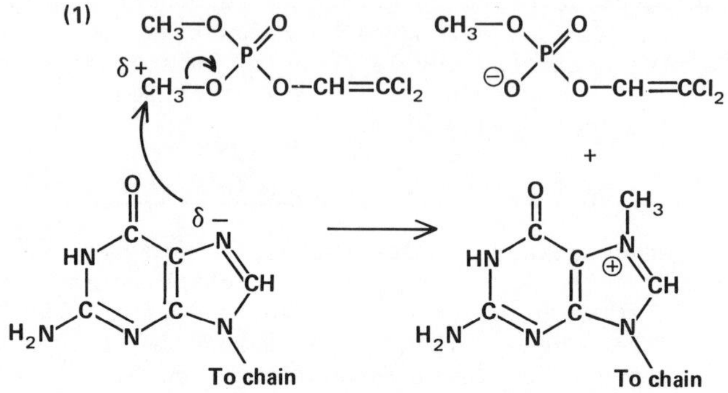

Figure 8.   Mechanism of the predominant reaction between dichlorvos and DNA in vitro

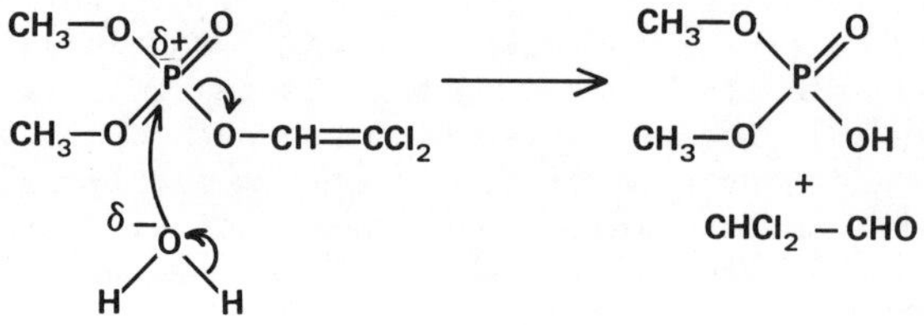

Figure 9.   Hydrolysis of dichlorvos

the possibility of 'spontaneous' alkylation reactions occurring
in vivo, particularly reactions with nucleophilic centres in
'shielded' targets such as DNA, is considerably reduced by
the marked susceptibility of the phosphoryl centre to esterase-
catalysed, nucleophilic attack by water (28, 29) (Figure 9).
    The results of very sensitive experiments conducted in
our laboratory have demonstrated that dichlorvos does not
give rise to detectable methylation of the DNA of mammalian
tissues when it is administered by the inhalation route at
practical use concentrations (30). This result provided direct
proof of the very efficient metabolism of dichlorvos in
mammals which directly leads to a loss of methylating
reactivity. The results also provided a scientific explanation
why dichlorvos induces mutation in bacteria but fails to
induce mutation or cancer in vivo.
    Very recent work conducted by Segerbäck (31) has shown
that the intraperitoneal injection of a high dose of dichlorvos
gives rise to detectable methylation of the DNA of the pooled
soft tissue organs of mice. However, Segerbäck has concluded
that, because of the low efficiency of dichlorvos as a
methylating agent in vivo, the genetic risk, posed by this
methylating reactivity, to humans receiving the maximal daily
intake recommended by FAO/WHO (4 μg.kg^{-1}.day^{-1}) is so small
as to be negligible.

Literature Cited

1. Gillette, J. R. Biochem. Pharmacol. 1974, 23, 2785.
2. Gillette, J. R. Biochem. Pharmacol. 1974, 23, 2927.
3. Miller, E. C. and Miller, J. A. in 'Chemical Carcinogens',
    Ed. Searle, C. S., ACS Monograph 173, Am. Chem. Soc.,
    Washington, 1976, p. 737.
4. Ehrenberg, L., Hiesche, K. D., Osterman-Golkar, S. and
    Wennberg, I. Mutation Res. 1974, 24, 83.
5. Lutz, W. K. and Schlatter, Ch. Arch. Toxicol. (Suppl. 2).
    1979, p. 411.
6. Lutz, W. K. Mutation Res. 1979, 65, 289.
7. Wilhelm, R. C. and Ludlum, D. B. Science. 1966, 153, 1403.
8. Ludlum D. B. Biochim. Biophys. Acta. 1970, 213, 142.
9. King, H. W. S., Thompson, M. H. and Brookes, P. Cancer
    Res. 1975, 35, 1263.
10. Thompson, M. H., Osborne, M. R., King, H. W. S. and
    Brookes, P. Chem. Biol. Interact. 1976, 14, 13.
11. Bigger, C. A. H., Tomaszewski, J. E. and Dipple, A.
    Biochem. Biophys. Res. Comm. 1978, 80, 229.
12. Kapitulnik, J., Wislocki, P. G., Levin, W., Yagi, H.,
    Jerina, D. M. and Conney, A. H. Cancer Res. 1978, 38, 354.
13. Grover, P. L., Hever, A., Pal, K. and Sims, P. Int. J.
    Cancer. 1976, 18, 1.

14. Osterman-Golkar, S., Ehrenberg, L., Segerbäck, D. and
     Hällström, I. Mutation Res. 1976, 34, 1.
15. Ehrenberg, L., Osterman-Golkar, S., Segerbäck, D.,
     Svensson, K. and Calleman, C. J. Mutation Res. 1977,
     45, 175.
16. Calleman, C. J., Ehrenberg, L., Jansson, B., Osterman-
     Golkar, S., Segerbäck, D., Svensson, K. and Wachtmeister,
     C. A. J. Env. Path. Toxicol. 1978, 2, 427.
17. Hamboeck, H., Fischer, R. W. and Winterhalter, K. H.
     Abstracts of 11 F.E.B.S. Meeting, Copenhagen, August
     1977.
18. Crawford, M. J., Hutson, D. H. and Stoydin, G.
     Xenobiotica. 1980, 10, 169.
19. Poirier, M. C. Santella, R., Weinstein, I. B., Grunberger,
     D. and Yuspa, S. H. Cancer Res. 1980, 40, 412.
20. Poirier, M. C., Yuspa, S. H., Weinstein, I. B. and
     Blobstein, S. H. Nature. 1977, 270, 186.
21. Poirier, M. C., Dubin, M. A. and Yuspa, S. H. Cancer
     Res. 1979, 39, 1377.
22. Miller, R. and Rajewsky, M. F. Cancer Res. 1980, 40, 887.
23. Löfroth, G., Kim, C. H. and Hussain, S. Environ. Mutat.
     Soc. New. Lett. 1969, 2, 21.
24. Löfroth, G. Nuturwissenschaften. 1970, 57, 393.
25. Lawley, P. D., Shah, S. A. and Orr, D. J. Chem. Biol.
     Interact. 1974, 8, 171.
26. Wennerberg, R. and Löfroth, G. Chem. Biol. Interact.
     1974, 8, 339.
27. Wild, D. Mutation Res. 1975, 32, 133.
28. Wright, A. S., Hutson, D. H. and Wooder, M. F. Arch.
     Toxicol. 1979, 42, 1.
29. Bedford, C. T. and Robinson, J. Xenobiotica. 1972,
     2, 307.
30. Wooder, M. F., Wright, A. S. and King, L. J. Chem. Biol.
     Interact. 1977, 19, 25.
31. Segerbäck, D. Toxicol. Appl. Pharmacol. In press.

RECEIVED February 2, 1981.

Problems and Pitfalls in Biochemical Studies for Pesticide Toxicology

T. BILL WAGGONER

Analytical Development Corporation, Monument, CO 80132

Biochemical investigations as related to pesticide toxicology consists of many types of studies depending on the objectives which may range from investigating the fate of a chemical in plants, animals, and the environment, such as soil, water, and air, to the investigation of the biochemical reactions in living systems and the effect of the chemical on these reactions. Further studies may be designed for detailed investigations of reaction mechanisms involved when the chemical interacts with biological systems either in vitro or in whole animals or mechanisms of toxic action.

This discussion will be related to the fate of pesticides in animals and plants where humans may be exposed to intake of metabolite residues. This area of pesticide biochemical research still has problems that involve identification of metabolites, recognizing the significance of major and minor metabolites, interpretation of data, and the significance of the results as they relate to the toxicity of the compound.

All pesticide chemists have their own objectives and ways to accomplish them. The ultimate objective on a broad basis is to seek the truth and to understand the nature of chemicals and their interaction with living systems, the most important of which is the human being. This broad ultimate goal is represented by a governmental (people) effort to regulate chemicals. This has been applied to chemicals used as drugs, chemicals used as pesticides, and now, all chemicals. The problems that are discussed for pesticides exist for all chemicals, and the solutions to solve problems in the biochemistry of pesticides will apply to other chemicals as well.

Traditional Approaches

The efforts of pesticide chemists and biochemists have been pretty much consistent during the past 15 years. Metabolism

0097-6156/81/0160-0305$05.00/0

studies are usually designed to give pharmacokinetic data and identifications of major metabolites by those scientists actively engaged in the governmental registration process. The changes in these studies have been in the hardware and manipulations that one can carry out to purify the unknown metabolite. The successful investigator is the one who uses specific separation techniques and, on a case-by-case basis, appropriate instrumentation.

Objectives. The objectives of metabolism studies which were concerned previously with identifying organosoluble metabolites now include characterizing and identifying water-soluble and insoluble fractions which previously were discarded but are now saved for further investigation. Concepts to which many chemists were exposed in their formal and on-the-job training are now either lost or being questioned. For example, conversion from apolar to polar compounds (metabolites), either through hydrolysis or oxidation, and subsequent excretion was and still is recognized as a detoxification procedure. However, the concept of a reactive intermediate and its role in the toxicity of certain chemicals should remind the chemist that formation of polar metabolites does not eliminate the chances of toxic effects. This is readily observed when dealing with organophosphorous compounds, such as parathion, an acutely toxic chemical. It was previously mentioned by Gillette in this symposium that parathion is believed to inhibit cytochrome P-450 enzymes via a reactive metabolite. Water-soluble residues in tissues are now evaluated closely, and our techniques have not been able to answer the questions that have been asked regarding their significance and toxicity. Insoluble residues that are difficult to extract are now questioned and suspect, even if they are not absorbed by the gut (bioavailable) in so-called relay toxicity studies. Interpretation of these studies is a big problem. Who knows what to do with insoluble residues?

An approach to determine the significance of metabolite residues which are not extracted by normally-used organic solvents, such as methylene chloride and ethanol from plant tissues, is feeding the extracted solids fraction to a test animal and observing the excretion and distribution pattern. Typical results as obtained by Sutherland (1) and Bakke (2) show the major route of excretion via feces and much smaller amounts via urine. Metabolite residues remaining in tissues of the animals are generally non-detectable. The results from a typical bioavailability study are represented in Table I, where plant solids from rice previously treated with phenyl-^{14}C-propanil were administered to rats. Aproximately 63% of the total radioactivity in the plants was unextractable. The major stumbling block regarding the bioavailability approach has been interpretation of results. Even if all the unextractable residue was excreted in the feces and was shown not to be

absorbed, some toxicologists have still rejected this approach
to demonstrate safety. The primary need to conduct the bioavailability study is because the identity of the unextractable
residue is unknown. This situation represents a frontier of
metabolite identification and will probably require years to
solve.

TABLE I

BIOAVAILABILITY OF UNEXTRACTABLE ^{14}C-RESIDUES

Excretion Route	Percent Recovery	
	Immobile	Mobile
Feces	76	89
Urine	2.4	11
Total	78.4	100

Tissues showed   0.05 ppm

All chemists and toxicologists are influenced directly by
governmental regulations, and a scientific meeting like this
does not operate in a vacuum and ignore this fact. We all wish
to leave this meeting with more knowledge and understanding of
our common scientific concerns. To this end, we eagerly
participate. To many of us, however, the bottom line is
regulatory needs and how to fulfill them. Does this major
concern cause a gap in our work? Does it place blinders on us
to prevent discovery and development of new ideas and theories?
Does it prevent interaction of disciplines, such as chemistry
and toxicology?

The answer to all these questions is "yes" to a certain
extent. Based upon the presentations of the previous speakers,
searching and discovering are part of their activities. The
result is the development of new ideas and theories. In the
author's experience, the biochemical investigations in the
industrial sector of the pesticide industry are somewhat
restricted in their approach to problems. For example, the fate
of a chemical in a test animal could be determined above and
below the pharmocokinetic threshold, but how many studies have
included this aspect in the past? Any pesticide chemist that
conducts biochemical studies for registration purposes is
restricted to a degree in the approach to completing the
objective. Other pesticide chemists are not as restricted.
Yet, all pesticide chemists will generally experiment, and
because of this, new theories will come. Biochemical investigations by all chemists and toxicologists on any chemical will

most likely result in an interaction of the two fields due to the interdisciplinary nature of the investigations, and this will hopefully open new avenues.

A major pitfall in present research in metabolism and toxicology is the lack of coordinated planning. Typical metabolism studies have little or no input from toxicologists. This may not be correct for all laboratories, but from our experiences, it appears to be the case. Nothing is lost by integrating metabolism and toxicology and there may be something to gain. There is a gap between the two fields, and there may be answers that both could provide by working together.

The chemists proceed down one path and the toxicologists proceed down another. The only time the two meet is usually to resolve an unusual problem or question. This frequently is a result of a pesticide product in the field showing properties that had not been previously observed. The approach to resolving the problem then is uncertain, and the chances of a successful outcome are low. This is expected since investigators in different fields normally work independently and therefore have not established any basis on which to handle not only the major problems jointly, but also to plan the original research.

The lack of a joint effort could be due to separate objectives that must be met. The chemist wants to know "what it is", and the toxicologist wants to know "how safe it is". However, what about the other questions that neither one appears to address to a significant extent? What is the mechanism or mode of action? This is a lower priority question to many investigators because it does not readily provide the answers that are needed, now. It may be that ignoring the mechanistic aspects of toxic effects is the major pitfall that widens our gap of understanding. There must be a way that this pitfall can be eliminated. An integration of efforts between chemists and toxicologists may provide a way.

For example, consider bromobenzene as a cause of liver necrosis. Identification of metabolites formed from bromobenzene and their relationship to observed necrosis has been investigated, and a reactive intermediate of bromobenzene was implicated. Pretreatment with inducers and inhibitors of bromobenzene metabolites were used experimentally. This is represented by a general relationship shown in Figure 1 (3). In the case of bromobenzene, the reactive intermediate is 3,4-bromobenzene oxide. These studies require the efforts of biochemists and toxicologists, and the interdisciplinary nature of the investigation is readily seen. The joint effort promotes understanding.

Comparative Metabolism. Comparative metabolism has always been a concern to chemists. Is the metabolic product of a chemical identical in the rat vs the dog vs the pig vs the dairy cow vs the goat, etc.? Extrapolation from one species to the next is a common problem to both chemists and toxicologists.

Basic metabolism of natural substances in all species that are studied by pesticide chemists and toxicologists is essentially identical. The pathway from glucose to carbon dioxide is a universal one. All mammals need oxygen. Yet xenobiotics have various effects on the system, and in most cases in our experience of observing metabolic pathways of organophosphorous compounds and carbamates, little difference exists among the common species that are investigated, such as the rat, dog, pig, cow, and chicken. However, the slight differences in rates of biotransformation may have profound influences on the observed toxicity.

Toxic responses to aflatoxin levels are significantly different depending on the species. A summary is shown in Table II (4,5).

TABLE II

SPECIES VARIABILITY OF
TOXIC EFFECTS OF AFLATOXIN

	Acute, Oral LD_{50} (mg/kg)
Rat (F)	16.0
Rabbit	0.3
Rainbow Trout	0.5
Channel Catfish	15.0

	Tumor Induction (ppm)
Mouse	150
Rat	0.1

Mice did not show tumor induction when fed levels up to 150 ppm for 85 weeks, whereas rats showed tumor induction when fed levels of 0.1 ppm in their diets. Biochemical studies have shown that metabolic conversions of aflatoxin B_1 in liver microsomal fractions to less toxic metabolites occurred in addition to formation of a possible reactive intermediate, aflatoxin B_1-2,3-epoxide. The latter may account for mutagenic and carcinogenic properties and is suggested from hydrolysis of an RNA-aflatoxin B_1 adduct.

This illustrates a major pitfall in many metabolism studies with pesticides, where potentially toxic effects may go undetected when determining the fate of a chemical and not relating the fate with the resultant observed toxicity. It may be a general occurrence that toxic effects are more susceptible to species variation than the different metabolites that may be formed.

Testing Systems. Administering test substances to whole animals according to proposed exposure does not provide all the metabolic information that is needed. It complicates the attempts for identifying metabolites. Due to the multiple dosing and relatively long exposure time, metabolism of parent and its metabolites that are initially formed results in a distribution of radioactivity into just about every tissue and in many forms: organosoluble, water-soluble, and insoluble. The distribution is so diffuse that little if any information can be obtained to indicate a metabolic pathway or provide a mechanism of metabolic transformation. Our experience in using this approach has resulted in a considerable amount of time in extraction and purification without a significant amount of metabolite identification beyond organosoluble metabolites. The results are useless in supporting toxicity studies. Perhaps the metabolism studies are too limited in their design and should be expanded to support toxicity studies.

Another approach would be in vitro studies, which have features that would overcome the pitfalls of the whole animal approach. The major advantages are a better-defined reaction system and a better opportunity to investigate the formation of individual metabolites. The chances of observing mechanistic features are also increased. The only problem is that the effect of the whole animal system cannot be observed, but this could be overcome by conducting toxicity feeding studies. In some situations, the advantage of metabolite identification from in vitro studies may be considerable compared to the metabolic effects of the whole system.

In vitro and in vivo studies are both utilized in a routine manner, but in vitro studies are generally not investigated. Pesticide chemists have focused on the total system, and after approximately forty years of using radioisotopes, are still "administering chemicals in the front end of the animal and collecting the products in a bucket at the other end". In vitro techniques have been indispensible in developing the metabolic picture of natural substances from biosynthesis to biological degradation.

In many situations where the metabolic pathway has not been clearly defined from identification of excretion products and tissue metabolites, the in vitro approach has provided an

insight. An example is the formation of metabolites within
minutes after oral administration that are only tentatively
identified in the liver due to either the low levels found or a
rapid conversion to final degradation products. Purification
and identification could be completed using a liver microsomal
fraction and thereby would confirm what was implied from the in
vivo study.

However, the danger of using in vitro techniques either
alone or in conjunction with in vivo studies is the interpretation of data and the practical consequences. So-called
"toxic" metabolites discovered under in vitro conditions may
eliminate a potential pesticide from further development
regardless what further work is done. In vitro studies have
their place in biochemical studies but should be used in proper
balance with other approaches.

Chronic feeding toxicity studies are certainly involved
with the total system. Interpretation is based on many observations ranging from gross effects to cellular effects. It
seems that all effects could be reduced to the chemistry, the
molecular level, and the problem is how to approach this aspect
to bridge the vacuum between metabolism and toxicity.

The route of exposure is a consideration for metabolism
studies and is generally via the oral route. Few studies are
conducted to investigate metabolism of compounds from dermal and
inhalation routes of exposure. Therefore, the techniques which
have been developed to study each of the routes are relatively
new, and to our knowledge, have never been standardized. A
major problem is to reduce the exposure to a single route. For
example, a whole body exposure of an animal to an aerosol,
smoke, gas, etc., could represent all three: inhalation,
dermal, and oral. Inhalation toxicity studies now represent
significant efforts, and metabolism studies from inhalation
exposure are still relatively few in number.

The problems associated with conducting inhalation excretion studies with test animals are demonstrated by Langard
and Nordhagen (6), where rats were exposed to zinc chromate
aerosol with a respirable fraction of 76%. The rats were
exposed to the aerosol using two different techniques: the
whole body was exposed in the first situation, and the fur coat
was protected from direct exposure to the aerosol in the second
situation. The results are shown in Table III.

TABLE III

INHALATION STUDY OF
RATS EXPOSED TO ZINC CHROMATE

	Total Cr Excretion	
	Exposed Coat	Protected Coat
Urine	653 nmol	118
Feces	97 μmol	11
Urine/Feces	0.7	1.0

Total excretion of chromium was calculated 60 hours after exposure. The total amount excreted by rats entirely exposed was approximately eight, and five times greater in feces and urine, respectively, than rats with their coats protected. The urine represents uptake in the respiratory organs and gastro-intestinal tract. The excretion in feces arises mainly from unabsorbed ingested chromates, bronchial dust clearance, and excretion in the bile. The authors concluded that licking of the coat was a significant factor for the exposure time of less than one hour, and therefore, ingestion and gastrointestinal absorption are important factors in such studies. Other factors were suggested for consideration, such as grooming patterns with different compounds, water solubility of the compound, and time of exposure (once or multiple exposures).

Metabolite Identification.
    Dosage. Identification of metabolites in animals is more difficult when the compound is administered in multiple doses compared to a single dose. A typical response curve of a conjugate is shown in Figure 2. A single dose will give a minimum level of radioactivity in the solids fraction (solid line). After the last dose of multiple treatment, the level of radioactivity in the solids (dotted line) will be considerably higher and cause major problems in isolating, purifying, and identifying significant metabolites.
    The slow accumulation of insoluble residues during or after the dosing regime raises questions: What is the nature of the accumulated residue? How is it extracted and purified? Is there enough to work with? Is it toxic? Is it significant? How do accumulated insoluble residues from multiple five-day doses compare with those from a chronic feeding study? What could be gained from administering a low-level, radioisotopically-labeled compound during the chronic feeding study?
    The overall problem now is a lack of new breakthroughs to investigate insoluble residues. Perhaps the combined effort

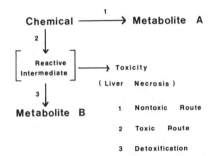

Figure 1.   Metabolic pathways

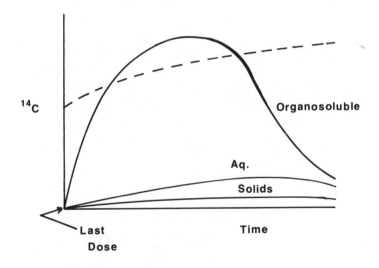

Figure 2.   Multiple vs. single dose

of chemists and toxicologists could resolve the questions. Up until now, little cooperative effort has been given to this topic.

A single administration of a radiolabeled compound is easier, since purification and identification of metabolites are more straightforward. Is it enough to know the nature of major metabolites from single doses as it relates to the toxicity of the compound? It appears that no real answers are evident, now, and probably will not be in the foreseeable future unless new approaches are investigated. Interweaving of toxicity and metabolic studies would also be a different approach with new problems of interpretation.

The present approach for metabolism studies of pesticides and animal health drugs is to administer the compound to plants or animals in a manner that represents normal use conditions in the field. This approach prevents successful identification of metabolites and restrains what should be the proper way to identify metabolites. Little attempt is made to correlate metabolic results and toxicity. The chemist already knows that insoluble, unidentifiable residues will most likely occur from multiple dosing and that the problem of identifying metabolites and determining their toxicity is still present. For example, the industrial situation, in general, is not conducive to doing it any other way.

Here is a situation that has existed for many years. Scientists are attempting to standardize metabolism studies to determine fate of chemicals and to assess the toxicity of the chemical and sometimes the metabolites in selected test animals. Studies, such as those for developing new ideas, testing new theories, and determining mechanism of the toxic effect, cannot be standardized. These studies should be designed on an individual basis after certain properties of the chemical have already been observed. In contrast, standardized studies are designed without prior knowledge of the properties of chemicals that potentially will be tested.

The pitfalls between the choice of a standardized and individual study is a bias toward the standardized study with terrible consequences. For example, if the standardized metabolism study to determine the fate of the chemical is multiple dosing to a dairy cow because the chemical will be administered that way as a growth promotant, then how will 90% of the metabolic residue in the animal ever be identified? The chances are very low, and in most cases are zero, based on the current state-of-the-act. A better approach to this fate study is administering a single dose to eliminate interferences due to distribution of non-metabolic residues, that is, radioactivity that has been incorporated into naturally-occurring substances such as the extracted solids. The point is that standardized tests should not prevent one from looking at a single dose. A

standardized test to simulate multiple dosing usually gives
little if any metabolite significant identification.

Isotope Labeling. Heavy isotope labels, alone, would
very likely not produce better results than radioisotope labels.
Problems of isolation and purification would be more difficult
than radioisotope studies due to the nature of the detection
system which would be needed, mass spectrometry. However, the
combined use of heavy isotope and radioisotope labeling can
provide very definitive information on confirming the identification of suspected metabolites in the same manner as using
dual radioisotopes, such as ^{14}C and ^3H, or ^{14}C and ^{32}P. However, reactions that split the molecule into two parts, one with
the radioisotope and the other containing the heavy isotope,
destroy the usefulness of the dual labels, due to purification
problems and interpretation. Intact metabolites containing both
labels are the easiest to confirm. Therefore, parent molecules
that tend to split into multiple metabolites complicate isolation, purification, and identification, and contribute to a
complicated interpretation.

Stability. A traditional problem in metabolite
identification has been stability of the unknown metabolite
during isolation and storage. Independent synthesis will
indicate storage stability of the pure metabolite under various
conditions, and recoveries may be determined during workup of
the sample. But what about the metabolite that may easily
undergo reaction during workup prior to its identification?
This will probably always plague the pesticide chemist. Perhaps
the only way to determine its significance, if formed at all, is
to determine the toxicity of the parent compound and assume its
formation by metabolic conversion of the parent. The problem,
however, still exists as to the levels in the tissue which could
probably not be assessed due to the absence of its detection
during metabolism studies.

Metabolites and Effects. The problem remains for us
to determine the metabolic products and their effects on the
system, whether those products remain in tissue (animal or
plant) or whether they are excreted. Even though the excretion
of metabolic products may be rapid, a gap still exists regarding
the effects of the parent or its metabolic products. It appears
that a combined effort from toxicologists and chemists may give
information to fill this gap.
        For example, organophosphorous compounds as a class
are excreted fairly rapidly due to hydrolysis and oxidation, and
generally do not accumulate in tissues. Yet, the compound may
be neurotoxic -- it may cause paralysis in some species and not
in others. What causes the neurotoxicity -- parent or metabolite? What concentration is required to allow the mechanism of

this effect to be indicated? As a routine investigation of this
toxicity in the industrial sector, I'm sure that little, if any,
investigations are underway.

Developing Studies

   Metabolism of Metabolites. Feeding radiolabeled metabolites isolated from plant tissue to animals can be a complex
experiment causing a problem of interpretation. A major
question should concern the need for such studies, and what one
could learn that would be different from feeding the parent
compound. Various types of plant metabolites could be fed:
organosoluble (generally unconjugated compounds), water-soluble
(generally conjugates), and insoluble residues.

   Organosoluble Residues. Identified metabolites in the
organosoluble fraction, labeled with the appropriate radioisotope, have been administered to animals traditionally. In
most cases, the metabolite fed from the plant was also formed in
the animal from the parent. The major objective for such
studies is confirming the metabolic pathway in the animal. This
approach would be useful if the metabolite, when formed from the
parent in the animal, is so rapidly transformed that it would
normally have only a fleeting existence and therefore would be
relatively difficult to detect having the characteristics of a
reactive intermediate. In addition to confirming the similarity
of metabolic pathways in plants and animals, a plant metabolite
not previously detected in animals should be studied in the
animal system. Its fate and toxicity relative to the parent
compound could provide essential information.

   Water-Soluble Metabolites. Administering unidentified
metabolites from plants to animals is difficult to conduct due
to practical considerations. The concentration of enough
radioactive material in a sufficiently small volume to administer is difficult to achieve. Enough radioactivity is
needed to provide sufficient radioactivity to determine excretion patterns and rate and for identifying metabolites. The
difficulty of too little activity is enhanced if metabolites in
tissues, eggs, or milk must be identified. Unless the specific
activity is in the 20-25 mCi/mmole range, the identification of
further metabolites in animals would be difficult due to losses
normally encountered during extraction and purification
procedures.

   Insoluble Residues (Metabolites). Administering
unidentified insoluble metabolites isolated from plant tissue to
animals is probably the most difficult experiment to conduct.
Again, the need for sufficient radioactivity to produce a
sufficient level of metabolite in the animal for subsequent

identification is a major problem. Interpretation of results is
the biggest problem area. Generally, the insoluble residue is
not identified, and the concentration of any single metabolite
is usually too low for identification. In many cases, the
amount of radioactivity absorbed in the gut of the animal is
very low, and therefore, the level of radioactivity in tissues
is frequently below detection. The questions from these types
of studies are unanswered to the satisfaction of both toxicologists and chemists: How does lack of absorption contribute to
the evaluation of whether insoluble metabolites are significant
to the parent compound's toxicity?

New Approaches

Mechanisms.
        Reactive Intermediates. Intermediates in organic
chemical reactions may be long- or short-lived and are frequently difficult to isolate. Isolation is best accomplished by
a trapping experiment or by conducting the study at very low
temperatures. The same techniques used in studying organic
chemical reactions could be applied to biochemical reactions
under in vitro conditions.
        The reactivity of a suspected reactive intermediate is
illustrated in Figure 3. Reaction of the intermediate to give
products must proceed through a pathway requiring an activation
energy, E_{A1}. If a trapping agent is added that reacts with the
intermediate through a pathway requiring a lower activation
energy, E_{A2}, then this pathway would be favored to give a
product that could be identified. Examples of trapping experiments are shown in Figure 4. The same concept could be applied
to biological reactions to confirm the identity of a suspected
reactive intermediate. This is identical to the detoxication
pathway of bromobenzene with glutathione to produce the corresponding conjugate.
        One would be limited in trapping intermediates under
in vitro conditions: (1) temperature range small; (2) a
selective trapping agent needed; (3) interpretation. The
temperature range would most likely have to be near 37°C,
especially since enzymic reactions would be involved. A
trapping agent that is selective only for the reactive intermediate would be needed, and it should give little or no
interference with the normal biochemical reactions under study.
Interpretation would always depend on the nature of the system
and the understanding of what was occurring. Carbonium ion free
radical intermediates could be defined in terms of their organic
reaction mechanisms under controlled conditions and extrapolation of this information to biological conditions.
        The ten steps outlined by Gillette to determine
whether a reactive intermediate is responsible for a toxic
response is time-consuming and probably expensive. As a
scientific activity for identifying the toxic entity and

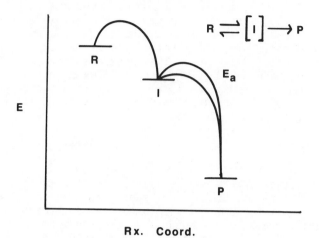

Rx.   Coord.

Figure 3.   Energy relationships: E, energy; R, reactants; I, intermediate; E_a, activation energy from intermediate to product; P, product

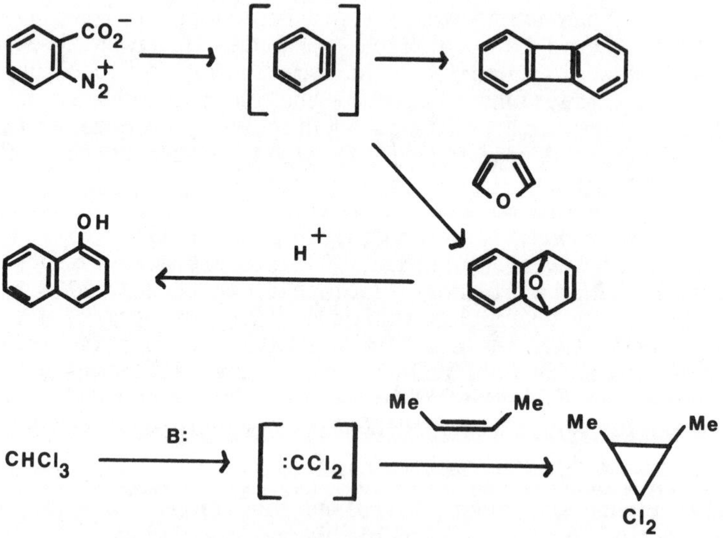

Figure 4.   Trapping intermediates

possibly an explanation of ^{14}C-bound activity which is found, the approach most likely has a useful place in pesticide research -- but where?

One cannot object to investigations of this nature, because the frontiers of biochemistry and toxicology are being expanded. The benefits are most likely long-term for evaluating pesticide safety. Many investigators in pesticide biochemical research do not have the facilities for these types of investigations. The problem is one of justifying the effort. For example, what is the time needed to carry out the 10 steps in identifying a reactive intermediate? An estimate, based on our experience in pesticide research, is given in Table IV.

TABLE IV

EFFORT NEEDED TO DETERMINE THE PRESENCE OF
A REACTIVE INTERMEDIATE

	People - Months
Species toxicity and dose response (1,2)	2 - 3
Development of analytical methods (3)	18 - 24
Pretreatments and toxicity (4)	2 - 3
Pretreatments and metabolism (5)	4 - 6
Target tissue binding and dose response (6)	4 - 6
Pretreatments, rates of metabolism, binding (7)	(8 - 12)
In vitro, K_m and V_m, several tissues (8)	6 - 12
Identify decomposition products (9)	18 - 24
Supportive evidence (10)	6 - 12
Total	68 -102

(5-8 people-years)

What type of toxicity should be observed when identifying a reactive intermediate? For acetaminophen, hepatic necrosis was observed and depended upon the species. Most animals were affected at a 150 mg/kg dose to hamsters; whereas, less than 10% of the animals were affected when rats were given 1,500 mg/kg. How can this concept be applied to the variety of classes of pesticides? The question must be resolved on a

case-by-case basis.   Toxicities of various pesticides are shown
in Table V.

TABLE V

OBSERVED TOXICITIES OF VARIOUS PESTICIDES

	Acute, Oral LD_{50} Rats (mg/kg)
Captan	10,000
Imidan	300
Parathion	3.6-13

	No Observable Effect Level
Organophosphorous compounds	<1-2 ppm in diet
Carbamates	<50 ppm in diet

Compounds with higher acute toxicities, such as parathion, would
be more difficult to investigate than captan which is less toxic
by a large factor.   Parathion inhibits cytochrome P-450 enzymes
via reactive intermediates, but is the level in the tissue sufficiently high to identify degradation products?   What specific
activity of parathion is needed?   No observable effect levels
for organophosphorous compounds and carbamates are frequently
based on the level that causes a significant depression of an
enzyme level of activity which can be less than 0.05 mg/kg for
the rat.   It doesn't appear that extrapolation of the investigative technique for reactive metabolites with drugs can be
readily undertaken for pesticides because of the wide range of
pesticide toxicities.   The question of the significance of
reactive intermediates in pesticide toxicity can be raised.
However, for all chemicals and their behavior and effect on
biological systems, if a general concept is possible, all compounds should be treated the same, but it's probably not possible
to use the same investigative technique for all chemicals.

Acceptance.   New ideas are slowly accepted in science.
However, the tendency of scientists to accept new ideas too
readily may result in long-term problems.   For example, the in
vitro studies, such as microbial mutation to give preliminary
evaluations of a chemical's toxicity, are readily accepted by
some.   The assessment and critical evaluation of theories is a
time-consuming process and requires flexibility to modify one's
own ideas.

All people resist change to a certain degree and are unwilling to try a new approach, unless, of course, it is their own idea. An example was FDA's attempt to publish the SOM (Sensitivity of Method) document to guide the registration process of animal health drugs (7). FDA might have just as well said that the earth is flat, judging from the response of industrial scientists. The scientific merits of the SOM are not the major concern here, but it represents a new approach compared to EPA's registration process for pesticides. It met with considerable resistance. Probably if scientists outside governmental regulation had presented the concept first, FDA would have rejected it with the same vigor.

The problem in pesticide metabolism research is the acceptance of a non-traditional approach. People don't want to change unless they are forced to do so, or unless it is safe. Integration of metabolic and toxicity studies is a new approach that will face the problem of acceptance. Maybe chemists and toxicologists are even opposed to it. However, all must be willing to seek new ventures and hopefully achieve the answers to our current questions.

Literature Cited

1. Sutherland, M.L. in "Bound and Conjugated Pesticide Residues" (Kaufman, D.D.; Still, G.G.; Paulson, G.D.; Bandal, S.K., Eds.); American Chemical Society: Washington, DC, 1976; p. 153.

2. Bakke, J.E.; Shimabukuro, R.H.; Davison, K.L.; Lamoureux, G.L. Chemosphere, 1972, 1, 21.

3. Jollow, D.J.; Kocsis, J.J.; Snyder, R.; Vainio, H., Eds. "Biological Reactive Intermediates"; Plenum Press: New York, NY, 1977; p. 42.

4. Gall, E.A.; Mostofi, F.K., Eds. "The Liver"; The Williams & Wilkins Company: Baltimore, MD, 1973; p. 161.

5. Uraguchi, K.; Yamazaki, M., Eds. "Toxicology Biochemistry and Pathology of Mycotoxins"; John Wiley & Sons: New York, NY and Kodansha Ltd.: Tokyo, Japan, 1978; p. 121.

6. Langard, S.; Nordhagen, A. Acta pharmacol. et toxicol., 1980, 46, 43.

7. FDA "Carcinogenic Residues in Food-Producing Animals"; Federal Register, 1979, 44, 17070.

RECEIVED February 2, 1981.

Biochemical Aspects: A Summary

GINO J. MARCO

CIBA–GEIGY Corporation, Biochemistry Department, 410 Swing Road, Greensboro, NC 27409

Historically, cancer has been associated with some kind of chemical exposure as was first documented from human expo- sure to environmental contaminants. Over 2 centuries ago, Percival Pott reported a high increase of scrotal cancers in young chimney sweeps. However, there have always been various incidences of certain cancers that had no clear association with a specific environmental contaminant. When I was in graduate school, biochemical studies devoted to understanding the mechanism of cancer were considered the "graveyard" for biochemists. It seems that many biochemical interpretations were based on minimal, if any, knowledge of the biological processes involved in cancer development. In the session's first paper, Dr. Laishes pointed to highly significant advances, armed with hard data, and in certain instances, left us with some exciting potential directions leading out of the "graveyard." The biochemistry of cancer development is far from understood and may differ in each target tissue. In seeking to solve this mystery, the bio- chemist, as a chemical Sherlock Holmes in this detective game, has been offered some interesting clues, for example, the demonstration that carcinogenesis can be divided, at least in some instances, into two qualitatively different biological processes, that is, initiation and promotion. Focus is now on new efforts into understanding, not only the molecular defects in target cells, but also the physiologic milieu necessary for the "promotion" of early altered cells to the development of frank, invasive, and even metastatic carcinoma.

The generality of the initiation-promotion, two-step system is one of our most readable directional signs. The fascinating sequence relationships, that is, the need to apply the initiator prior to the promoter, provided a remark- able clue in our understanding. Not only has the irrevers- ible, additive concept surfaced again, but commonality of mechanisms in many chemicals was seen. The common formation

0097–6156/81/0160–0323$05.00/0
© 1981 American Chemical Society

of electrophilic reactants from structurally
diverse chemical carcinogens remains as one of the most
powerful contributions of the past 20 years. With the
concept of covalent binding to form carcinogenic adducts, we
truly were in the realm of the chemist. Chemicals, with
antagonistic effects on the process of carcinogenesis,
provided fuel for the discovery of enzyme induction in
mammals. As the concept of precarcinogenic compounds
yielding proximate and finally ultimate carcinogens developed
as metabolic processes, the biochemist moved closer to his
areas of interest and expertise. Finally, what was certainly
logical, but required evidence, was the concept of repair
mechanisms of carcinogenic DNA-adducts.
     We are now acquiring the tools to allow us to delineate
those alterations in cellular regulatory molecules, induced
by chemical carcinogens, that are essential to the
biochemistry of cancer development. With our realization
that AAF, as a "complete" carcinogen, has both initiating and
promoting properties, the possibility of a simple
straightforward mechanism of carcinogenesis is more remote.
     In attempting to unravel some of these biochemical
pathways having toxicological consequences, Dr. Gillette
indicated that toxic potential between parent compound and
its chemically stable metabolites was relatively simple.
Isolation, identification, synthesis and testing of these
metabolites for their toxic effects have been an effective
way to deal with them. However, other strategies must be
used for that elusive, often speculative, short-lived
chemically reactive metabolite. Here, the detective in the
chemist must again surface. While nucleic acids are targets
leading to potential serious consequences, other targets are
equally of concern. Intracellular enzymes, proteins in
general, cell membranes, and locations of repair processes
are some of the more important additional interaction sites.
However, direct covalent binding of a chemical is not the
only alteration possible. As one example, indirect attack
can occur by free radicals generated by the chemically
reactive intermediate. We now have truly elusive mechanisms
to sort out. The parent, stable metabolites, reactive
intermediates and indirect reactive entities provide us with
the concepts to elucidate the processes of all toxicological
phenomena, not solely carcinogenesis. A complicating factor
is that ability of the chemically reactive metabolites to
react with multiple cell components, and proceeding at
varying rates. Yet, the chemically reactive metabolite may
be scavenged by cellular components with large numbers of
nucleophilic groups leading to their preferential attack.
Also, these reactions might be developed after the toxic
effects of the parent have been expressed, possibly leading
our detective down a wrong trail.

By use of various inducers and inhibitors of the metabolism of the toxicant and the emphasis on covalent binding to
protein, as an indirect measure of the concentration-time
exposure of the reactive metabolite to the target, we are
provided with a strategy permitting the use of kinetics
already well defined in the protein-enzyme field. By conducting a sequence of in vivo and in vitro experiments using the
stated strategy, a case can be made for the involvement of a
chemically reactive metabolite in a given toxicity. This
approach certainly highlights the multi-disciplined and
diverse methodological approaches requiring the philosophy of
the scientific generalist and interdisciplinary action in
experimental design.

The repair mechanism ideas developed in the previous
papers encourage one to believe that an absolute threshold
exists for chemical induction of cancer. It is the concept
of one irreversible molecular event leading to the induction
of cancer that provides the stimulus arguing against a threshold. Implied in the concept of threshold is the ability to
quantitatively estimate the carcinogenic risk at low levels
of exposure. Dr. Ramsey addressed this question which brings
to bear the elucidation of the shape of dose-response curve.
He discussed the pharmacokinetic characteristic of a chemical
since they are intricately linked to its toxic response. The
clues to unravel in this detective's game are how biological
rate processes vary with chemical concentration. We saw that
the transition from linear to non-linear kinetics, as the
dose level increased, constitutes the pharmacokinetic threshold; and this transition was a gradual one. The extrapolation made from data obtained above or below the pharmacokinetic threshold dose were shown to be of major concern
when attempting to predict toxic effects at low exposure
levels. In assessing risk by use of models, the main differences are in the rapidity with which zero exposure is
approached. But, few make provision for an absolute
threshold for carcinogenic response. There is absolute
dependence on concentration of the carcinogenic entity being
directly proportional to dose or parent chemical. Thus,
predictions based on dose levels alone can lead to a
fallacious conclusion. The speaker indicated that the
relationship between steady state concentrations and administered dose levels to be crucial in interpreting and predicting any toxic response as a function of exposure level.
Thus, we have added mathematical tools to our chemical ones
in order to expose a chemical of toxicological concern.

There is another complicating threshold concept. The
cytotoxic threshold is considered to be the result of the
finite capacity of the cell to tolerate injury to the multiplicity of cellular components other than critical parts of
the genome before the cell itself sustains injury. Recurring

cytotoxic injury may lead to an increased rate of tumor production. While the cytotoxic threshold is not easily
expressible in the formal mathematical terms of the pharmacokinetic threshold, it nevertheless comprises a range of
dose levels above which the rate of chemically induced cancer
may be disproportionately much greater than that at lower
levels.

Dr. Ivie proceeded to show the complexities seen in how
biological systems deal with a xenobiotic, with emphasis on
pesticides. Since pesticides by design are meant to be toxic
and all living things have much in common biochemically, the
toxicological consequences to man must be considered in their
use. Yet, the judicious use of pesticides contributes in
positive ways to human welfare. Studies in pesticide metabolism not only show the role in the expression of pesticide
toxicity but aid in the evaluation of toxicological significance of these metabolic products. Studies leading to an
understanding of the mechanisms of pesticidal action aid in
appropriate selection of pest control agents with minimal
environmental consequences. The evaluation of toxicological
significance of pesticides must include its metabolites since
much of human exposure is related to the decomposition
products of the pesticide. Yet, pesticide metabolism studies
cannot be considered as an end to themselves, but rather are
a means to an end. That is, they are intended to gain data
of value toward assessment of the toxicological significance
of the pesticide.

As the speaker showed, the metabolism is composed of
complex, multi-stepped reactions, leading to complicated,
difficult to identify metabolites, often present in extremely
small concentrations. This places the chemist working in
metabolism research at the knife-edge of modern technology
and at the razor's edge of interpretation. For example, what
is a major or minor metabolite and of what significance is a
bound residue? An added wrinkle is the impact of regulatory
requirements on the why, what and how of metabolism research.
As posed in this paper, can pesticide metabolism studies be
more effectively used in the safety evaluation process,
especially with more toxicological relevance to the ultimate
biological system, man? Possibly, the direct use in man of
relatively safe techniques, such as heavy isotopes, could be
one of several ways in future studies.

Dr. Wright continued the look into newer biochemical
strategies for understanding pesticide toxicology. The focus
was on approaches to improve the quality of human risk
assessment based on quantitative dose-response data generated
in experimental animals. The nature and magnitude of the
target dose can be a prime determinant of the nature and
amount of key lesions. In this approach, DNA is considered
the key (i.e., primary, critical) target of most chemical

mutagens. Thus, the target dose of such chemicals is DNA-
dose estimated by determining the nature of the adducts of
reaction of the ultimate mutagen with DNA and measuring the
amounts of these adducts. Reaction rate constants and biological half-lives of the adducts as well as duration of
exposure are needed for the calculation of target dose.
Provided that target dose can be accurately measured in
humans and in the experimental model, then the exposure
values in the extrapolative models for risk assessment can be
substituted by estimates of target dose. Substitution of
exposure values by target dose should improve the quality of
risk assessment by emphasizing the factors that influence the
nature and the concentration of the toxicant at its critical
target. It was emphasized that the appropriateness of the
metabolizing system of the model to find these adducts needs
to be experimentally established, not simply assumed. However, human primary cell cultures could effectively mimic in
vivo metabolism.
      The major practical and conceptual problems associated
with the target dose approach center around the determination
of target dose (DNA-Dose) in humans. While radiation techniques are useful in experimental animals, this approach is not
applicable to humans. The use of haemoglobin as a dose-
monitor for DNA-adducts was discussed with indications that
this technique may not always be appropriate. The use of
immunochemistry in assessing target dose was suggested with
ample opportunity for chemists and immunologists to develop a
joint endeavor by the detection and assay of protein and
nucleic acid adducts. The target dose approach is designed
to take into account differences between the biological model
and human factors for determining the rate of formation of
key lesions. But, the risk model takes no account of differences between test system and humans in factors to determine
progression of key lesions into overt biological effects.
Future work should identify the relevant species differences
in factors determining the progression phase. The target
dose approach for assessing genetic risk in man is yet to be
applied to the pesticide field except for determining the
relevance of bacterial mutation test data for prediction of
genetic risk in mammals. However, this is a new strategy
worth considering.
      Every scientific discipline has its problems and pit-
falls and biohemistry is no exception. Most of those
discussed by Dr. Waggoner may not seem new. Maybe that is a
reason they are still problems and pitfalls. One pitfall is
the often lack of coordinate planning, especially between the
fields of chemistry and toxicology. An example of this is
the minor effort in defining the mechanism of toxicological
action. Is the pressure to meet regulatory needs a cause of
this? Such pressure could even be the reason for minimal

amounts of comparative biochemical studies. While animal
metabolic studies are a way to quick data, is this detracting
from more intensive development of in vitro biochemical
studies? The route of exposure certainly provides problems
and pitfalls in all aspects (methodologically and interpre-
tatively), especially in oral vs. dermal vs. inhalation
treatments. Separation of the three in a meaningful way is a
real challenge. Metabolite identification is continually in
a state-of-the-art world simply because new tools and techniques open approaches not previously accessible. As highly
complex metabolites are isolated at lower concentrations,
problems magnify, including those of contamination and
stability. What do the studies of metabolism of unknown
metabolite mixtures really show? Since all the metabolites
in a tissue may not be ingested, how are the results interpreted, especially when low levels generally must be fed?
Possibly the largest problem and pitfall, related to all of
science as well, is the resistance to change and opposition
to non-traditional approaches.

   So where do we now stand in the biochemical area? Biochemical studies to understand carcinogenic action certainly
are no longer a graveyard. In fact, the area is very much
alive with new tools, methodology and concepts. New information is continually surfacing, providing many new ideas about
the carcinogenic processes. Rather than a graveyard, it is
more like a six-lane expressway with results and conclusions
speeding, with direction, toward a rational elucidation of
these complex interactions.

RECEIVED February 9, 1981.

ANALYTICAL ASPECTS

Analytical Aspects: An Introduction

WILLIAM HORWITZ

Bureau of Foods, Food and Drug Administration, HFF-101, 200 C Street SW, Washington, DC 20204

We frequently hear statements regarding the fantastic advances in analytical chemistry over the last several decades. These statements tell us how the analytical chemist has increased his ability to detect and measure many chemicals from parts per thousand to parts per million, then to parts per billion, and now even parts per trillion. Such statements are finding their way into the morning newspaper, complete with 3 and 4 significant figures. Rarely, however, do we see a discussion as to whether or not these figures are correct. Even if they are correct, what is their reliability or its converse, its uncertainty. Even among scientists, a report from a laboratory showing the presence of several parts per trillion of a toxic chemical is accepted without question. What is worse, the accompanying uncertainties of analytical measurements are not recognized by many analytical chemists.

The first two lectures in this session on the Analytical Chemist and Modern Toxicology will introduce some of the marvels which have been accomplished with the aid of analytical chemistry.

Our first two speakers are from the National Center for Toxicological Research of the Food and Drug Administration at Jefferson, Arkansas. The National Center for Toxicological Research has been one of the first institutions that has been using good laboratory practices both in toxicology and in analytical chemistry. The original concept of this institution was to handle what was termed a "mega-mouse" study. Sampling statistics tell us that the probability of determining a very low incidence of cancer in animals, say at levels of 0.1% or 1%, requires tens of thousands if not hundreds of thousands of test animals to be sure to differentiate between a tenth of a percent induction and the background. The sheer logistics of such an operation soon scaled down the original version to a 25,000 mouse version on a known

THE PESTICIDE CHEMIST AND MODERN TOXICOLOGY

standard carcinogen as a model substance in a test run. The
results of this first gigantic experiment were recently published
in the Journal of Environmental Pathology and Toxicology.

   In conducting this study, numerous problems and side issues
have been handled in setting up, maintaining, and interpreting the
experiment. But one of the main tools was analytical chemistry,
maintaining the integrity and purity of the diet, water, and air
without interruption for more than a two-year period, assuring the
presence of the anticipated amount of the test substance as well as
its purity and integrity, providing the chemical information
required to determine the toxicity profile such as metabolism and
pharmokinetics, and finally interpreting the results of all of the
analyses. Along the way, additional peripheral but important
chemical operations were necessary to guarantee the safety of the
personnel and animals from contamination, and the disposal of all
exposed and contaminated experimental materials. Many new analytical methods had to be developed and validated to assure their
applicability to the problem at hand. In addition, specifications
had to be developed to protect the experimental animals and the
controls from exposure to materials which might perturb the
responses sought.

   In order to interpret the results from chemistry and toxicology programs, it is necessary to apply statistics. Dr. Tiede
will point out the major statistical tools required in this area.
An important thing to remember in statistics is that to measure
small effects or small quantities you need large samples. Also, if
you wish to be more confident of your results, you need a larger
sample. Only if you wish to be sloppy and not be very confident,
can you get along with a small number of samples. This applies
whether you are measuring toxicological effects or physical
amounts of substances in micrograms, nanograms, or picograms.
Another important fact to remember is that it is impossible to
design, conduct, or interpret any work in these areas without a
working knowledge of statistics. At the very least the scientist
must become an amateur statistician in order to tell the statistician what is wanted and to understand the answer that is
provided.

   In my lecture I am warning you that the analytical chemist is
nowhere near as good as he thinks he is or that he makes it appear
that he is. For those who do not wish to be confused by
variability, the analytical chemist will give you a number. But
the variability is still there. To properly interpret chemical
values in terms of biological phenomena, the analytical variability must be removed to assure that the final results are truly
of toxicological significance and not merely the analytical error
of the chemist. The toxicologist must be particularly wary when
the chemist operates near the limits of measurement. It appears

that false positives and false negatives are inherent in the data when a method is pushed beyond its capabilities. The toxicologist and chemist, not the statistician, must stay clear of that precipice.

Finally, Dr. McKinney, from his vast experience in environmental chemistry will point out some very practical matters, such as how to handle the sample to protect it from things which will mimic the compound sought. To emphase this point even further, much of the trace element analyses in biological materials in literature today is invalid because the investigator was unaware that metal tools have a sufficient solubility or fragility to impart significant amounts of trace elements into a biological sample from mere contact.

The intriguing advertisements of instrument manufacturers suggest that they have the true salvation for analytical problems. Very often the instrument manufacturer will apply his equipment to the ideal situation of a pure compound in a pure solvent. The results are strictly true, but may be grossly misleading if applied to a biological matrix. The same applies to recovery studies even in biological matrices. The native compound may be tightly bound by reaction or by absorption and may be lost to your determination, although additions above this point are recovered satisfactorily. Under other conditions, these materials which do not respond to a normal analysis are released to give an unanticipated effect. Furthermore, never forget blanks and controls. Any irregularity in their values requires further investigation. Blanks are also critical for proper statistical interpretation of calibration functions, recoveries, and limits of measurement. Only by keeping in mind all of these various factors which tend to subvert or mislead the investigator, can the exquisite results described by Dr. McKinney, be obtained.

RECEIVED March 18, 1981.

The Increased Role of Chemistry in Toxicology

THOMAS CAIRNS

Department of Health and Human Services, Food and Drug Administration, National Center for Toxicological Research, Jefferson, AR 72079

In this century society has by its proliferation of synthetic organic and inorganic chemicals including pesticides and herbicides given birth to a whole host of new interdisciplinary sciences. Toxicology has obviously been a product of this evolutionary process and is continuing to be a complex martrix of many basic sciences neatly and somewhat conveniently blended with the skill and creativity of a gourmet chef in the presentation of a great classic and outstanding dish for immediate consumption and praise. Indeed, it is perhaps presumptuous for a humble chemist to attempt to delineate his particular professional role in this new technological development. The myriad of sub-specialities that are the building blocks of modern toxicology have largely contributed to a lack of understanding of a clear definition of the exact science of toxicology. Nevertheless the strengths of modern toxicology must be considered a direct synergistic result of the various component disciplines from which its comprises. Fundamental to modern toxicology is the role that analytical chemistry can play and has honestly derived from various quantum leaps in instrumental technologies through a continuing process of aggressive pioneership by the profession itself and related disciplines. In reflection, developments in capabilities have somehow managed to be in a kind of synchronous step with solving evolving problems that those enhanced capabilities have introduced to society. Chemical technology has greatly advanced our standard of living with concurrent threats to the health of society. In particular, further generations must be fully protected by a rational policy on chemical utilization. It is interesting to ponder the question: "Was it growing knowledge of modern toxicology that compelled a re-examination of public policy regarding human exposure to toxic substances, or was it increased public concern that forced science into greater participation?".

0097–6156/81/0160–0335$05.00/0

Experiments in which suspect carcinogens are administered to animals in massive quantities over relatively short periods of time have been challenged on scientific grounds and are confusing and often distrusted by laymen who myopically can only see the differences between the experimental process and real life. It is true to say that the "800 cans of diet soda" perception has been a serious impediment to the credibility of modern toxicology and the resultant regulatory processes. There can be no doubt that in the environment and in certain occupational situations, there are chemical agents which can increase the likelihood or threat of human cancer.

As an analytical chemist fortuitously transposed to oversee the scientific and administrative direction of the National Center for Toxicology Research (NCTR) for the last two and one half years I consider I have had more than ample time to soak up the essences of modern toxicology and break that barrier of presumptiveness as a chemist to discuss authoritatively the demanding role that chemistry contributes to modern toxicology by illustrating a few selected examples from experiences at NCTR and then advancing future trends and ideas of current research topics.

Nonclinical Laboratories Studies - Good Laboratory Practices

At the present time, three principle sources of evidence exist for the identification and removal of a chemical substance that might pose a carcinogenic threat to public health:
1. Epidemiologic evidence from exposed human populations;
2. Long-term chronic bioassays from animal studies;
3. Short-term or other tests that suggest carcinogenic activity.

Of these three options, a properly conducted long-term chronic bioassay has been accepted as the definitive model for estimating the carcinogenic risk for humans. Having mammalian tumor-induction as its end-point, the chronic bioassay is the only source of direct evidence (other than in humans) of chemically induced tumors in the mammalian species. Of all test systems it comes the closest to mimicking human routes of exposure and metabolic/pharmacological processes which activate and distribute chemicals.

In testing for carcinogenicity via such a chronic bioassay protocol, the implications on chemistry placed by the recent FDA regulations, "Good Laboratory Practice", can be summarized as follows
1. Identity, purity, chemical properties and stability of the test substance.
2. Handling and storage of the test substance.
3. Analysis of the bioassay supplies for essential and/or deleterious ingredients.

4. Homogeneity, stability and proper concentration of the test substance in the dosage form.
5. Safety surveillance of personnel and work areas.
6. Safe disposal of the chemical and contaminated experimental materials.

Obviously, the integrity of a long-term study is therefore highly dependent upon a number of the above factors. For instance, the compound 2-acetylaminofluorene (2-AAF) was a known model carcinogen selected for a 33-month study at NCTR involving 24,192 female BALC/c mice fed 30, 35, 45, 60, 75, 100 and 150 ppm plus a control (1). Initially, the first 10kg batch of 2-AAF acquired was 85-90% pure and hence was purified in house to the desired level. However, a later shipment of 2-AAF received assayed at 16.2% pure. Had this single fact gone undetected the entire investment in the experiment might have resulted in erroneous data being reported.

Test substances are usually administered to the animals in either diet or the drinking water. For very obvious cost-effective reasons (both manpower and choice of method of analysis), the drinking water is the preferred route if the test compound is both soluble and stable enough. A good example can be illustrated from the stability studies on 4-aminobiphenyl at pH7 and pH2 (Table I).

Table I. Stability of Aqueous Solutions of 4-Aminobiphenyl.HCl

Sampling intervals, days	Conc of 4-aminobiphenyl.HCl solns indicated[a]			
	1.0 ppm[b]	100 ppm[b]	1.0 ppm[c]	100 ppm[c]
0	0.989±0.003	98.9±0.35	1.01 ±0.001	93.8±0.00
1	0.973±0.012	99.2±0.42	0.781±0.001	79.3±0.02
2	0.968±0.005	97.9±0.90	0.649±0.019	73.5±0.17
4	0.976±0.021	98.6±1.0	0.459±0.020	60.4±2.3
8	0.950±0.001	98.3±0.31	0.365±0.023	62.4±0.66
16	0.936±0.002	98.9±0.50	0.282±0.011	57.4±1.1

[a] Mean and standard error from triplicate assays.
[b] Aqueous HCl solution (0.01N, pH2), samples adjusted for control.
[c] Deionized water solution, samples adjusted for control.

With recent emphasis on conducting chronic bioassays sometimes at low concentration ranges of the test compound, the question of toxicant and nutrient variability of commercial laboratory animal diets has been extensively examined at NCTR over the last five years (2,3). The animal diet must be considered an important source of variation since the relative proportions and/or source

THE PESTICIDE CHEMIST AND MODERN TOXICOLOGY

of different ingredients may well vary depending on the availability and cost of raw materials. Respecting this possible variability, commercial rodent feed has been analyzed at NCTR for the past five years and the results of 148 lots are displayed in Table II. As anticipated, the variability of nutrient concentrations was much less than that of the trace pesticides found or the heavy metals. Using this data base, the specification limits indicated in Table II were strictly adhered to and several three ton lots of feed had to be discarded. This type of survey has also provided the information to select the specifications based on what the market place could produce. In general, however, the annual average Cu and vitamin A concentrations were at least 12% lower than the approximate concentrations listed by the manufacturer whereas Ca, protein and vitamin B were within +5% and fat and Zn within +8% of the manufacturer's specifications. Frequently, Se was found at concentration levels at which it has been shown to interact with the process of chemical carcinogenesis. Occasionally DDT,

TABLE II. The Twenty Parameters Used in Animal Feed Surveillance*

Parameter	Specification Limitation		Mean	Std. Dev. (n=148)
	Min.	Max.		
Aflatoxin, ppb (B₁, B₂, G₁, G₂)	-	5		N.A.
Lindane, ppb	-	100	1.67	3.6
Heptachlor, ppb	-	20	1.07	2.2
Malathion, ppm	-	5	0.33	0.52
DDT (Total), ppb	-	100	27.72	48.4
PCB, ppb	-	50	8.7	15.1
Dieldrin, ppb	-	20	2.4	4.6
Cadmium, ppb	-	250	87.3	33.2
Arsenic, ppm	-	1.0	0.31	0.18
Lead, ppm	-	1.5	0.47	0.38
Mercury, ppb	-	200	0.024	0.02
Selenium, ppm	.05	0.65	0.34	0.15
Calcium, %	0.75	-	1.16	0.18
Copper, ppm	8	-	15.0	2.8
Zinc, ppm	75	-	108.2	9.7
Vitamin A, I.U./g	15	75	41.6	36.9
Vitamin B₁, mg/100g	7.5	12.5	9.1	1.25
Estrogenic activity, ppb	-	5	5	N.A.
Total Protein, g/100g	21.0	23.0	24.2	2.4
Total Fat, g/100g	4.3	6.7	5.54	0.58

*148 lots of Purina autoclavable Rodent Laboratory Chow 5010 analyzed prior to autoclaving.

dieldrin, Cd and Pb were present close to the concentration levels
known to have biological effects.   In this monitoring program,
animal supplies were examined prior to use to provide assurances
that acceptable levels of nutrients were present and to prohibit
the entrance of unacceptable levels of nutrients of contaminants
such as pesticides and heavy metals.

In addition to quality control of the diet, chemical
surveillance has always been employed to assure accurate dosages
of the test compounds in animal diets.   The principle requirements
to prepare a dosed animal diet include a series of sterilizing,
screening, blending and packaging operations within enclosed
safety cabinetry.   Autoclaving of dosed animal feed is normally
necessary to ensure microbiological integrity and the effect of
such autoclaving on both trace nutrients and contaminants must of
necessity also be closely monitored (Table III).   The data exhibit
the expected decrease in the concentration levels of vitamin A and
vitamin B.   The sharp reduction in malathion after autoclaving
reflects this particular pesticide's known thermal instability.

TABLE III.   Effects of Autoclaving on Selected Feed Components

Analyte	Units	No. of samples	Concentration (mean ± SD) Before Autoclaving	After Autoclaving
DDT (total)	ppb	24	7.1 ± 8.6	9.9 ± 16.5
Dieldrin	ppb	23	1.2 ± 1.6	2.0 ± 3
Lindane	ppb	24	2.3 ± 0.9	1.7 ± 1.0
Matathion	ppm	24	0.8 ± 0.9	0.1 ± 0.1
As	ppm	24	0.3 ± 0.2	0.6 ± 0.1
Cd	ppb	24	73 ± 43	108 ± 54
Ca	%	22	1.3 ± 0.2	1.3 ± 0.2
Cu	ppm	22	13.1 ± 2.5	14.2 ± 1.9
Se	ppm	24	0.4 ± 0.1	0.4 ± 0.1
Zn	ppm	22	110 ± 9	121 ± 9
Fat	%	20	5.3 ± 0.6	4.8 ± 0.7
Protein	%	22	24.3 ± 1.0	24.4 ± 0.8
Vitamin A	IU/g	24	44.2 ± 14.7	23.2 ± 15.3
Vitamin B1	ppm	24	89 ± 10	65 ± 16

During normal operations of conducting nonclinical studies
using mice, rats, monkeys, etc. a tremendous amount of
contaminated experimental materials is accumulated.   In
decontaminating animal cages large volumes of water are used and
the resultant waste water contains trace levels of all test

compounds. This burning issue was addressed at NCTR by devising
an adsorptive system to remove trace quantities of chemical
carcinogens and other test compounds (4). The success of the
pilot study has culminated in the construction and operation of a
waste water treatment plant at NCTR at a cost of 1.5 million
dollars to handle 100,000 gallons of waste water per day. A
schematic layout of the plant (Figure 1) illustrates the tandem
arrangement of filters, activated carbon and non-ionic polymeric
resin (XAD-2) to achieve a highly efficient and low-cost
operation. For the moment, treated samples are analyzed for all
carcinogens known to be present from experimental operations.
This method of monitoring is a costly procedure and attempts are
currently under way to develop a series of model marker compounds
(non-polar, semi-polar, and polar) to deliberately add to the
influx of the waste water from the facility and monitor only these
three to ensure the efficiency of the entire system.

Biochemical Mechanisms of Carcinogenesis

The lay public and many fellow scientists have long been
bitter critics of the currently accepted dose level studies of
in-vivo carcinogen testing. Extrapolation of test results of high
dose to low dose levels and to the genetically diverse human
population is an accepted regulatory posture (5). At this
juncture, it should be emphasized that the major advantage of
animal toxicity versus human epidemiology is that the toxicity can
be predicted before human exposure (e.g. asbestos). Attempts to
explore the shape of the dose response curve at one order of
magnitude lower than that previously performed were conducted at
NCTR, the so-called ED_{01} study to determine the dosage necessary
to produce a 1% tumor rate (1). The price tag of such extensive
explorations precludes their repeat with other chemicals and has
directly led to the dilemma of concurrent investment in basic
mechanism studies to seek out biochemical indicators of
carcinogenicity at extremely low doses rather than conventional
pathological indicators.

Carcinogenesis can be properly defined as a change in the
regulatory mechanism of a target cell which gives rise to a
progeny of altered cells constituting the basis of the neoplastic
disease. Therefore the initial molecular insult inflicted by a
specific carcinogen may be limited to only a few cells. Such
molecular events are the focal point for many inquiries into the
biochemical aspects of carcinogenesis. In very simple terms
certain compounds have the structural ability to become
electrophilic or electron deficient moieties via metabolic
activation, and then bind covalently to informational
macromolecules (DNA, RNA, proteins). These molecular events, for
example, result in residues or adducts to the base pairs of

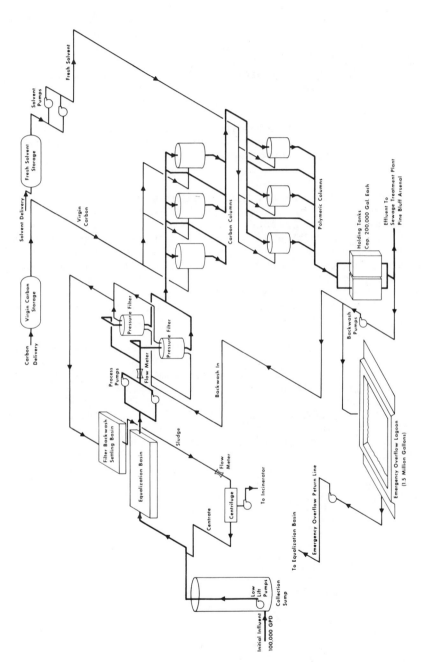

Figure 1.   Schematic of industrial waste water treatment facility at the National Center for Toxicological Research

DNA and sophisticated analytical techniques are often required to identity them (6). With the recent availability of high resolution NMR the exact site of interaction can be determined (Figure 2). Extensive studies at NCTR devoted to this scientific probe have resulted in a data base or compilation of the various sites of interaction on the bases of DNA (Figure 3).

It is optimistic to predict that this line of inquiry and the wealth of information on site attachment will replace the more coventional bioassay as a regulatory tool. However, it is realistic to assume that this line of attack on the biochemical mechanisms of carcinogenesis that has been initiated will yield up in several years some clues or guides to unravel the secrets of the basic mechanisms of carcinogenesis.

Future Trends in Chemistry and Toxicology

Health-oriented government agencies responsible for the protection of the public from possible adverse effects, such as chemical residues in the food supply, must somehow attempt to establish priorities on regulation as well as manpower to conduct monitoring programs. In 1975 in the United States alone, the pesticide industry used approximately 1400 active ingredients formulated by 4600 companies at 7200 plants to produce an estimated 35,000 - 50,000 separate products for an annual volume of 1.6 million pounds (approximately 45% of total world production) with a retail value of about three billion dollars. Staggering as these statistics of 1975 might sound regulatory agencies are faced with increasing problems of how best to serve and protect the public health. In an attempt to assist in this monumental task of providing maximum protection to the consumer while using the limited resources that are available, a risk assessment procedure has been constructed (7) as a possible technique to evaluate toxic materials that are potential candidates as residues in the food chain and to assign an index number that identifies a relative hazard. This procedure, developed to accomplish a possible ranking of the potential risks amongst the various chemical residues, is called the Surveillance Index (SI). The SI, which consists of three terms, can be expressed mathematically as follows:

$$\text{Surveillance Index (SI)} = TF + EF + BSF$$

where

    TF = Toxicity factor
    EF = Environmental factor
    BSF = BioSafety factor

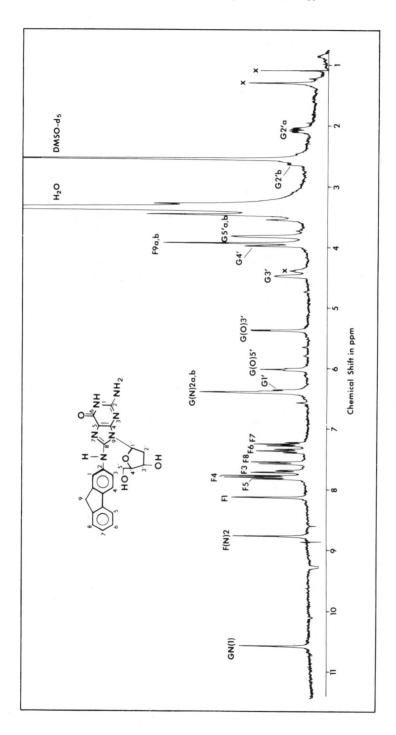

Figure 2   270-MHz proton NMR spectrum of N-(deoxyguanosin-8-yl)-2-aminofluorene

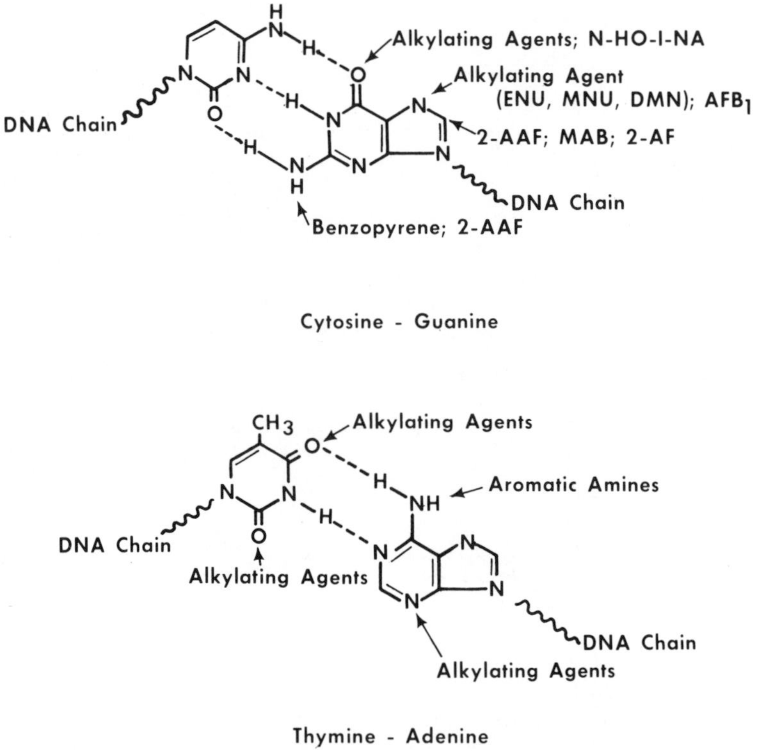

Cytosine - Guanine

Thymine - Adenine

Figure 3. Summary of the experimental investigations into the various sites of adduct formation on DNA by various known carcinogens

Toxicity Factor (TF) = KT x RTR

where
    KT = Kilotons of the compound released into the environment
annually
    RTR = Relative toxicity ratio = LD_{50} dieldrin/LD_{50} for the
compound (oral in rats)

Environmental factor (EF) = CV x $t_{1/2E}$
where
    CV = the sum of the crop values

Crop	Value
Cotton	1
Grains	3
Legumes	7
Vegetables	10
Fruits	10
Tobacco	5
Milk	20

$t_{1/2E}$ = Effective half-life = $(t_{1/2p} \times t_{1/2b})/(t_{1/2p} + t_{1/2b})$

where

$t_{1/2p}$ = physical half-life

$t_{1/2b}$ = biological half-life

BioSafety Factor (BSF) = PB x S x PAR/NOEL

where
    PB = Propensity to biomagnify
     S = Specificity (reactive sites in man)
    PAR = Population at risk
    NOEL = (Presumed) no observable effect level (ppm)

Applying this procedure to five selected environmental
pollutants (Table IV) has provided numerical values as potential
indicators of high risks. This equation is by no means set in
stone and work is continuing to refine and provide an exponential
term to encourage graphic displays for management purposes.

TABLE IV. Surveillance Indices for Selected Pollutants

Compound	1971	1978
p,p'-DDT	2,476	Banned
Toxaphene	1,019	1,025
Methyl parathion	270	310
Carbaryl	160	160
Aminotriazole	27	Banned

Currently, pesticide residues are monitored by a wide variety of chemical extraction and identification schemes. There is no single chemical multi-residue procedure available nor under development that can determine the entire spectrum of pesticide residues in a given sample. At present, even the most sophisticated procedures can only monitor at trace levels for several compounds in a few classes of pesticides. The problem is that assays performed by such procedures are labor intensive and sometimes employ expensive equipment and personnel. Therefore, what is sadly needed is a rapid, sensitive and relatively inexpensive multi-residue procedure to monitor for toxicants in the food chain. An investigation of bioassay systems employing four species of arthropods, Daphnia, Hyalella, Culex and Palaemonetes was initiated in response to the need for such an assay system (8). The evaluation of inherent toxicities related to types and amount of organic solvents commonly used in such systems indicated that dimethyl sulfoxide (DMSO) and methanol (MeOH) were least toxic in the aqueous test media. These solvents were then used in 18 hr. tests to determine sensitivities of the four organisms to a representative compound from six classes of pesticides. Stress factors such as amount of organic solvent and volume of test medium were adjusted to determine their effects on three of the organisms tested against dieldrin and parathion. The highest sensitivity obtained with dieldrin (50% mortality with 2 ng in a 25 ml test medium) was with Culex stressed with 2% of MeOH in a reduced test volume. Hyalella stressed with 2% of MeOH were most sensitive to parathion (50% mortality with 85 pg in a 100 ml test medium); further stress imposed by reducing the volume of test medium diminished sensitivity. These very preliminary experiments with various extracts of animal feed indicated that an extensive effort would be required to develop a method that could provide extracts compatible with the bioassay systems.

Conclusions

With the continuing increased knowledge and emphasis on modern toxicology the demands on the component disciplines such as chemistry must inevitably increase not as passive supporters but as aggressive partners demanding greater participatory roles in the design and research management areas of conceived experiments. Chemistry must assume its proper role in the hierarchy of modern toxicology and through application of its fundamental discipline contribute to major breakthroughs as well as continue to provide integrity of animal experiments.

Disclaimer

The views expressed are those of the author and do not necessarily reflect the policy of the U.S. Food and Drug Administration.

Literature Cited

1. Cairns, T.; J. Environmental Pathology & Toxicology, 1980, 3(3), 1.

2. Greenman, D.L.; Oller, W.L.; Littlefield, N.A.; and Nelson, C.J.; J. Environmental Pathology & Toxicology, 1980 6, 235.

3. Oller, W.L.; Gough, B.; and Littlefield, N.A.; J. Environmental Pathology & Toxicology, 1980, 3(3), 203.

4. Nony, C.R.; Treglown, E.J.; and Bowman, M.C.; Science of the Total Environment, 1975, 4, 155.

5. IRLG Report, J. Nat. Cancer Inst., 1979, 63, 245.

6. Kadlubar, F.F.; J. Nat. Cancer Inst., 1980, in press.

7. Oller, W.L., Cairns, T., Bowman, M.C.; and Fishbein, L.; Archives of Environm. Contamin. and Toxicology, 1980, 9, 483.

8. Bowman, M.C., Oller, W.L., Cairns, T., Gosnell, A.B., and Oliver, K.H.; Archives of Environm. Contamin. and Toxicology, 1981, January.

RECEIVED February 2, 1981.

Aspects of Analytical Toxicology Related to Analysis of Pesticidal Trace Contaminants: An Overview

LAWRENCE FISHBEIN

Department of Human and Health Services, Food and Drug Administration, National Center for Toxicological Research, Jefferson, AR 72079

We are all becoming increasingly aware of the potential adverse effects induced by trace levels of a spectrum of chemicals (primarily pesticides and industrial chemicals) in the environment. For example, the cause for public health concern over Mirex, Kepone, DBCP, HCB, PCBs, PBBs, nitrosamines and the widespread use of chemicals contaminated with polychlorinated dibenzo-p-dioxins and dibenzofurans are well documented. This of necessity has placed an increasing focus and pressure on both the analytical chemist and toxicologist. There is a primary need for the analytical chemist to develop and refine techniques relating to the detection, determination and confirmation of trace impurities (often at parts-per-billion or lower), in consumer products, in the workplace and in the environment. Toxicologists are increasingly confronted with an equally difficult array of problems relating to the elaboration of techniques and methodologies that will enable them to detect biological and toxicological events at what is increasingly recognized to be the major exposure problem, continuous low-level exposure at the sub parts-per-million or parts-per-billion level of trace impurites or trace levels of the toxicant per se.

In the forefront of chemicals of potential environmental and human toxicological concern are the pesticides both from the spectrum of agent and their use patterns as well as potential degree of population exposure. The latter includes those involved in the preparation, formulators, applicators, pickers, processors and finally the consumers. The major objectives of this overview are to highlight several of the newer advances in the analysis of trace impurities in and of pesticides per se.

Detection by the Thermal Energy Analyzer (TEA) and Electro-chemical Detection

It is recognized that other newer areas that deserve increasing recognition in pesticide and trace analysis include:

P^{31}-Fourier Transform NMRM, radioimmunoassay, pulse-polarography and low and room temperature phosphorescence analysis.

Eight distinct steps are recognized in trace organic analysis. These are: (a) collection, (b) storage, (c) extraction, (d) concentration, (e) isolation, (f) identification, (g) quantification, and (h) confirmation. The instrumental facilities for carrying out the three basic activities of analytical chemistry are separation, identification and measurement are shown in Table 1. We additionally recognize the fact that the power of analytical techniques can be increased by combining several analytical techniques, or what can be referred to as synergism between methods (1,2). Thus we can combine high discriminating power in one technique with a high separating power in the other. For example, gas chromatography's excellent quantitation and relatively poorer qualitation can be well matched to the good qualitation and relatively poorer quantitation of infra red or mass spectrometry. Table 2 illustrates the synergism and the strengths and weaknesses of analytical techniques which can be achieved between GLC, LC, TLC and MS and Fourier NMR. The various analytical systems can be ranked in the order of their usefulness for trace organic analysis. Mass spectrometry provides sufficient sensitivity for trace analysis and is easily interfaced to a gas chromatograph. It is generally acknowledged that combined GC/MS is currently the most powerful and useful technique for the identification of trace levels of organic compounds. It can provide qualitative information with nanogram quantities of single compounds present in the sample and in addition it provides a mass spectrum of each peak eluting from the GC. Hence, the GC/MS data can be plotted in the form of mass chromatograms as an additional interpretive aid (3). While gas chromatography is still the most widely utilized technique in trace organic analysis, it should be recognized that recent advances in HPLC have made HPLC comparable to GC in speed, convenience and efficiency (3-6). LC or HPLC with detectors such as MS, electrochemical, UV, and fluorescence is hence of increasing utility. Coupled to the various detectors, the minimal detectable quantities for LC are: UV, 10^{-9} g; electrochemical, 10^{-10} g; and fluorescence, 10^{-12} g. Sample sizes must be in the sub-ppm range (4). The UV detector is almost universal for organics while the electrochemical detector is selective and the fluorescence detector is even more selective (3,4). For example, with fluorescence spectroscopy it is possible to vary both the excitation wavelength and the wavelength at which the emission is observed thus providing additional spectrometric information (3).

Chemiluminescent Detectors (Thermal Energy Analyzers) in Nitrosamine Analysis

It is well recognized that humans may be exposed to N-nitroso compounds in a variety of ways, viz., (1) formation in the environment with subsequent absorption from air, water, food

TABLE 1

The Three Pillars of Analytical Chemistry (1)

Separation	Identification	Measurement
Instrumental separation by discriminating detection: Nmr (by chemical shift dispersion), Selective potentiometry, Ms (by single or multiple ion detection)	Physical Methods; Nmr, Ir/Raman, Ms and gc-ms, Uv, Comparison with properties of a standard	Dependent on physical properties: Fluorescence, Thermal analysis, Microscopy Sedimentation, UV absorbance, Atomic absorption, Nmr
Physical separation: Phase extraction, Chemical separation, Chromatography (lc,tlc,gc)	Chemical methods: Functional group analysis, Spot tests, Elemental analysis, Atomic absorption	Dependent on chemical properties: Polarography, Potentiometric titration Radiochemistry, Gc-ms

TABLE 2

Synergism of Analytical Techniques (1)

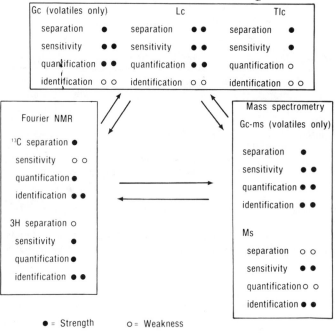

Gc (volatiles only)		Lc		Tlc	
separation	●	separation	● ●	separation	●
sensitivity	● ●	sensitivity	● ●	sensitivity	●
quantification	● ●	quantification	● ●	quantification	o
identification	o o	identification	o o	identification	o o

Fourier NMR

¹³C separation ●
sensitivity   o o
quantification ●
identification ● ●

3H separation o
sensitivity   ●
quantification ●
identification ● ●

Mass spectrometry
Gc-ms (volatiles only)

separation   ●
sensitivity   ● ●
quantification ● ●
identification ● ●

Ms

separation   o o
sensitivity   ● ●
quantification o o
identification ● ●

● = Strength   o = Weakness

and/or industrial and consumer products; (2) from the consumption or smoking of tobacco; (3) from naturally occurring compounds (considered to be an exceedingly minor contribution) and (4) formation in the human body from precursors ingested separately in food, water or air. The latter category is acknowledged by many to be of increasing concern but aspects of the potential risk are as yet to be unambiguously defined. The carcinogenicity, mutagenicity, and teratogenicity of a broad spectrum of nitrosamines has been increasingly and exceedingly well documented (7-11).

The occurrence of the nitrosamines, whether as direct emissions of N-nitroso compounds or via localized release of large amounts of precursor compounds (e.g., secondary amines, nitrogen oxides, nitrate, nitrites), effluent discharges from sewage treatment plants or runoff from feedlots or croplands treated with amine pesticides, ammonium fertilizers or nitrogenous organic materials, or accidental products in food processing and use, tobacco smoke, or via the body burden contributed by in vivo nitrosation reactions, has sparked ever increasing intensive investigations as to the overall scope of the potential sources, mechanism of in vitro and in vivo formation, body burdens as well as to the need to develop a proper scientific foundation for a human health risk assessment (7-14).

In order to best develop a proper scientific basis for the assessment of human risk associated with potential nitrosamine exposure, it is of course vital that we possess the requisite sensitive and selective analytical methodologies primarily for the detection and determination of exceedingly low levels (ppb-ppt) of nitrosamines, particularly in environmental samples.

A sensitive and selective chemiluminescent detector that has made an appreciable impact on the analysis of nitrosamines in environmental samples in the last several years is the thermal energy analyzer or (TEA) (15-19). This detector utilizes an initial pyrolysis reaction that cleaves nitrosamines at the N-NO bond to produce nitric oxide. Although earlier instrumentation involved the use of a catalytic pyrolysis chamber (15,17,19), in current instruments, pyrolysis takes place in a heated quartz tube without a catalyst (20). The nitric oxide is then detected by its chemiluminescent ion react with ozone. The sequence of reactions can be depicted in Figure 1. A schematic of the TEA is shown in Figure 2 (17). Samples are introduced into the pyrolysis chamber by direct injection or by interfacing the detector with a gas chromatograph (15,17,21,22) or a liquid chromatograph (22-25).

Chemiluminescence detectors possess considerable selectivity for nitrosamines because the light emitted from the NO-ozone reaction is in the near infrared region, whereas other known chemiluminescent reactions with ozone emit light in the visible or near UV region (17,20,26,27). An optical filter eliminates response to emissions occurring below 600 nanometers. Selectivity is additionally provided by a cold trap between the pyrolysis chamber and the NO-ozone reaction chamber which removes all but

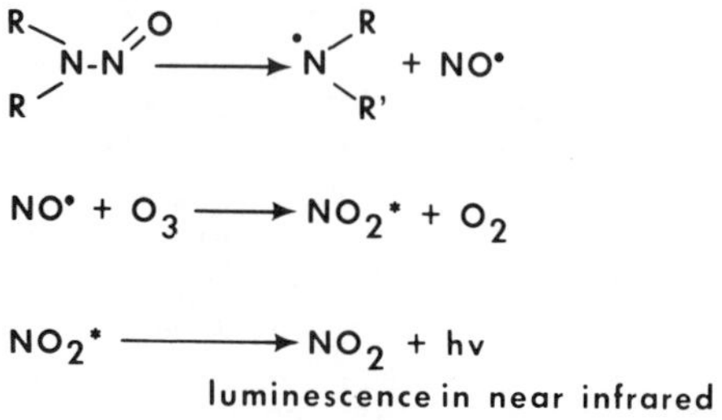

$$R\diagdown \quad \diagup O$$
$$\underset{R\diagup}{\overset{}{N\text{-}N}} \diagdown \quad \longrightarrow \quad \overset{\bullet}{N}\diagup\overset{R}{\diagdown}_{R'} \quad + \text{ NO}^{\bullet}$$

$$\text{NO}^{\bullet} + \text{O}_3 \longrightarrow \text{NO}_2^{\bullet} + \text{O}_2$$

$$\text{NO}_2^{\bullet} \longrightarrow \text{NO}_2 + h\nu$$
luminescence in near infrared

Figure 1.   Basis of chemuluminescent detection with a TEA

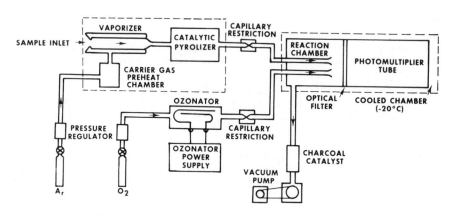

Analytical Chemistry

Figure 2.   Schematic of the TEA (17)

the most volatile compounds eluting from the pyrolysis chamber
(20). The TEA analyzer is sensitive to picogram quantities of
N-nitroso compounds (15,16,22-26) with a linear response extending
over five orders of magnitude.

While the chemiluminescence detectors have considerable
selectivity for nitrosamines it must also be recognized that the
possibility exists that any compound that can produce NO during
pyrolysis will produce a signal (20). For example, TEA responses
have been observed from organic nitrites, C-nitro and C-nitroso
compounds (17,28) and nitramines (29). In the routine analysis of
N-nitroso compounds, possible TEA analyzer responses to compounds
other than N-nitroso derivatives normally do not represent a
problem since the the identity of a compound can be readily
established by co-elution with known standards on GC-TEA and/or
HPLC-TEA systems (30-34). Additional confirmation could be
provided when the sample can be chromatographed on both GC-TEA and
HPLC-TEA (30,33). The technique accepted as the most reliable for
the confirmation of N-nitrosamines is based on mass spectrometry
(22,35,36). Low-resolution mass spectrometry is satisfactory for
the analysis of relatively simple mixtures and in those instances
in which extensive clean-up of samples has been performed.
However, complex samples require more sophisticated GC and MS
procedures (e.g., high resolution-MS).

Farrelli et al (37) described the determination of volatile
N-nitrosamines as pesticide contaminants utilizing gas chromato-
graph-mass fragmentography. Quantitation was accomplished by a
GC/MS (Finnigan Model 300) equipped with a programmed, multiple
ion detection system used in the E.I. mode. Trifluralin was found
to contain 34 ppm of dipropylnitrosamine by this technique.
Figure 3 shows a mass fragmentogram obtained by analyzing a
solution of trifluralin where a peak at m/e 130 can be observed
with the same retention time as dipropylnitrosamine (DPN). The
presence of DPN in the trifluralin sample was confirmed taking a
full mass spectrum of the contaminant (Figure 4).

Krull et al (30) recently described rapid and reliable
confirmatory methods for the thermal energy determination of
N-nitroso compounds at trace levels. These approaches utilize
minor modifications in the normal operation of the analyzer, GC
and HPLC interfaced with the analyzer, UV irradiation of the
sample and wet chemical procedures. Comparisons were made between
these analyzer associated methods of confirmation and other
approaches for the determination of N-nitroso compounds at trace
levels. Figure 5 illustrates the analysis scheme by Krull et al
(30) to distinguish N-NO compounds from C-NO, O-NO, N-NO$_2$, C-NO$_2$,
and O-NO$_2$ compounds utilizing the TEA analyzer.

There is recognized widespread concern about the possibility
of both false positive and false negative findings at low ppm to
low ppb concentration levels of the N-nitrosamines generally
reported. Such artifacts could arise during sample preparation,
extraction and/or subsequent chromatographic analysis (38). The

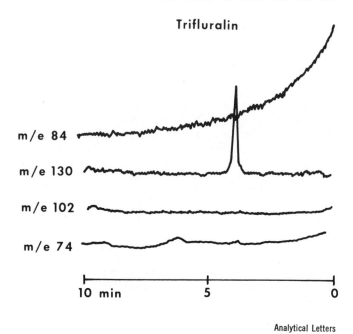

Analytical Letters

Figure 3.   Mass fragmentogram of trifluralin (37)

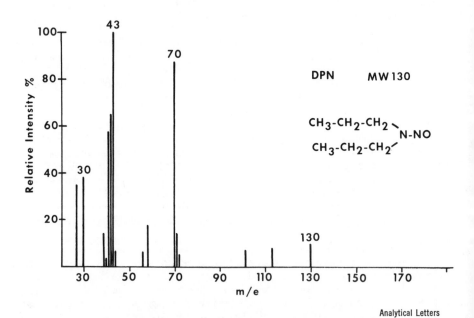

Analytical Letters

Figure 4.   MS of N-dipropylnitrosamine (37)

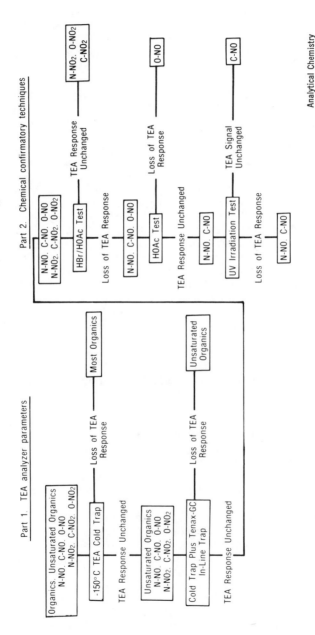

Part 1. TEA analyzer parameters

Part 2. Chemical confirmatory techniques

Figure 5. Analysis scheme to distinguish N-NO compounds from C-NO, O-NO, C-NO₂, and O-NO₂ compounds using the TEA (30)

Analytical Chemistry

source of nitrosating agent which could be responsible for a positive artifact, has included nitrite contamination of the sample itself, open column chromatography on nitrite contaminated packing materials for GC and LC columns, absorption of nitrogen oxides from ambient air, and nitrosamine contaminated deionized water and organic solvents. Precautions are also required to prevent the accidental destruction of N-nitroso compounds which can occur in sunlight and even under conventional fluorescent lightings (37). N-nitroso compounds can be destroyed during GC or HPLC. It is possible, as in the case of N-nitroso compounds with free OH groups such as N-nitrosodiethanolamine and N-nitrosamino acids that these compounds may give a sub-molar response by TEA detection.

While the utility of the thermal energy analyzer for the estimation of nitrosamines in air and water has been previously demonstrated by Fine and his co-workers (15-19,23-26), it is particularly relevant to consider its utilization in the determination of nitrosamines as trace impurities in pesticides as well as nitrosated pesticides. There are two major rather distinct problem areas that can lead to human exposure in this area and hence potential risk to consider. One area focuses on the concern that certain nitrogen-containing pesticides (e.g., carbamates, ureas, triazines, amides, anilides), as residues in soil, water, plants, etc., may be nitrosated by exogenous nitrite or by other nitrosating agents, e.g., nitrogen oxides from automobile, tractor or truck exhausts or other fuel consumption. The other area concerns the possibility that a variety of pesticides which are applied to soil and plants may contain nitroso compounds as impurities (39). These impurities may arise from the three most probable routes of N-nitroso contamination, e.g., (a) formation in the manufacturing process; (b) formation during storage and (c) contamination of amines used in the manufacturing process (39-47).

It was initially reported by Fan et al in 1976 that four of seven herbicides purchased in retail outlets had contained measurable concentrations of nitrosamines as detected with a thermal energy analyzer (43). Three of the herbicides consisted of polychlorobenzoic acids formulated as dimethylamine salts and contained dimethylnitrosamine as a contaminant in concentrations ranging from 0.3 to 640 ppm. It was postulated that nitrite used as a corrosion inhibitor in the metal containers reacted with dimethylamine during storage. The fourth herbicide is a formulation containing trifluralin (α, α, α-trifluoro-2,6-dinitro-N,N-dipropyl-p-toluidine) (Treflan   ), which is a dinitroaniline derivative rather than an amine salt. It was found to contain 154 ppm of dipropylnitrosamine and it was speculated that during the manufacturing process which employs sequential nitration and displacement of chloride by dipropylamine, nitrogen oxides or nitrous acid from nitric acid, can be carried along and react with excess dipropylamine used in the final step (39).

Bontoyan et al (40) examined over 90 technical and commercial pesticide formulations used in agriculture, hospitals and homes for the presence of N-nitroso compounds.   Tables 3-6 list the 18 triazines, 16 dinitroaniline and related derivatives, 28 amine salt and 12 miscellaneous herbicides (and other pesticides) respectively screened for nitrosamine contamination by GLC-TEA, LC-TEA, LC-UV, and GLC-Hall detection techniques.   Also examined were 4 alkyl amines used in manufacturing of the amine salt herbicides (Table 7).   Figure 6 illustrates typical instrument operating parameters for GLC and LC analysis for nitrosamines and Figure 7 depicts the procedure for screening and identifying non-volatile nitrosamines by LC-TEA analysis showing different solvent systems used.   Of the 91 pesticides and starting materials screened, 25 contained nitrosamines at or above 1 ppm.   Fourteen of these were the dinitroaniline formulations, seven were amine salts, three were amines used in the manufacturing process and one was the sample containing N-nitrosodiethanolamine.   The results indicate that the higher levels of N-nitrosamines are primarily found in substituted amine, dinitroaniline and amine salt formulations whereas the triazine compounds were free from nitrosamine contamination (40).   As might be expected, the nitrosamines in the amine salt formulations corresponded to the amine used therein, and those in the dinitroanilines corresponded to the dialkylamino group on the aromatic ring (40,41).   The finding of N-nitroso compounds in dinitroaniline products is suggested to result from a reaction of residual nitrous acid that is left from the nitration of the chlorobenzene and the excess secondary amine used in the amination step, while the N-nitroso compounds in formulations of amine salts of phenoxy herbicides probably results from the reaction of the corrosion inhibitor nitrite and the corresponding secondary amine used to form the amine salt (40).

Ross et al (42) employed TEA for detection of dimethylnitrosamine and dipropylnitrosamine in several herbicide formulations after separation by GC or HPLC.   With additional chromatographic cleanup, the identity of the compounds was confirmed by high resolution mass spectrometry.   These results further indicated that formulations of amine salts can form nitrosamines on storage and nitrosamines can be formed in preparations of nitroaniline based herbicides.   The results obtained for the determination of nitrosamines in seven technical herbicides are shown in Table 8.

The determination of volatile nitrosamines in crops and soils treated with dinitroaniline herbicides was reported by West and Day (47).   Measurement was accomplished by means of a gas chromatograph-thermal energy analyzer.   The sensitivity of the methods was 0.2, 0.05 and 0.01 ppb for dipropylnitrosamine in crops, soil and water respectively.

A recent paper by EPA reviewed the results of analysis of about 300 pesticides for nitrosamines.   A large number of amide, carbamate, organophosphate, triazine, urea derivatives and miscellaneous nitrogen-containing pesticides did not contain

Table 3   TRIAZINE HERBICIDES SCREENED FOR
NITROSOAMINE CONTAMINATION^a (4Q)

Sample	Ingredient
N-1	hexahydro-1,3,5-triethyl-S-triazine
N-2	4,6-dichloro-N-(2-chlorophenyl)-1,3,5-triazin-2-amine
N-3	metribuzin
N-4	hexahydro-1,3,5-tris(2-hydroxypropyl)-s-triazine
N-5	2-[(4-chloro-6-(ethylamino)-s-triazin-2-yl)amino]-2-methylpropionitrile
N-9	2-chloro-4-cyclopropylamino-6-isopropylamino-s-triazine
N-10	2-(tert-butylamino)-4-chloro-6-(ethylamino)-s-triazine
N-11	2-chloro-4,6-bis(ethylamino)-1,3,5-triazine
N-12	2-chloro-4,6-bis(isopropylamino)-1,3,5-triazine
N-13	2-(ethylthio)-4,6-bis-(isopropylamino)-1,3,5-triazine
N-15	2-methoxy-4,6-bis(isopropylamino)-1,3,5-triazine
N-16	2-methylthio-4,6-bis(isopropylamino)-1,3,5-triazine
N-17	2-chloro-4-ethylamino-6-isopropylamino-1,3,5-triazine
N-18	2-ethylamino-4-isopropylamino-6-methylthio-1,3,5-triazine
N-26	same as N-17
N-27	same as N-17
N-48	same as N-17
N-49	same as N-17
N-50	same as N-17
N-62	same as N-17
N-73	same as N-5

^{a}No nitrosamine was detected over 1 ppm.

Table 4   DINITROANILINE AND SIMILAR HERBICIDES SCREENED FOR NITROSAMINE CONTAMINATION[a] (40)

Sample	Ingredient	GLC-TEA,ppm	LC-TEA,ppm	LC-UV,ppm	GLC-Hall, ppm
N-20	2,6-dinitro-N,N-dipropyl-4-trifluoromethylaniline	121 DPNA[b]			150 DPNA
N-21	3,5-dinitro-N⁴,N⁴-dipropylsulfanilamide				
N-22	4-isopropyl-2,6-dinitro-N,N-dipropylaniline	54 DPNA			39 DPNA
N-32	2,6-dinitro-N,N-dipropyl-4-trifluoromethylaniline	11 DPNA			13 DPNA
N-37	4-(1,1-dimethylethyl)-N-(1-methylpropyl)-2,6-dinitrobenzenamine			74 nitros-amine of parent compd	
N-38	N-butyl-N-ethyl-2,6-dinitro-4-trifluoromethyl-aniline	38 BENA[c]			28 BENA
N-40	3,5-dinitro-N⁴,N⁴-dipropylsulfanilamide	sub 1, 1.5 DPNA; sub 2, neg	neg[f]		
N-51	N³,N³-diethyl-2,4-dinitro-6-trifluoromethyl-1,3-phenylenediamine	sub 1, 153 DENA[d]; sub 2, 100 DENA	146; 85		
N-63	3,5-dinitro-N⁴,N⁴-disopropylsulfanilamide				
N-64	N-butyl-N-ethyl-2,6-dinitro-4-trifluoro-methylaniline	8 BENA	3		
N-65	N-(1-ethylpropyl)-3,4-dimethyl-2,6-dinitro-benzenamine (technical)		102 nitros-amine of parent compd; neg	104 nitros-amine parent compd	
N-66	N-(1-ethylpropyl)-3,4-dimethyl-2,6-dinitro-benzenamine (anal. std)		present		
N-75	2,6-dinitro-N,N-dipropyl-4-trifluoromethyl-aniline	16 DPNA	present		
N-80	4-isopropyl-2,6-dinitro-N,N-dipropylaniline	9 DPNA			
N-87	2,6-dinitro-N,N-dipropyl-4-trifluoromethylaniline	6 DPNA			
N-91	N-(cyclopropylmethyl)-α,α,α-trifluoro-2,6-dinitro-N-propyl-p-toluidine	4 CMPNA[e]			

[a] A blank in the table means the sample was not analyzed by that method.
[b] N-Nitrosodi-n-propylnitrosamine.
[c] N-Nitroso-n-butylethylnitrosamine.
[d] Diethylnitrosamine
[e] N-Nitrosocyclopropylmethyl-n-propylnitrosamine.
[f] Less than 1 ppm.

Table 5 AMINE SALT HERBICIDES SCREENED FOR NITROSAMINE CONTAMINATION (40)

Sample	Ingredient	GLC-TEA. ppm	LC-TEA. ppm	LC-UV. ppm	GC-Hall ppm
N-6	dimethylamine salt of 2,4-D and 2,4,5-trichlorophenoxypropionic acid	neg a	c		
N-8	dimethylamine salt of 2,4-D	6 DMNA b			
N-23	dimethylamine salt of 2-methyl-4-chlorophenoxyacetic acid	neg	neg		
N-24	dimethylamine salt of 2-(2-methyl-4-chlorophenoxy)propionic acid	2 DMNA			
N-33	dimethylamine salt of 2,3,6-trichlorobenzoic acid	2 DMNA			
N-34	dimethylamine salt of 2,3,6-trichlorobenzoic acid	253 DMNA			
N-35	diethanolamine salt of 2,4-D	neg	neg		
N-36	dimethylamine salt of 2,4-D	neg	neg	neg	
N-42	diethanolamine salt of 6-hydroxy-3(2H)-pyridazinone	neg	neg	neg	
N-43	dimethylamine salt of 2,4-D	neg			
N-44	dimethylamine salt of 2-methyl-4-chlorophenoxyacetic acid	2.5 DMNA	6		
N-45	dimethylamine salt of 4-(2,4-dichlorophenoxy)butyric acid	neg	neg		
N-46	diethanolamine salt of 2-(2-methyl-4-chlorophenoxy)propionic acid		neg		
N-47	diethanolamine salt of 2,4-D		neg		
N-52	diethanolamine salt of 2,4-D	neg	neg		
N-54	dimethylamine salt of 2,3,6-trichlorophenylacetic acid	18 DMNA	19. 24		
N-55	morpholine salt of 2,4-D	neg			
N-56	dimethylamine salt of 2,4-D	neg			
N-57	diethanolamine salt of 2-methyl-4-chlorophenoxyacetic acid	neg	neg		
N-58	diethanolamine salt of 2,4-D		neg		
N-59	diethanolamine salt of 2-(2-methyl-4-chlorophenoxy)propionic acid	neg			
N-61	diethanolamine salt of 2,4-D				
N-71	diethanolamine salt of 6-hydroxy-3(2H)-pyridazinone				
N-72	dimethylamine salt of 2-(2,4,5-trichlorophenoxy)propionic acid	neg	neg		
N-78	dimethylamine salt of 2,4-D	1 DMNA			
N-79	dimethylamine salt of 2-(2-methyl-4-chlorophenoxy)propionic acid	neg			
N-82	dimethylamine salt of 2,3,6-trichlorobenzoic acid	neg			
N-83	diethanolamine salt of 3-trifluoromethylsulfonamido-p-acetotoluidide	neg			

a Less than 1 ppm. b Dimethylnitrosamine. c A blank indicates that the sample was not run by that method.

INSTRUMENT OPERATING PARAMETERS

	Column	T°C	Flow Rate
GLC-TEA	14 ft-1/8″ 10% Carbowax-20M & 0.05% KOH on Chromosorb WHP-80/100	175	30 ml/min
GLC-HALL DETECTOR	6 ft-1/4″ 3% Carbowax-20M on Chromosorb W-80/100	120	30 ml/min
HPLC-TEA HPLC-UV (254 nm)	2 - 3.9 mm ID x 30 cm μ Porasil connected in Tandem		1.5 ml/min

HPLC Solvent Systems

<u>HPLC-UV</u>

Volatile Nitrosamines	15% Isopropanol in Iso-Octane
Non-Volatile Nitrosamines in:	
Triazine Herbicides Prowl Butralin	3% Dimethoxyethane in Iso-Octane plus 0.02% 75/25 (Isopropanol-Water)
Diethanolamine Salts	50% Dimethoxyethane in Iso-Octane plus 0.02% 75/25 (Isopropanol Water)

<u>HPLC-TEA</u>

Volatile Nitrosamines Non-Volatile Nitrosamines	10% Acetone in Iso-Octane
Diethanolamine Salts	40/60 Acetone Iso-Octane

Journal of Agricultural and Food Chemistry

Figure 6.   Typical instrument operating parameters for GLC and LC analysis for nitrosamines (40)

Table 6 MISCELLANEOUS HERBICIDES SCREENED FOR NITROSAMINE
CONTAMINATION; OTHER PESTICIDES SCREENED FOR NITROSAMINE CONTAMINATION (40)

Sample	Ingredient	GLC-TEA, ppm	LC-TEA, ppm	LC-UV,ppm	GLC-Hall, ppm
	Miscellaneous Herbicides				
N-19	3,6-dichloro-o-anisic acid	neg[b]	[d]		
N-68[a]	diethanolamine salt of 2-sec-butyl-4,6-dinitrophenol		233 DELNA[c]	217 DELNA	
N-70	2-sec-butyl-4,6-dinitrophenol	neg			
N-81	3-(3,4-dichlorophenyl)-1,1-dimethylurea	neg			
N-86	3-(3,4-dichlorophenyl)-1,1-dimethylurea	neg			
N-88	2-ethoxy-2,3-dihydro-3,3-dimethyl-5-benzofuranyl methanesulfonate	neg			
N-89	same as N-88	neg			
	Other Pesticides				
N-41	bis(dimethylthio)carbamoyl)disulfide	neg			
N-60	diphenylamine	neg			
N-76	sodium [4-(dimethylamino)phenyl]diazene sulfonate	neg			
N-77	same as N-76	neg			
N-84	bis(dimethylthiocarbamoyl) disulfide	neg			
N-7	2,3,5-triiodobenzoic acid	neg			

[a]Presumed to be diethanolamine salt. [b]Less than 1 ppm. [c]Diethanolnitrosamine
[d]A blank indicates that the sample was not analyzed by that method.

Table 7 Alkyl Amines Used in Manufacturing Screened
for Nitrosamine Contamination (40)

Sample	Ingredient	GLC-TEA, ppm	LC-TEA, ppm	LC-UV, ppm	GLC-Hall, ppm
N-28	dimethyl-amine	34 DMNA	[b/]	26 DMNA	
N-31	dimethyl-amine	28 DMNA		29 DMNA	
N-67	triethanol-amine		neg[a/]		
N-69	diethanol-amine		neg		
N-85	dimethyl-amine	4 DMNA	6 DMNA		

[a/]Less than 1 ppm. [b/]A blank indicates that the sample was not analyzed by that method.

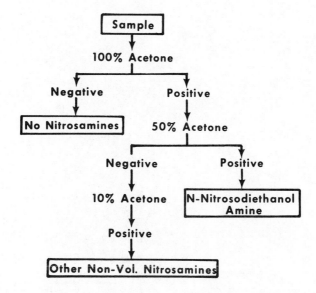

Journal of Agricultural and Food Chemistry

Figure 7. Procedure for screening and identifying nonvolatile nitrosamines on LC-TEA showing the different solvent systems used (40)

Table 8

DETERMINATION OF NITROSAMINES IN TECHNICAL HERBICIDES (42)

Sample	Herbicide formulation	EPA registration no	Compounds determined	mg/L	Determination procedure
1	2,4-dichlorophenoxyacetic acid 2-(2-Methyl-4-chlorophenoxy)-propionic acid as DMA salt 3,6-Dichloro-o-anisic acid as DMA salt	1386-569	NDMA	0.30	GC-TEA HPLC-TEA
2	2,4-Dichlorophenoxyacetic acid as DMA salt 3,6-Dichloro-o-anisic acid as DMA salt	539-226	NDMA	ND	GC-TEA
3	2,4,5-Trichlorophenoxopropionic acid as DMA salt 2,4-Dichlorophenoxyacetic acid as DMA salt	5887-92AA	NDMA	ND	GC-TEA
4	2,3,6-Trichlorobenzoic acid as DMA salt	352-250AA	NDMA	187	GC-TEA HPLC-TEA GC-MS
5	2,3,6-Trichlorobenzoic acid as DMA salt	352-250AA	NDMA	195	GC-TEA HPLC-TEA
6	2,3,6-Trichlorobenzoic as DMA salt	264-92AA	NDMA	640	GC-TEA HPLC-TEA
7	Forumation of α,α,α-trifluoro-2,6-dinitro-N,N-dipropyl-p-toluidine		NDPA	154	GC-TEA HPLC-TEA GC-MS

*ND = <0.05 mg/L

detectable levels (<1 ppm) of N-nitrosamines (46). Oryzalin (3,5-dinitro-N,N-dipropylsulfamilamide) did not contain detectable levels of di-n-propyl nitrosamine, while most dimethylamine salts of phenoxyalkanoic acids contain low or non-detectable levels of dimethylnitrosoamine. Although levels of dialkylnitrosamines in dinitroaniline herbicides were originally found to be high (e.g., up to 115 ppm), these levels have been decreased considerably due to process changes in manufacture. It was stated that "most N-nitroso contaminants in pesticides can be avoided by simple process changes and the elimination of nitrite salts in the formulation" (46).

The most important products thus far shown to contain nitrosamines are the dinitroaniline herbicides. The greatest focus has been on trifluralin, a pre-emergent soil incorporated herbicide that is widely used on cotton and soybeans as well as on several other field crops, fruits and vegetables to control broadleaf weeds and annual grasses. The scope of utility of trifluralin in the U.S. alone could be gleaned from the following. Currently 38% of the soybeans and over 60% of the cotton are grown in trifluralin-treated soil. It should be noted that the current levels of di-n-propyl nitrosamine in trifluralin are at least an order of magnitude lower than the 150 ppm originally discovered and it is anticipated that further decreases will result from further modifications of the production process (49).

It is important to know the possible extent of human exposure to pesticides containing trace amounts of nitrosamines or agricultural residues of nitrosamines. We are concerned with the potential hazard of exposure to applicators, field workers, as well to those using herbicides formulated for home use. It has also recently been suggested that trace nitrosamine residues may occur on crops for human consumption. Data relating to the above are meager. Studies by Fan and co-workers in 1976 (50,51) revealed no di-n-propylnitrosamine (or NDPA) in air samples (limit of detection 1 ng/m^3) above a California field before and immediately after spraying the field with a trifluralin formulation containing 154 ppm of the nitrosamine. Additionally, no NDPA was found in any irrigation water from the field (limit of detection 0.02 ppb) (52). Another study disclosed that cottonseed, soybeans and carrots grown in trifluralin-treated fields contained no nitrosamines (at sensitivities of 0.2-0.3 ppb) (53). Samples of water from ponds and wells located near fields with histories of heavy trifluralin usage did not contain nitrosamines (sensitivity 0.01 ppb). Of 24 samples of soil from a trifluralin-treated cotton field taken just before, and one week after, planting, only four were positive, containing 0.12 to 0.19 ppb NDPA.

In six separate studies comprising a total of 80 air samples over fields in three states before, during and after trifluralin application, NDPA was detected at levels of 0.005 to 0.007 ppb in five samples (54). No detectable nitrosamine residues (at a test sensitivity of 0.2 ppb) were observed in any crops treated with the herbicides trifluralin (Treflan), Balan [benefin; N-butyl-N-ethyl-2,6-dinitro-4-(trifluoromethyl)benzenamine] and Surflan (oryzalin, 3,5-dinitro-N,N-dipropylsulfanilamide) (Table 9) (55). Ross et al (44) reported similar findings on tomatoes harvested from a Treflan treated field. These results were consistent in view of the absence of detectable amounts of nitrosamines in the soils from which these crops were harvested. It is of interest to examine aspects of the stability of nitrosamines in soil as it may impact on subsequent crop uptake and human exposure. The half-life of dimethyl-, diethyl-, and di-n-propyl nitrosamines in aerobic soils is on the order of three weeks (56) with most of the initial losses being due to volatilization following surface application. Following soil incorporation, degradation to CO_2 predominated over volatilization (57). The most likely mechanism of dissipation of the nitrosamine contaminant is volatilization followed by vapor phase photolysis (58).

While laboratory studies indicate that low molecular weight nitrosamines including NDPA can volatilize rather rapidly after application to the surface of warm soil, incorporation into the soil of the nitrosamine co-applied with a dinitroaniline herbicide, decreases both the rate and extent of volatilization (49,59). However, in either case volatilization observed occurred within 3 or 4 days after application. No uptake of ^{14}C into the stems, leaves and beans was found when soybeans were grown in soils treated with 100 ppb of NDPA-^{14}C, or N-nitrosopendimethalin-[14]C [N-(1-ethyl propyl)-N-nitroso-3,4-dimethyl-2,6-dinitro-benzenamine (60). It should be noted that N-nitrosopendimethalin (a contaminant of the pesticide pendimethalin) (61) was relatively stable in soil and significant quantities could be recovered unchanged after several months.

Nitrosation of Pesticides in Soil, Water and Plants

The second focal point of concern as noted earlier is that certain pesticides as residues in soil, water, plants, may be nitrosated in situ. While it is readily recognized that nitrite is an intermediate in the soil nitrogen cycle and in the presence of an acid or other suitable catalyst would be a potential nitrosating agent in soil, however, under normal conditions, nitrite concentration is extremely low (49). Also, frequently cited is the possible circumstance that formation of nitrite from nitrate or ammonium fertilizers might promote the nitrosation of a pesticide (62). While with appropriate catalysts nitrosations can occur under neutral or even mildly alkaline conditions (63) there is sparse evidence of the reaction indeed occurring under these soil conditions (49).

It should be noted, however, that under model experiments, nitrosamines could be formed when high levels of amines and nitrite are added to soils (64). For example, dimethylnitrosamine was formed in soils such as spodosol (pH 3.8), a silty clay loam

Table 9. VOLATILE NITROSAMINE RESIDUES IN CROPS AND PLANTS FROM FIELDS TREATED WITH DINITROANILINE HERBICIDES (42)

Herbicide	Rate. kg/ha	No. of Applic.	Crop	Part	No. of Samples	Residue. ppb
Treflan	0.56-1.1	5-13	cotton	seed	5	NDR a/
Treflan	0.56-0.84	1	cotton	seedlings	10	NDR
Treflan	0.56-2.2	1-10	soybeans	seed	6	NDR
Treflan	0.56-1.1	2	carrots	roots	4	NDR
				tops	4	NDR
Treflan	1.1	1	cauliflower	fruit	1	NDR
				leaves	1	NDR
				alfalfa (volunteer)	1	NDR
Treflan	0.84	1	cotton	seed	6	NDR
Surflan	0.56-1.1	1	soybeans	leaves	3	NDR
Balan	1.7	2	lettuce	nuts	1	NDR
Balan	1.7	1	peanuts	shells	1	NDR

a/NDR   no detectable residue at a test sensitivity of 0.2 ppb (pH 5.8) and a silt loam (pH 6.5) when trimethylamine or dimethyl-amine were added to a soil to a concentration of 50 and 500 ppm as nitrogen. Dimethyl nitrosamine also was produced from the di- or trimethylamine in the spodosol soil in the absence of supplemental inorganic nitrogen. It was also found that the fungicide thiram could be converted to dimethylnitrosamine in the spodosol treated with nitrate or nitrite (64).

Pesticides N-nitrosated under in vitro and in vivo conditions in the laboratory have included the fungicide ziram (zinc dimethyldithiocarbamate), the insecticides carbaryl (1-naphthyl methyl carbamate) and propoxur (o-isopropoxy phenyl methyl carba-mate) and the herbicides benzthiazuron [N-(2-benzothiazolyl)-N'-methylurea], simazine [2-chloro-4,6-bis(ethylamino)-s-triazine] and atrazine [2-chloro-4-(ethylamino)-6-(isopropylamino)-s-triazine).

There is a paucity of information regarding the environmental formation or stability of N-nitrosopesticides. For example, the extensive use of atrazine in crop production programs utilizing heavy application of nitrogen fertilizers has raised the possibility of its N-nitrosation in soils (65).

Both atrazine and butralin [4-(1,1-dimethylethyl)-N-(1-methylpropyl)-2,6-dinitrobenzenamide] (66) are nitrosated per se (I and II) in soil but only in the presence of high levels of sodium nitrite (e.g., 100 ppm nitrogen as NO_2. However, no nitrosoatrazine or nitrosobutralin were observed when ammonium nitrate was substituted for sodium nitrite.

I
Nitrosoatrazine

II
Nitrosobutralin

While both nitrosoatrazine and nitrosobutralin formed rapidly in the above studies, nitrosoatrazine also disappeared rapidly (65) whereas traces of nitrosobutralin were still detectable in soil after 6 months (66). More definitive studies have been reported by Kearney and co-workers (65) on the distribution, movement, persistence and metabolism of N-nitrosoatrazine in soils and a model aquatic ecosystem. The results would suggest that the possibility of nitrosoatrazine formation is extremely remote in good agricultural soils (pH 5.0-7.0) receiving normal applications of atrazine (2 ppm) and even high rates of nitrogen fertilizer

(100 ppm nitrogen). In both soil and water, synthetic nitroso-atrazine is unstable and is degraded, usually by denitrogation to atrazine and polar products. High concentrations of NO_2 produce transient amounts of nitrosoatrazine in acid soil. It was concluded by Kearney et al (65) that based on past failures to detect nitrosoatrazine in a number of systems (67) and the noncarcinogenic response observed for the structurally related nitrosated herbicide nitrososimazine (68), nitrosoatrazine seems to pose no environmental threat.

It should be noted that although many pesticides (e.g., amides, ureas, carbamates) that are potentially nitrosatable, no nitrosamides appear to have been formed or detected in soils even in the laboratory. However, at this stage, it is not certain whether this reflects limited investigation, lack of formation or the general instability of nitrosamides (49). The most recent U. S. Environmental Protection Agency (EPA) assessment of nitros-amines in pesticides (April, 1980) stated that pesticides that contain nitrosamine levels higher than 1 ppm will be subject to a "full-scale risk assessment" under the EPA's rebuttable presump-tion against registration RPAR review process. Under an "interim policy" the agency will conduct spot checks of pesticides that, because of their chemical structure, are likely to contain nitros-amines. Those pesticides found to have nitrosamines at levels greater than 1 ppm will thus be subject to RPAR review (69). The proposed policy would (a) establish new data requirements for both existing registrants and future applicants whose products are contaminated with N-nitroso compounds; (b) propose risk criteria which will guide the agency in deciding whether to allow registration or to immediately review the compounds in the RPAR process or other regulatory action; (c) describe ways in which applicants can reduce the risks associated with human exposure to N-nitroso compounds in pesticides and (d) establish regulatory priorities and processing requirements.

Electrochemical Detectors

A major limitation in the increased application of modern liquid chromatography is the sensitivity of the detector system (70,71). For example, although a few UV detectors have been described which are usable at low nanogram levels, these involve special operating conditions or compounds with higher molar absorptivities (70,72,73). With ordinary UV absorbing species, quantitative LC is usually limited to a few tenths of micrograms (71).

Recent studies of Lores et al (74,75) described the utility of HPLC with electrochemical detection for the determination of halogenated anilines and related compounds. The halogenated anilines are an extremely important category of environmental toxicants which can enter the environment via a variety of sources including: (a) pesticides degradation and transformation, (b)

industrial discharge, (c) dyes and reduction of nitrosubstituted
aromatic compounds, and (d) combustion of plastics and urethane
products (74). One of the most common of these pathways is pesticide degradation and metabolism. Table 10 lists of the possible
origins of various anilines. Carbamates, ureas and anilides can
be metabolized or degraded in the environment to yield halogenated
anilines which in some cases are more toxic than the parent
compound. The polar nature of substituted anilines makes analysis
by GC difficult without prior derivatization of the compounds. It
should be noted that even after derivatization, it has been found
difficult to separate some isomers by GC (76). The procedure of
Lores et al (74) utilized an HPLC method that allows separation
and detection of sub-nanogram quantities of halogenated anilines
without derivatization. The separation of these compounds was
accomplished on a 15 cm Zorbax C-18 column, utilizing an inexpensive electrochemical detector. For the detection of quantities
ranging from 10 nanograms to several micrograms a UV detector (254
nm) was used. The separation of seven anilines listed in Table 12
was accomplished in this case by programming the mobile phase from
80% phosphate buffer/20% acetonitrile to 40% phosphate buffer/60%
acetonitrile at 10%/min (Figure 8). For HPLC chromatography of
anilines for electrochemical detection (which is extremely
sensitive to any changes in parameters) solvent programming was
not possible. Aniline, 2-amino-4-chlorophenol and o-chloroaniline
(compounds I, II, III, Table 10) were chromatographed with a
mobile phase of 80% 0.15 M phosphate buffer/20% acetonitrile and a
flow rate of 1 ml/min. For the separation of later eluting
anilines, e.g., o-bromoaniline; m-chloroaniline; p-chloroaniline
and 3,4-dichloroaniline (compounds IV-VII, Table 10), the mobile
phase was changed to 60% buffer/40% acetonitrile and the flow rate
was increased to 2 ml/min. The detector oxidation potential was
maintained at + 1.1 V throughout the experiment. Two different
solvent systems were required with the electrochemical detector
since no single solvent system would permit the elution of the
seven anilines in a reasonable period with sufficient separation.
Figures 9 and 10 illustrate chromatograms obtained from these two
solvent systems used with an electrochemical detector (operated at
an oxidative potential of + 1.1V with a CP-W graphite paste
electrode) (75). These chromatograms were obtained using the same
column that was used (a 15 cm Zorbax C-18 column) with solvent
programming and UV detection for the separation of seven anilines
as shown in Figure 8.

The percent acetonitrile and the maximum flow rate that can
be used are limited by the electrochemical detector. For example,
buffer solutions that contain more than 50% acetonitrile do not
provide enough electrolyte for proper performance of the detector.
Use of quaternary ammonium salts allows higher concentrations of
acetonitrile in the mobile phase. Flow rates greater than 2.5
ml/min can erode the surface of the carbon paste electrode. The
limits of detection for the seven anilines (based on a quantity

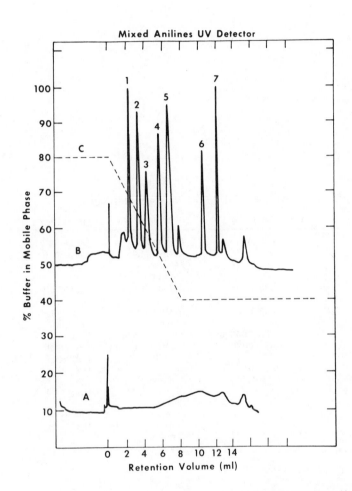

Figure 8.   Chromatograms obtained using a 15-cm Zorbax C-18 column with solvent programming and UV detection which show: (A) a blank profile run; (B) a chromatogram of a mixture of standards containing: (1) 40 ng aniline; (2) 70 ng 2-amino-5-chlorophenol; (3) 40 ng p-chloroaniline; (4) 75 ng p-bromoaniline; (5) 60 ng m-chloroaniline; (6) 90 ng 2-chloroaniline; (7) 180 ng 3,4-dichloroaniline; (C) the gradient profile showing the percent buffer in the mobile phase (75)

Table 10

POSSIBLE ORIGINS OF VARIOUS ANILINES (75)

#	Metabolite	Origin	Type Pesticide
I	aniline	Propham	Carbamate
		Carbetamide	Carbamate
		Fenuron	Urea
		Siduron	Urea
II	2-amino-4-chlorophenol	Barban	Carbamate
III	o-chloroaniline	Drazoxolone	-isoxalone
		Dyrene	Triazine
IV	o-bromoaniline	Metabromuron	Urea
V	m-chloroaniline	Chloropropham	Carbamate
		Barban	Carbamate
VI	p-chloroaniline	Monuron	Urea
		Monolinuron	Urea
		Urox	Urea
		Dimilin	Urea
VII	3,4-dichloroaniline	Propanil	Anilide
		Diuron	Urea
		Linuron	Urea

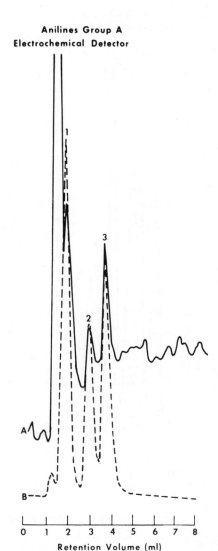

Anilines Group A
Electrochemical Detector

Retention Volume (ml)

Figure 9. Chromatograms of the anilines in Group A obtained using an electrochemical detector that show: A (1) .52 ng aniline; (2) .32 ng 2-amino-5-chlorophenol; (3) .64 ng p-chloroaniline; B (1) 65 ng aniline; (2) 40 ng 2-amino-5-chlorophenol; (3) 80 ng p-chloroaniline (75).

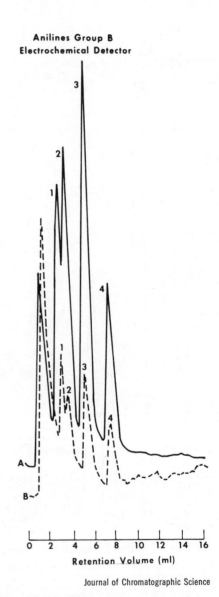

Figure 10. Chromatograms of the Group B anilines obtained using an electrochemical detector show: A (1) 15 ng p-bromoaniline; (2) 10.7 ng m-chloroaniline; (3) 18.7 ng o-chloroaniline; (4) 12.5 ng 3,4-dichloroaniline; B (1) 1.2 ng p-bromoaniline; (2) 0.86 ng m-chloroaniline; (3) 1.5 ng o-chloroaniline; (4) 1 ng 3,4-dichloroaniline (75)

which gives a signal/noise ratio ⪧2 using the conditions described above) were as follows (in ng): aniline, 0.23; 2-amino-4-chlorophenol, 0.28; p-chloroaniline, 0.33; p-bromoaniline, 0.38; m-chloroaniline, 0.27; o-chloroaniline, 0.28 and 3,4-dichloroaniline, 0.38.

The factors that can affect the sensitivity of the electrochemical detector should be noted. These include: applied voltage, flow rate, volume injected, and background current. Applied voltage, for example, can exert very dramatic effects on the sensitivity. The voltage can be used to "tune" certain compounds in or out. This can be very useful in cases when compounds cannot be separated and the compound of interest has a lower oxidation potential. For this purpose a scanning cyclic voltagram can be used for determining the oxidation potential. Figure 11 illustrates cyclic voltamograms of aniline, 2-amino-4-chlorophenol, p-bromoaniline and 3,4-dichloroaniline. The electrochemical detector appears to be linear over a wide range for the compounds tested. Figure 12 illustrates a graph of detector response versus quantity injected (constant volume 10 microliters) for aniline, o-chloroaniline, 2-amino-4-chlorophenol and p-bromoaniline.

The pH of the phosphate buffer as well as the percent acetonitrile in the mobile phase are the major factors affecting the separation of the anilines on the C-18 columns. By employing a low pH, the halogenated aniline compounds are separated not as the neutral free bases, but as their corresponding anilinium ions. Thus the use of a lower pH buffer will decrease the elution time for all the compounds (74).

The utility of HPLC with electrochemical detection for the determination of halogenated anilines in urine was also recently reported by Lores et al (75). The only previous existing method for the analysis of halogenated anilines in urine requires derivatization, silica gel cleanup and gas chromatography (76). HPLC eliminates the need for derivatization and makes the cleanup easier. The anilines studied were divided into two groups depending on which mobile phase was required. The mobile phase used for aniline, p-chloroaniline and p-bromoaniline was a mixture of 80% 0.1 M phosphate buffer adjusted to a pH of 3.0 and 20% acetonitrile. The mobile phase mixture used for m-chloroaniline, o-chloroaniline and 3,4-dichloroaniline was 60% 0.15 M phosphate buffer adjusted to a pH of 2.1 and 40% acetonitrile. The electrochemical detector was operated at an oxidative potential of +1.1V with a CP-W graphite paste electrode. The limits of detection for this method will depend on the sample size and the noise level of the detector. In this study with an injection volume of 70 ul and without concentration of the sample, levels below 5 ppb can be detected. Unsubstituted aniline could not be detected using either of the mobile phases since it was obscured by peaks eluting with the solvent front (75).

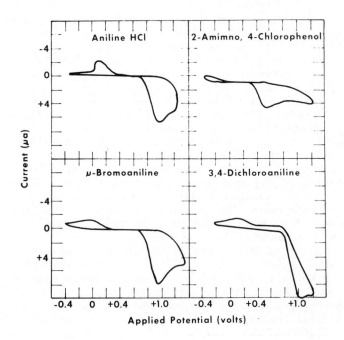

Journal of Chromatographic Science

Figure 11.   Cyclic voltamograms of some of the anilines. Sweep rate: 50 mV/s; graphite paste electrode, Ag/AgCl reference electrode. The solvent was 50/50 acetonitrile/phosphate buffer, and the concentration was ~ 0.1 mg/mL. Note that the oxidation potential of the aminochlorophenol is much lower than the anilines (75).

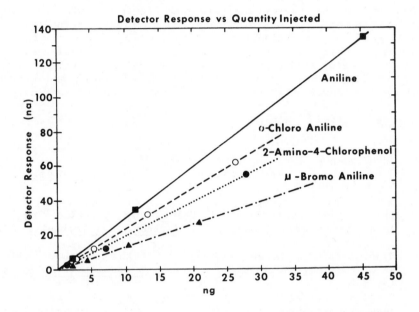

Journal of Chromatographic Science

Figure 12. Detector response vs. quantity injected (constant volume — 10 μL) for several anilines. The other anilines fall within the slopes of those indicated (75).

Summary

There is an increasing awareness of the potential environmental and health hazards of trace levels of pesticides, their
trace impurities as well as metabolites and/or degradation
products. Hence, there is a need to develop and refine requisite
analytical and toxicological methodologies for the detection and
estimation of these agents as well as their biological and
toxicological activity.
Although GC/MS techniques remain the basic hallmark of
analysis for pesticides and their derivatives, it is important to
stress the need and the specific role of additional methodologies
such as thermal energy analysis and electrochemical detectors.
The former technique was discussed with particular emphasis on its
utility of the determination of trace levels of nitrosamines and
nitrosated pesticides in agricultural products and residues.
The latter technique of electrochemical detection focused on
its utility (coupled with HPLC) for the determination of
halogenated anilines in environmental and biological samples.

Literature Cited

1. Selby, I. A., Analytical chemistry: The hub of pharmaceutical
   development, Chem. in Britain, 1978, 14 606-613
2. Hirschfeld, T., The hy-phen-ated methods, Anal. Chem., 1980,
   52, 297A-311A
3. May, W. E., Chesler, S. N., Wise, S. A., and Hertz, H. S.,
   Trace organic analysis, Anal. Chem., 1978, 50, 428A-436A
4. Anon, Trace organic analysis, Environ. Sci. Technol., 1978,
   12, 757-759
5. Majors, R. E., Recent advances in high performance liquid
   chromatography packings and columns, J. Chromatog. Sci., 1977,
   15, 334
6. Wise, S. A., and May, W. E., Res. Develop., 1977, 28, 54
7. IARC, Monographs on the Evaluation of the Carcinogenic Risk of
   Chemicals to Humans, Vol. 17, Some N-Nitroso Compounds,
   International Agency for Research on Cancer, Lyon, 1978
8. Magee, P. N., Montesano, R., and Preussmann, R., N-Nitroso
   Compounds and Related Compounds, In: "Chemical Carcinogens",
   (ed), Searle, E., ACS Monograph No. 173, Washington, DC, 1976;
   p. 491-625
9. Montesano, R., and Bartsch, H., Mutagenic and carcinogenic
   N-nitroso compounds. Possible environmental hazards, Mutation
   Res., 1976, 32, 179
10. Druckrey, H., Preussmann, R., Ivankovic, S., and Schmahl, D.,
    Organotrope carcinogene wirkungen bei 65 verschiedenen N-ni-
    trosoverbindungen an BD-ratten, Z. Krebsforsch., 1967, 69,
    103-201

11. WHO, Environmental Health Criteria No. 5, Nitrates, Nitrites and N-Nitroso Compounds, World Health Organization, Geneva, 1977; p. 59-61
12. Lijinsky, W., Health problems associated with nitrites and nitrosamines, Ambio, 1976, 5, 67-72
13. Lijinsky, W., and Epstein, S. S., Nitrosamines as environmental carcinogens, Nature, 1970, 225, 21-23
14. Fishbein, L., Overview of some aspects of occurrence, formation and analysis of nitrosamines, Sci. Total Environ., 1979, 13, 157-188
15. Fine, D. H., Rufeh, F., and Gunther, B., Group specific procedure for the analysis of both volatile and non-volatile N-nitroso compounds in picogram amounts, Anal. Lett., 1973, 6, 731
16. Fine, D. H., Rufeh, F., and Lieb, D., Group analysis of volatile and non-volatile N-nitroso compounds, Nature, 1974, 247, 309
17. Fine, D. H., Rufeh, F., Lieb, D., Rounbehler, D. P., Description of the thermal energy analyzer (TEA) for trace determination of volatile and non-volatile N-nitroso compounds, Anal. Chem., 1975, 47, 1188-1191
18. Fine, D. H., Rounbehler, D. P., and Oettinger, P. E., A rapid method for the determination of sub-part per billion amounts of N-nitroso compounds in foodstuffs, Anal. Chim. Acta., 1975, 78
19. Fine, D. H., Lieb, D., and Rufeh, F., Principle of operation of the thermal energy analyzer for the trace analysis of volatile and non-volatile N-nitroso compounds, J. Chromatog., 1975, 107
20. Hansen, T. J., Archer, M. C., Tannenbaum, S. R., Characterization of pyrolysis conditions and interference by other compounds in the chemiluminescence detection of nitrosamines, Anal. Chem., 1979, 51, 1526-1528
21. Fine, D. H., and Rounbehler, D. P., Trace analysis of volatile N-nitroso by combined gas chromatography and thermal energy analysis, J. Chromatog., 1975, 109, 271-279
22. Preussmann, R., Walker, E. A., Wasserman, A. E., and Castegnaro, M. (eds), Environmental Carcinogens-Selected Methods of Analysis, Vol. 1 "Nitrosamines", IARC Scientific Publications, No. 18, Lyon, 1978
23. Oettinger, P. E., Huffman, F., Fine, D. H., and Lieb, D., Liquid chromatograph detector for trace analysis of non-volatile N-nitroso compounds, Anal. Lett., 1975, 8, 411-414
24. Fine, D. H., Rounbehler, D. P., Rounbehler, A., Silvergeld, A., Sawicki, E., Krost, K. and DeMarrais, G. A., Determination of dimethylnitrosamine in air, water, and soil by thermal energy analysis: Measurements in Baltimore, MD, Environ. Sci. Technol., 1977, 11, 581-584

25. Fine, D. H., Rounbehler, D. P., Sawicki, E., and Krost, K., Determination of dimethylnitrosamine in air and water by thermal energy analysis: Validation of analytical procedures, Environ. Sci. Technol., 1977, 11, 577-580
26. Fine, D. H., "An Assessment of Human Exposure to N-Nitroso Compounds" In IARC Scientific Publication No. 19 (eds) Walker, E. A., Castegnaro, M., Gricuite, L. and Lyle, R. E., International Agency for Research on Cancer, Lyon, 1978
27. Clough, P. N., and Thrush, K., Mechanism of chemiluminescent reaction between nitric acid and ozone, Trans. Faraday Soc., 1967, 63, 915
28. Fan, T. Y., Vita, R. and Fine, D. H., C-nitroso compounds: A new class of nitrosating agents, Toxicol. Lett., 1978, 2, 5-10
29. Hotchkiss, J. H., Barbour, J. F., Libbey, L. M., and Scanlan, R. A., Nitramines as thermal energy analyzer positive non-nitroso compounds found in certain herbicides, J. Agr. Food Chem., 1978, 26, 884
30. Krull, I. S., Goff, E. U., Hoffman, G. G., and Fine, D. H., Confirmatory methods for the thermal energy determination of N-nitroso compounds at trace levels, Anal. Chem., 1979, 51, 1706-1709
31. Castegnaro, M., and Walker, E. A., In "Environmental Aspects of N-Nitroso Compounds" (eds) Walker, E. A., Castegnaro, M., Gricuite, L. and Lyle, R. E., IARC Scientific Publication No. 19, International Agency for Research on Cancer, Lyon, 1978; p. 53
32. Havery, D. C., Fazio, T., and Howard, J. W., In "Environmental Aspects of N-Nitroso Compounds" (eds) Walker, E. A., Castegnaro, M., Gricuite, L. and Lyle, R. E., IARC Scientific Publication No. 19, International Agency for Research on Cancer, Lyon, 1978; p. 41-52
33. Krull, I. S., and Fine, D. H., In "Handbook of Carcinogens and Other Hazardous Substances, Chemical Trace Analysis", Chapter 6 (ed) Bowman, M. C., Marcel Dekker, New York, 1979
34. Krull, I. S., Goff, U., and Wolf, M. H., In "Canadian Chromatography Conference-I" (ed) Bhatnager, V. M., Marcel Dekker, New York, 1978
35. Gough, T. A., Determination of N-nitroso compounds by mass spectrometry, Analyst, 1978, 103, 785-805
36. IARC, Monographs on the Evaluation of the Carcinogenic Risk of Chemicals to Humans, Vol. 17, Some N-Nitroso Compounds, International Agency for Research on Cancer, Lyon, 1978
37. Fanelli, R., Chiabrando, C. and Airoldi, L., Determination of volatile N-nitrosamines as pesticide contaminants by gas chromatography-mass fragmentography, Anal. Letters, 1978, A11, 845-854
38. Krull, I. S., Fan, T. Y., and Fine, D. H., Problem of artifacts in the analysis of N-nitroso compounds, Anal. Chem., 1978, 50, 698-701

39. Fishbein, L., Nitrosamines in Pesticides and Agricultural Residues, Proceedings of Toxicology Forum, Washington, DC, Feb. 28–March 1, 1980
40. Bontoyan, W. R., Law, M. L., and Wright, D. P., Jr., Nitrosamines in agricultural and home-use pesticides, J. Agr. Food Chem., 1979, 27, 631–635
41. Oliver, J. E., Nitrosamines from pesticides, Chem. Tech., 1979, 9, 366–371
42. Ross, R. D., Morrison, J., Rounbehler, D. P., Fan, S., and Fine, D. H., N-Nitroso compound impurities in herbicide formulations, J. Agr. Food Chem., 1977, 25, 1416–1418
43. Fan, S., Fine, D., Ross, R., Rounbehler, D. R., Silvergleid, A., Song, L., American Chemical Society 172nd National Meeting, San Francisco, CA, Sept. 2, 1976
44. Day, E. W., Jr., West, S. D., Koenig, D. K., and Powers, F. L., Determination of volatile nitrosamine contaminants in formulated and technical products of dinitroaniline herbicides, J. Agr. Food Chem., 1979, 27, 1081–1095
45. Anon, Draft pesticide nitrosamine policy sets maximum limits of exposure and risk, Pesticide & Toxic Chem. News, August 8, 1979, 7, 24–27
46. Anon, Pesticide nitrosamine levels. Ease of removal noted in EPA-er's paper, Pesticide Toxic Chem. News, October 24, 1979, 7, 18
47. Ridgeway, R. L., Pesticide use in agriculture, Environ. Hlth. Persp., 1978, 27, 103
48. West, S. D. and Day, E. W., Jr., Determination of volatile nitrosamines in crops and soils treated with dinitroaniline herbicides, J. Agr. Food Chem., 1979, 27, 1075–1080
49. Oliver, J. E., Nitrosamines from pesticides, Chem. Tech., 1979, 9, 366–371
50. Ross, R. D., Morrison, J., Rounbehler, D. P., Fan, S., and Fine, D. H., N-Nitroso compound impurities in herbicide formulations, J. Agr. Food Chem., 1977, 25, 1416–1418
51. Fan, S., Fine, D., Ross, R., Rounbehler, D. R., Silvergleid, A., Song, L., American Chemical Society 172nd National Meeting, San Francisco, CA, Sept. 2, 1976
52. Anon, EPA asked to suspend 4 herbicides with "intolerable" nitrosamine levels, Pesticide & Toxic Chem. News, Sept. 8, 1976, 4, 22–23
53. Zaldivar, R., Nitrate fertilizers as environmental pollutants: Positive correlation between nitrates ($NaNO_3$ and KNO_3) used per unit area and stomach cancer mortality rates, Experientia, 1977, 33, 264–265
54. Statement of E. F. Alder at the Environmental Protection Agency Hearing on Pesticide Products Containing Nitrosamines, Washington, DC, March 8, 1977
55. West, S. D., and Day, E. W., Jr., Determination of volatile nitrosamines in crops and soils treated with dinitroaniline herbicides, J. Agr. Food Chem., 1979, 27, 1075–1080

56. Ross, R., Morrison, J., and Fine, D. H., Assessment of dipro-
    pylnitrosamine levels in a tomato field following application
    of Treflan EC, J. Agr. Food Chem., 1978, 26, 455-457
57. Oliver, J. E., Kearney, P. C., and Kontson, A., Degradation of
    herbiciderelated nitrosamines in aerobic soils, J. Agr. Food
    Che., 1979, 27, 887-891
58. Hanst, P. L., Spence, J. W., and Miller, M., Atmospheric chem-
    istry of N-nitrosodimethylamine, Environ. Sci. Technol., 1977,
    11, 403-405
59. Oliver, J. E., American Chemical Society National Meeting,
    Anaheim, CA, March 17, 1978
60. Kearney, P. C., Oliver, J. E., Konston, A., Fiddler, W.,
    Pensabene, J. W., American Chemical Society, 175th National
    Meeting, Anaheim, CA, March 17 1978
61. Bontoyan, W. R., Law, M. L., and Wright, D. P., Jr., Nitro-
    samines in agricultural and home-use pesticides, J. Agr. Food
    Chem., 1979, 27, 631-635
62. Take, R. L., and Alexander, M., Stability of nitrosamines in
    samples of lake water, soil, and sewage, J. Natl. Cancer
    Inst., 1975, 54, 327-330
63. Challis, B. G., Edwards, A., Hunma, R. R., Kyrtopoulos, S. A.,
    and Outram, J. R., In: "Environmental aspects of N-nitroso
    compounds", (eds) Walker, E. A., Castegnaro, M., Gricuite, L.,
    and Lyle, R. E., IARC Scientific Publications No. 19, Lyon,
    1978; p. 127-142
64. Ayanaba, A., Verstraete, W., and Alexander, M., Formation of
    dimethylnitrosamine, a carcinogen and mutagen, in soils
    treated with nitrogen compounds, Soil Sci. Soc. Am. Proc.,
    1973, 37, 565-568
65. Kearney, P. C., Oliver, J. E., Helling, C. S., Isensee, A. R.,
    and Kontson, A., Distribution, movement, persistence and meta-
    bolism of N-nitrosoatrazine in soils and a model aquatic eco-
    system, J. Agr. Food Chem., 1977, 25, 1177-1181
66. Oliver, J. E., and Konston, A., Formation and persistence of
    N-nitrosobutralin in soil, Bull. Environ. Contam. Toxicol.,
    1978, 20, 170-173
67. Marco, G. J., Bora, G., Cassidy, J. E., Ryskiewich, D. P.,
    Simoneaux, B. J. and Sumner, D. D., 172nd National Meeting of
    American Chemical Society, Pest. Section 100, San Francisco,
    CA, August, 1976
68. Eisenbrand, G., Deutsches Krebsforsch ungs Zentrum, Heidel-
    berg, Federal Republic of Germany, personal communication
    (1977) quoted in reference 101
69. EPA, EPA Proposed Policy for Handling Registrations of Pesti-
    cides Contaminated with N-Nitroso Compounds, U. S. Environmen-
    tal Protection Agency, Federal Register, June 25, 1980, 45,
    42854-42858
70. Zweig, G., and Sherma, J., Chromatography, Anal. Chem., 1972,
    44, 42R

71. Kissinger, P. T., Refshauge, C., Dreiling, R., and Adams, R. N., An electrochemical detector for liquid chromatography with picogram sensitivity, Anal. Letters, 1973, 6, 465–477
72. Brooker, G., Effect of temperature control on the stability and sensitivity of a high pressure liquid chromatography ultraviolet flow cell detector, Anal. Chem., 1971, 43, 1095
73. Porcaro, P. J., and Shubiak, P., Detection of nanogram quantities of hexachlorophene by ultraviolet chromatography, Anal. Chem., 1972, 44, 1865
74. Lores, E. M., Bristol, D. W., and Moseman, Determination of halogenaetd anilines and related compounds by HPLC with electrochemical and UV detection, J. Chromatog. Sci., 1978, 16, 358–362
75. Lores, E. M., Meekins, F. C., and Moseman, R. F., Determination of halogenated anilines in urine by high-performance liquid chromatography with an electrochemical detector, J. Chromatog., 1980, 188, 412–416
76. Bradway, D. E., and Shafik, T., Electron capture gas chromatographic analysis of the amine metabolites of pesticides, J. Chromatog. Sci., 1977, 15, 322–328

RECEIVED February 2, 1981.

Statistical Considerations in the Evaluation of Toxicological Samples

JAMES J. TIEDE

Bristol Laboratories, Syracuse, NY 13201

"You can prove anything with statistics." Few people, if any, have not heard this statement. Statistics is the science which deals with the collection, evaluation, interpretation and presentation of experimental data. As a science, it is governed by a set of fundamental underlying assumptions which, if violated, can invalidate the results of a statistical analysis. Thus, while the opening statement of this paper is not correct, a relatively accurate statement can be made after a slight modification. "With an improper analysis, you can prove anything with statistics." This is certainly true. If one ignores the assumptions underlying a statistical method of analysis or employs an improper method of analysis for the experimental design, any desired conclusion can be obtained. However, is it not true of any scientific discipline that if the fundamental rules are violated, questionable results can be obtained? The same is true of statistics. An appropriate statistical treatment of experimental data, one which will withstand critical peer review will, in general, lead to unequivocal results.

The objective of a statistical evaluation of experimental data is to provide results which are meaningful to the experimentor. The most rigorous analysis may have less value than a simple graph if it does not aid the experimentor in the interpretation of his data. On the other hand, a simple analysis may prove to be misleading if unjustified assumptions about the data are made. Even the most appropriate analysis will not guarantee that the desired conclusion will be obtained.

"It's obvious that there is a difference in this data. Why don't the statistics prove it?" This is a comment which has been heard by all consulting statisticians. There are numerous factors influencing a statistical evaluation. Among these are:

0097–6156/81/0160–0387$05.75/0

inadequate sample size, biased sample, improper design and uncontrolled exogenous variables. If one or more of these are not addressed appropriately, the results of the statistical evaluation may not necessarily agree with the biological or physical interpretation of the data. Insufficient sample size is a particularly important factor. When there is a great deal of variability in a set of data (i.e. the signal to noise ratio is small), larger sample sizes are required if the experiment is to have a reasonable degree of sensitivity (the ability to detect differences among groups or between a sample and a reference standard) associated with it. The problem here is that samples are often expensive, both in terms of dollars and time. All of these factors (sensitivity, sample size and cost) must be considered before an experiment is conducted. This requires, however, that the statistician be consulted before the study is even designed and not just after the data is collected. Constant interplay (interface) between the statistician and the experimentor, from the beginning of a project until the end, will optimize the amount and quality of information which can be obtained.

"How can these data show significance? Its obvious that there are no real differences here." This is another frequent plea made to statisticians. This touches on a difficult problem; statistical significance versus clinical, biological or physical significance. Often, ones intuition or experience will suggest that data which show statistical significance may not be biologically significant. As stated previously, the causes of such differences (contradictions) are numerous; improper design, uncontrolled variables in the experiment, a sample which is not representative of the population at large are but a few. It is the existence of this statistical-biological contradiction which underscores the need for constant interaction between experimentor and statistician. With the continued interplay between statistician and biologist, the potential for contradiction can be minimized.

Statistics is a tool for scientists just as the brush is a tool for the painter. When used properly in the hands of an artist, the paintbrush can help transform a blank piece of canvas into a masterpiece. When improperly used, it can destroy that same piece of canvas. The same is true of statistics. When properly employed by a professional, the "picture" which the data conveys can be extracted. When improperly applied, questionable results can be expected.

The purpose of this paper is to present and discuss some of the more commonly used statistical methods. The emphasis will be on understanding the concepts behind the methods and on interpretation of the results of the analyses. Since this is to be an overview of the methods presented, detailed discussion will not be possible. Relevant references will be included for further details on the concepts presented.

Statistics Based on a Sample

A statistical population is the total collection of all possible values of the attribute with which one is concerned. For example, the blood pressure of American adult males or the body weight of Fisher rats are statistical populations. Generally, the population is so large, it is impossible to directly access the effect that a chemical agent (food additive, drug, etc) would have on the population. To administer the chemical agent to each member of the population and determine the effect, if any, would be impossible. As an alternative, the effect of the chemical agent on a small portion or sample from the population can be evaluated and from this evaluation, inferences can be made about the population at large. If the sample is characteristic of the population, it should provide good insight into the nature of the population. One of the functions of statistics is to make objective inferences about the population response based on the data obtained from the sample.

Confidence Intervals. In this section, some of the fundamental statistical concepts which relate to data collected from a sample will be discussed. Although only the one sample problem will be discussed in detail, many of the concepts can be extended to cases where there are more than one sample.

Consider the following (hypothetical) body weight data:

Change in body weights(g)

15.5	17.8	21.9	20.6	13.9
17.2	21.3	29.4	13.3	19.2
24.6	14.1	15.9	7.7	18.3
18.8	8.6	22.0	24.1	17.5

It is assumed that these data represent a random sample from the population, that is, a sample which was collected in such a manner that each member of the population had an equal chance of being chosen. A question which immediately arises is what inferences, based on the sample data, can be made about the mean and standard deviation of the population. The most common method of estimating the population mean is to use the average of the sample date, i.e. the sample mean, $\bar{x}$. The standard deviation is most commonly estimated by the sample standard deviation $s = \left[\sum_{i=1}^{n} (x_i - \bar{x})^2 / (n-1) \right]^{1/2}$ where x_i refers to the individual observations and n is the number of observations in the sample. Assuming that the data come from a normal (Gaussian or bell-shaped) distribution, the sample mean ($\bar{x}$) and the sample standard deviation(s), in addition to the obvious advantages in terms of familiarity and interpretability, have optimal statistical properties. Both of these statistics ($\bar{x}$ and s) are called point estimates because they provide a single number estimate of the population parameter of interest. For the sample data, $\bar{x}=18.1$ and $s=5.24$.

Consider, for the moment, the sample mean. Although x provides the "best" estimate of the population mean, it is based on the sample and hence, it is not exact. If another sample is taken from the same population, it is likely that a numerically different estimate of the population mean would result. A third sample would yield a third estimate of the mean. Thus, while x provides a point estimate of the population mean, it would also be of value to have an interval which, with a specified degree of assurance or confidence, would contain the population mean. A confidence interval provides such an estimate.

For a specified degree or level of confidence P (for example 90%), a one or two sided confidence interval for the population mean can be constructed from the sample data. The value of P can be altered depending on the desired level of confidence. The larger the value of P, that is, the greater the degree of confidence that is desired, the wider the corresponding interval will be.

For the example data, the 95% (2-sided) confidence limits for the population mean are (15.6,20.5). When interpreting these limits, it is not proper to say that there is a 95% probability that the population mean is in the interval (15.6,20.5). The mean either is (probability = 1) or is not (probability = 0) within the confidence interval. The correct statement is that one can be 95% sure that the population mean does lie between 15.6 and 20.5. That is, the probability that the interval (15.6,20.5) contains the population mean is 95%.

Another way of interpreting the confidence interval is as follows. Suppose that 100 random samples were taken from a single population and that 100 confidence intervals were computed. It could be expected that 95% of the 100 confidence intervals would encompass the population mean.

Confidence intervals serve another useful purpose. They can assist one in determining whether the population mean may equal a specific value. To illustrate, suppose prior to the collection of the example data, one wished to determine whether the population mean might be equal to 25. Since 25 does not fall within the confidence interval for the population mean, one could feel reasonably confident that, based on the data, the population mean did not equal 25.

To illustrate the application of this concept to the two-sample problem, consider the comparison of two population means. For example, suppose a chemical agent was added to the feed of one group of mice while a second group had a chemical free diet. Suppose further that one wished to assess the effect, if any, of the chemical on body weight. Based on the experimental data, one could construct a confidence interval for the difference of the two mean values. By virtue of the discussion above, if this confidence interval contained the value 0, one could conclude that there was no difference in the mean values.

While only confidence intervals for the population mean have been discussed, confidence intervals can be computed for the population standard deviation. In addition, as described in the last example, confidence intervals can be applied to more than one sample. The construction of these intervals is discussed in most statistical textbooks including those cited at the end of this paper.

Tolerance Intervals. In some situations, given a sample from a population, one is interested in constructing limits not on the mean or standard deviation, but limits wich will provide an idea of a range within which a certain percentage of the population will fall. Such limits are called tolerance limits. For example, one may wish to determine the tolerance limits which contain 95% of the population.

If the population characteristics (mean and statistical deviation) are known, tolerance limits can be precisely determined. However, since only the sample characteristics are generally available and, as previously discussed, these are not exact, tolerance limits can be determined only wthin a certain degree of confidence. For example, one could determine the limits which with 90% confidence, will contain 95% of the population.

To calculate tolerance limits, two values must first be specified; C the proportion of the population to be covered (the "coverage") and P the confidence coefficient. For given values of C and P, one or two sided tolerance limits for the population can be calculated.

Using the data from the discussion above, suppose one wished to obtain a 2 sided tolerance interval with C = 75% and P = 95%. The appropriate interval would be (9.7,26.5). Based on this calculation, one is 95% sure that 75% of the population falls between 9.7 and 26.5.

A more detailed description of the calculation of tolerance intervals can be found in 3, 7 and 8.

Relationships Among Variables

The methods which have been presented to this point are used in the evaluation of one or more samples. In many instances, however, one wishes to evaluate the relationship between two or more variables. The relationship between dose of a drug in the diet and body weight of mice, the plasma levels of a drug as a function of time or a comparison of a new analytical method compares to a standard method are examples of such problems. In this section, discussion will focus on the case of two variables X and Y. It will be initially assumed that the variable X (the independent variable) is measured with little or no error and is set prior to the time when the experiment is conducted. The second variable, Y, (the dependent variable) is the response variable which is dependent on the value of X. Y is measured with error. The evaluation of the relationship between X and Y is termed a <u>regression</u> <u>problem</u>. Although the concepts presented here apply specifically to the linear regression model, they can be extended to the cases of polynomial regression, regression with many X variables (<u>multiple</u> <u>regression</u>) and nonlinear regression models.

Least Squares Analysis. In the case of linear regression, the theoretical relationship between the two variables X and Y can be expressed as $y = A + Bx +$ error where A is the theoretical (population) intercept and B is the theoretical slope. Given a sample of n independent pairs (x_1,y_1), (x_2,y_2), ..., (x_n,y_n), the observed relationship between x and y is expressed as $y = a+bx$ where 'a' is an estimate of A and 'b' is an estimate of B.

Numerous methods exist for estimating A and B. The most common approach is the <u>least squares</u> method. The least squares method is based on minimizing the square of the distance between the observed value y_{obs} and the "fitted" value $y_{fit} = a+bx$. This distance is represented by the interval 'd' in Figure 1. Thus, the least squares method is based on minimizing

$$\sum d^2 = \sum (y_{obs} - y_{fit})^2.$$

Based on the assumptions that the data (y's) have a normal distribution and that the standard deviation of the y's is the same as each x, the least squares estimates 'a' and 'b' of A and B are unique and unbiased. It is these properties which make the least squares estimates of A and B attractive.

Since 'a' and 'b' are statistical estimates based on a sample, they have an error term (the standard error of the estimate) associated with them. Among all estimates A and B, the least squares estimates have the smallest error term. This is the third important property of the least squares estimates.

Consider the (hypothetical) data presented below.

Table 1	X	Y
	0	27.3
	5	40.5
	10	63.1
	20	91.6
	30	117.7

Inspection of a plot of Y versus X (Figure 1) suggests that a straight line might be a reasonable fit to the data. Using the method of least squares, the estimates of A and B are found to be a=28.3 and b=3.06. This means that, based on the data, the relationship between X and Y can be expressed as

$$y = 28.3 + 3.06x.$$

Prediction. The estimation of the parameters of a regression line is often the first step in an analysis. Frequently, the regression line is used to predict a value of Y for a new value of X. (The opposite problem, predict an X for a new Y will be discussed later.) When the new value x* falls between the X's used to estimate the regression line (the regression limits), the

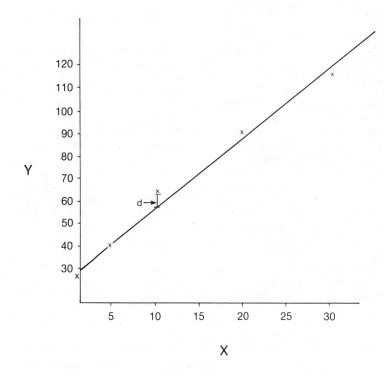

Figure 1. Sample data for regression analysis

prediction of the corresponding Y value, y*, is called
interpolation. When x* falls beyond the regression limits, the
prediction of Y is termed extrapolation.

Consider the problem of interpolation. Given a new
observation x*, both point and interval estimates of Y can be
calculated. The point estimate of Y is called the predicted value
and is easily calculated as y* = a+bx*. The interval estimate
for the predicted value is called the prediction interval. The
calculation of a prediction interval is essentially the same as
the calculation of a confidence interval. Both require the
specification of the confidence coefficient P and both (overlaps)
intervals have the interpretation; the interval contains the true
(population) value with probablilty P.

There are two types of prediction intervals which can be
constructed in the regression problem, prediction intervals for a
population mean (the mean response y* for a given x*) and
prediction intervals for individual observations (i.e. the
prediction interval for a particular patient). Conceptually, the
difference between the two intervals is subtle. In the first
case, one is interested in an interval for the population mean at
a given value of X. In the second case, an individual
observation from the population is of interest. In practice the
difference between the two can be substantial since the
prediction interval for the population mean is more narrow than
that for an individual. To illustrate this difference, consider
the data in Table 1. The 95% prediction interval for the
(population) mean response at x*= 15 is (74.0,84.4). If,
however, one were interested in obtaining a prediction for a
particular patient who had x*=15, this would be (66.7,91.7).

Extrapolation, the prediction of values beyond the range of
the independent variable, is highly dependent on an important
assumption. It is assumed that the relationship between the two
variables remains contant for all values of X, up to and
including the value of interst. G. Hahn (6) presents an
excellent example of the potential danger involved with
extrapolation. Consider the data in Figure 2 . Inspection of
the plot of X versus Y suggests a linear relationship between the
two variables. Fitting a straight line to the data (Figure 3)
and extrapolating out to x*=50 produces a predicted value of
y*=138. The plot in Figure 2 is a hypothetical plot of the
height of a random sample of males between the ages of 8 and 14.
The extrapolation of x*=50 suggests that a 50 year old male would
have a height of 11+ feet. The assumption that the linear
relationship between X and Y will continue to hold for x*=50 is
obviously erroneous. Thus when one extrapolates beyond the
regression limits, one should do so very cautiously, especially
as one get further away from the regression endpoints. The
concepts of point and interval prediction estimates as discussed
in the case of the interpolation can also be applied to the
extrapolation problem.

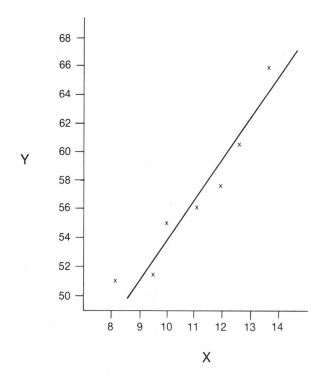

Figure 2.   Linear relationship between variables X *and* Y

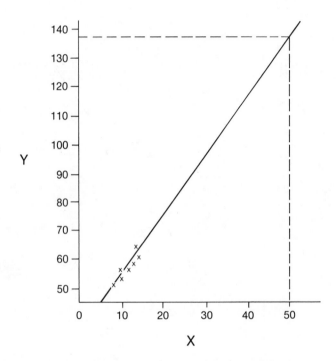

Figure 3. *Extrapolation of data in Figure 2 to* x = 50. *The variable* X *is age in years and the variable* Y *is height in inches.*

Tests for goodness of fit. While a linear model can be fit
to any set of data, a straight line may not be the best model.
It is possible in the regression analysis to check for lack of
fit, that is, the inability of the model to adequately describe
the relationship between X and Y. This statistical test requires
that two or more independent observations (measurements) must be
made at each level of X. Independent here means that unique
responses must be obtained. Determining the white blood cell
count from the same sample three times does not provide three
independent observations. Although the test for lack of fit
cannot indicate what the appropriate model would be, it can
enable the experimentor to assess the validity of the assumed
model. This is why it is frequently requested by statisticians
that multiple observations be obtained for the various levels of
the X variable.

Logrithmic Transformations. It is common in the biological
sciences to find that while the relationship between X and Y is
not linear, a logrithmic transformation of one or both of the
variables will produce a straight line relationship. The
regression methods as described above can be applied to the
transformed data in order to estimate the parameters of the
model, to make predictions of future values and to obtain the
corresponding confidence limits. Caution must be used, however,
when interpreting the results of analyses based on transformed
data, particularly when discussing the results in the original
scale of measurement. All results (parameter estimates,
confidence limits, etc.) pertain to the transformed data.
Expressing the results of the statistical evaluation in the
original scale of measurement (i.e. by taking anti-logrithms)
does not preserve the statistical interpretation. For example,
the 95% confidence limits for a predicted value of log Y are not,
after taking anti-logs, the 95% confidence limits for the value
of Y in the original scale of measurement. This fact can be
illustrated using the following set of data:
15, 17, 10, 22, 13, 15, 18.
The mean and 95% confidence limits for these data are 15.71 and
(12.18,19.24), respectively. Taking the logrithm of each of the
values, the mean and 95% confidence limits are 2.73 and
(2.50,2.96). Taking anti-logs of these values, one obtains that
the mean and 95% confidence limits in the original scale of
measurement are 15.31 and (12.14,19.30), respectively.
Comparison with the first set of statistics reveals distinct
differences.
The discussion above applies to most other transformations
which are used to linearize a set of data, i.e. exponentiation,
taking roots, raising to a power, probits, logits, etc. Only for
those transformations which themselves are linear (that is, are
of the form y=b+mx) will the statistical interpretations be
preserved before and after transformation.

Statistical Calibration. The discussion, to this point, has dealt with the prediction of y* for a given x*. In many problems such as radioimmunoassay, one wishes to predict X from a given (observed) value of Y. Such problems are called inverse prediction or calibration problems. A general outline of the calibration problem can be described as follows. Two variables, X (the independent variable) and Y (the dependent variable), are such that X is difficult (or impossible) to measure directly while Y is relatively easy to measure. In the first part of the experiment, the corresponding Y's are obtained for n known values of X. These data are frequently called the calibration or standards data. Later, m additional values of Y are (i.e. responses from m patients) obtained and the objective is to estimate the corresponding values for the X's.

The first step in the calibration analysis is to determine the relationship between X and Y by fitting a model to the calibration data using the method of least squares . When estimating the parameters of the calibration line, it is not correct to reverse the role of X and Y and then to use the procedure for prediction in the regression analysis. The theory of least squares is based on the assumption that the X's are error free and that the Y's are measured with error. To regress X on Y violates this fundamental assumption. The correct procedure for the calibration problem is to regress Y on X as in the regression problem in order to estimate the calibration line. Predicted values for X can be obtained as follows.

The point estimate of X (x') for a new value y' is easily calculated as x'=(y'-a)/b. The calculation of an interval estimated for X' is more difficult than calculation of prediction limits for y* in the regression problem. In the calibration problem, there are 2 error terms which must be considered; the error in establishing the calibration line and the error associated with measuring y'. Since the calibration line is based on a random sample of observations, it is not exact. A different set of x's (at the same levels as before) would have resulted in a numerically different estimate of A and B. This lack of precision must be accounted for in the calculation of the interval estimate for x'. In the same way, if a second sample were taken from the same patient, two numerically different Y's would likely result. Thus, this error or lack of precision must also be considered in the calculation of the interval estimate for x'. The procedure for incorporating these two error terms in the calculation of the interval estimate for x' is illustrated in Figure 4 . Although such intervals may appear to be relatively large, it would be inappropriate to ignore one of the error terms in order to reduce the interval width.

To illustrate these concepts, suppose one wished to obtain point and interval estimates for X when y=50 in the data in Figure 1. The point estimate of X is calculated to be 7.1. Using the procedure outlined above, the interval estimate for X is found to be (2.74,11.18).

Errors in Both Variables.   Until now in the discussion of the
relationship between two variables, it has been assumed that the
X variable is measured without error or with negligable error.
If X is measured with non-negligable error, one solution to the
problem of evaluating the relationship between X and Y is called
a correlation analysis. Given two variables X and Y, a measure
of the linear association between them is given by the
correlation coefficient r . For a given problem the value of r
can fall anywhere in the interval (-1, 1). A value of r=-1 is
reflective of a "perfect" negative linear relationship between X
and Y(Figure 5a). A value of r=+1 is reflective of a perfect
positive linear relationship between X and Y(Figure 5b). The
absence of a linear relationship between X and Y(Figure 5c) is
suggested by r=0. It is important to note that this does not
mean that X and Y are not related. To illustrate, consider the
hypothetical data in Figure 5d. For this data, r would equal 0.
However, it is rather apparent that there is a (nonlinear)
relationship between the two variables. Thus, when determining
the correlation coefficient between two variables, it is
important to keep in mind that r is providing a measure of the
linear association between the two variables.

Suppose that, in the evaluation of two variables X and Y,
both of which are measured with non-negligable error, one wishes
to determine the functional relationship between the two
variables and not just the correlation coefficient. To
illustrate, suppose the theoretical relationship between the dose
D of a drug and the response metameter R is given by the model
R=A+B*D+error and that one wished to estimate the parameters of
the model. Suppose further that both D and R are measured with
error so that what are actually observed are $d=D+error_1$ and
$r=R+error_2$. Since both D and R are measured with error, the
regression methods previously described cannot be used to
estimate the parameters A and B of the model. The solution to
this problem requires that additional assumptions must be made
about the data.

There are two particular approaches which have been proposed
for addressing this problem which has been frequently referred to
as the error in variables problem. One is to assume that the
ratio of the error in D to the error in R is constant, that is,
to assume that k=error(R)/error(D). Even though the experimentor
does not know the exact values for the standared deviations of D
and R, he may feel, for example, that the error in R is of the
same order of magnitude as that for D (i.e. k=1). Or, the
experimentor may be able to obtain a reasonable estimate of k
based on prebvious experience. With the determination of k,
estimates of A and B can be obtained.

The second approach is called the controlled-independent-
variable approach. In this approach, the experimentor decides
before the experiment, what values of d will be observed (hence
the name controlled-independent-variable). For example, the
experimentor may choose to obtain responses at d=5, 10, and 20

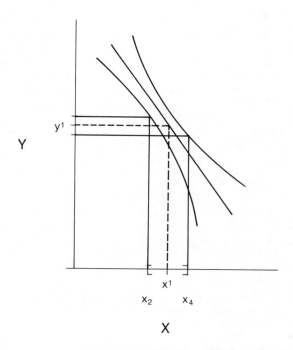

Figure 4. *Calculation of prediction intervals in calibration analysis: y^1 is the observed value of Y; x^1 is the predicted value of X corresponding to y^1; x^2 and x^4 are the lower and upper prediction limits, respectively.*

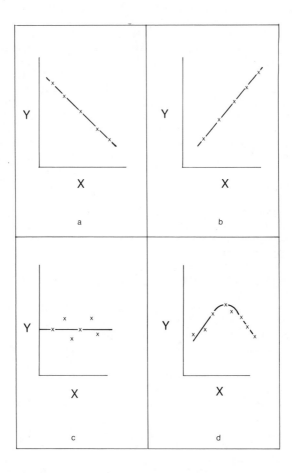

Figure 5. Possible outcomes from correlation analysis: (a) r = 0, *negative linear relationship; (b)* r = 0; *positive linear relationship; (c)* r = 0, *no linear relationship; (d)* r = 0, X *and* Y *are related but not linearly.*

mg. Although the true dose D will not equal 5, 10 or 20 mg, the fact that responses are obtained at the observed doses(d)=5, 10 and 20 mg is what is important. Assuming the data are collected under these conditions, estimates of A and B can be subsequentially obtained.

To summarize, when one wishes to evaluate the relationship between two variables, one of which is fixed (the independent variable) and one which is allowed to vary (the dependent variable), the analysis is termed a regression problem. A special type of regression problem is called calibration. In the calibration problem, one wishes to predict (future) values of the independent variable for given values of the independent variable. If, on the other hand, both variables are measured with error, correlation analysis and error in variables analyses are two approaches which one can use in the evaluation of the data. More extensive details of regression and correlation analysis are found in the references cited at the end of this paper. The calibration problem is discussed in 2, 8, 9 and 10.

Data Smoothing

The concepts which have been previoulsy presented dealt specifically with the analysis of experimental data. An equally important aspect in the evaluation of experimental data is how the data are presented and, in particular, the graphical display of data. One of the purposes of graphing data is to illustrate trends or cycles which may exist in the data. If, however, the data are "noisy"(i.e. there exist large variations to to random or experimental error, it is often difficult to observe the important trends or patterns in the data. The elimination of the "noise" from a graph is called data smoothing.

There are many methods which are used for data smoothing. Three methods which are particularly useful because of their of simplicity are; the method of averages, the method of medians and the method of differences. In the discussion to follow, these three methods will be described and applied to the data presented in Figure 6a and Table 2.

Smoothing by averages involves the replacing of an observed value by the average of that value and surrounding observations. To maintain symmetry, an equal number of observations are generally taken on either side of the value to be smoothed. The calculation of the smoothed value can be based on assigning equal weights to all data in the average(i.e. the arithmetic average) or by computing a weighted average of the data. 1-2-1 smoothing, smoothing in which the center value receives twice the weight of the outside values is an example of the use of a weighted average. Figure 6b presents a smoothed plot of the data in Figure 6a in which each observation is replaced by the arithmetic average of the observation and one value to either side of it. Thus, for example, the value -0.28 is replaced by -0.76 (see Table 2). It should be noted that in the smoothed curve, 2 points are "lost", the first and the last. This is due to the fact that a three point smoothed value could not be obtained for these data.

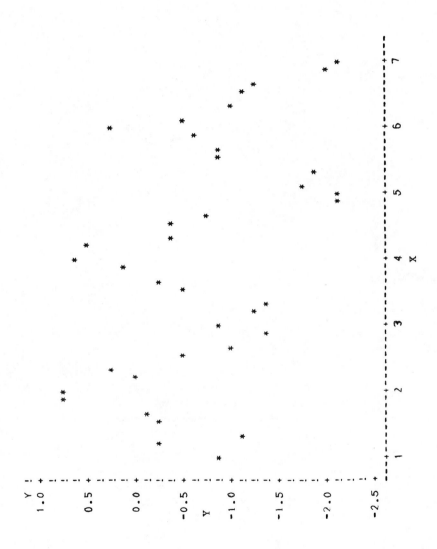

Figure 6a. Sample data to be smoothed

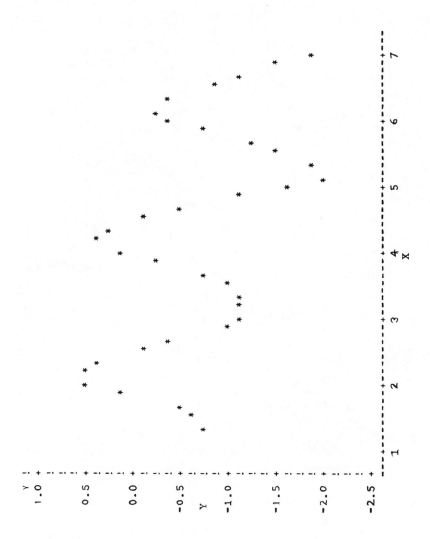

Figure 6b.   Sample data smoothed by method of averages

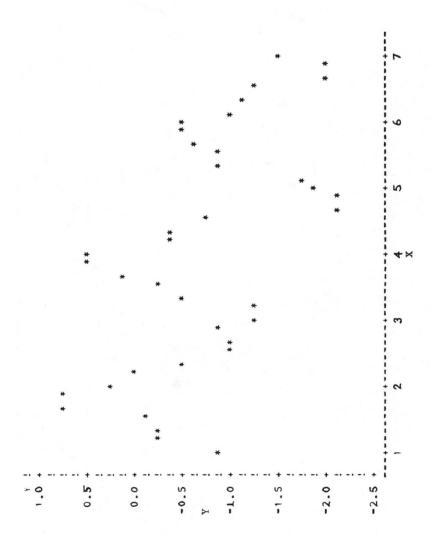

Figure 6c.   Sample data smoothed by method of medians

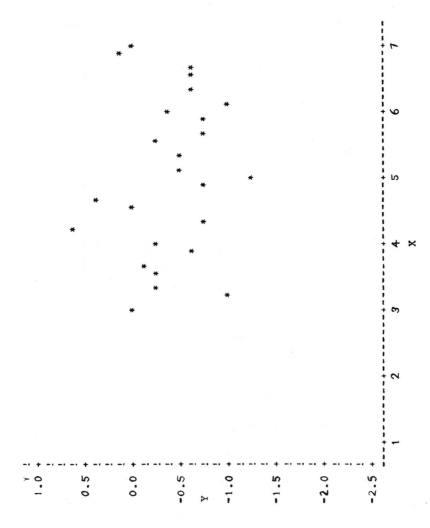

Figure 6d.   Sample data smoothed by taking the 12th differences

Table 2

Data Used to Illustrate Methods of Smoothing

Original Data	Smoothed by Averages	Smoothed by Medians	Smoothed by 12th Differences
-0.82	.	.	.
-0.28	-0.76	-0.82	.
-1.17	-0.56	-0.28	.
-0.24	-0.51	-0.24	.
-0.12	0.13	-0.12	.
0.76	0.48	0.76	.
0.81	0.51	0.76	.
-0.04	0.35	0.27	.
0.27	-0.08	-0.04	.
-0.46	-0.41	-0.46	.
-1.05	-0.97	-1.05	.
-1.41	-1.11	-1.05	.
-0.89	-1.17	-0.89	-0.06
-1.22	-1.17	-1.22	-0.94
-1.41	-1.06	-1.22	-0.24
-0.55	-0.73	-0.55	-0.31
-0.23	-0.21	-0.23	-0.11
0.16	0.17	0.16	-0.60
0.59	0.43	0.53	-0.22
0.53	0.23	0.53	0.54
-0.42	-0.10	-0.40	-0.69
-0.40	-0.51	-0.42	0.05
-0.70	-1.09	-0.70	0.35
-2.16	-1.66	-2.10	-0.76
-2.10	-1.99	-2.10	-1.22
-1.71	-1.92	-1.93	-0.49
-1.93	-1.49	-1.71	-0.52
-0.83	-1.23	-0.83	-0.27
-0.92	-0.77	-0.83	-0.69
-0.57	-0.42	-0.57	-0.73
0.25	-0.26	-0.46	-0.34
-0.46	-0.41	-0.46	-0.99
-1.03	-0.85	-1.03	-0.61
-1.07	-1.14	-1.07	-0.67
-1.31	-1.48	-1.31	-0.61
-2.06	-1.81	-2.06	0.11
-2.08	.	.	0.03

Smoothing by medians is very similiar to smoothing by averages. In this method, an observation is replaced by the the median of itself and the 2 neighboring(adjacent) points. (The number 2 is arbitrary. Other values, 4 or 6 for example could have been used. However, this may cause too much smoothing and the overall character of the data may be lost.) To illustrate, in the example data, the value -0.28 is replaced by -0.82, the median of -0.82, -0.28 and -1.17. The advantage of smoothing by medians is that if there is an occasional outlier observation (excessively large of small value) in the data, the smoothed plot will not be effected by it. When smoothing by averages, the existance of outliers will still be apparent in the smoothed plot. Figure 6c and Table 2 illustrate the effect of smoothing the data in Figure 6a using the method of medians.

Smoothing by differencing is generally used when there are cycles in the data which might mask underlying trends or when observations are dependent on previous values. The diurnal variations in blood pressures is an example of a cycle dominating a set of data. If the blood pressures of a patient treated with an antihypertensive drug were taken hourly for three days, the effect of the drug would probably not be evident in a graphical presentation of the data because of the diurnal variations. The removal of this cycle by differencing would reveal the overall decreasing trend in the data.

Consider for the moment, data which is collected hourly. To obtain the smoothed data using the method of differencing, each observation is replaced by the difference of itself and the observation obtained x hours previously. First order differences involve obtaining the difference between the "current" observation and the previous observation; second order differencing involves the difference between the "current" observation and the observations 2 hours previously, etc. The level of differencing will depend and the period of the cycle. For example, if a set of data has a six hour cycle and the data are collected hourly, sixth order differences would be appropriate.

In Figures 6d, the data in Figure 6a are smoothed by taking twelfth order differences. These "de-cycled" data can now be examined/evaluated for the existance of trends. Note that the first 12 observations are "lost" in the smoothing.

Averages, medians and differences are but three methods which can be used for data smoothing. The advantage of these methods over other methods such as exponential smoothing is that these methods are easily applied to most sets of data.

Summary

To summarize, a number of statistical methods have been briefly presented in order to establish a conceptual framework for the reader. The discussions have not been detailed since such an in depth accounting of each concept would not serve our purposes here. Detailed discussions of each statistical method are available and references have been cited for those who seek such depth.

Statistics is a very viable tool in modern scientific research. With continued interface between the scientist and the statistician, the resulting research and only be enhanced.

Literature Cited

1. Bennett, Carl A. and Franklin, Norman L. Statistical Analysis in Chemistry and the Chemical Industry. 1954. John Wiley and Sons, Inc. New York.
2. Box, G. E. P., Hunter, W. G. and Hunter, J. S. Statistics for Experimentors. 1978. John Wiley and Sons, Inc. New York.
3. Dixon, Wilfred J. and Massey, Richard H. Introduction to Statistical Analysis. Third Edition. 1975 McGraw-Hill Book Company. New York.
4. Draper, N. R. and Smith, H. Applied Regression Analysis. 1966 John Wiley and Sons, Inc. New York.
5. Graybill, F. A. An Introduction to Linear Statistical Models. Volume I. 1961. Mc Graw-Hill Book Company, Inc. New York.
6. Hahn, Gerald, J. The Hazards of Extrapolation. 1978. Chemical Technology. 8. pp 699-701.
7. Natrella, Mary G. Experimental Statistics. National Bureau of Standards. Handbook 91. 1966. U.S. Government Printing Office Washington D.C.
8. Ostle, Bernard and Mensing, Richard W. Statistics in Research. Third Edition. 1975. The Iowa State University Press. Ames.
9. Snedacor, George W. and Cochran, William G. Statistical Methods. Sixth Edition. 1967. The Iowa State University Press Ames.
10. Sokal, Robert R. and Rohlf, F. James. Biometry. 1969. W. H. Freeman and Company. San Francisco.

RECEIVED April 20, 1981.

Analytical Measurements:
How Do You Know Your Results Are Right?

WILLIAM HORWITZ

Bureau of Foods, Food and Drug Administration, HFF-101, 200 C Street, SW, Washington, DC 20204

The scientist these days has a new partner--the auditor. He is not a financial auditor, but rather an examiner of knowledge. He is a verifier of accounts, as the dictionary puts it. In this case, he intends to verify that the public's trust in science is well founded.

The presence of the science auditor is the result of revelations that some laboratories were submitting false or faulty data to government agencies as the basis for obtaining permission to expose the public and the environment to potentially hazardous materials, such as pesticides, food and color additives, and more recently, "toxic substances." In granting permission to use toxic chemicals to control agricultural pests, to construct protective food-contacting polymers, and to fabricate foods and articles useful to consumers, Congress required its public servants to assure themselves that no harm would occur to the ultimate users of the products. Congress did not require that the tests to assure absence of harm be performed by the presumably neutral government; on the contrary, they accepted the common portrait of a scientist as the altruistic individual whose main desire was to satisfy his thirst for knowledge. As a practical matter, however, we have discovered that there is a long road between the laboratory data and a regulatory petition that leads through the office of the laboratory manager, the vice president in charge of research and development, the chief legal counsel, and apparently more often than not, the director of public relations.

It appears that we have placed so much emphasis on certainty that we are uncomfortable with uncertainty. In handling data, we tend to avoid and hide the uncertainties in our obsession to produce "clean" data. But clean data are more a matter of judgment than of actuality. Raw data are frequently disorderly in the sense that they are full of perturbations resulting from the many outside influences on the particular property we are measuring. The value which we obtain at any given moment is equivalent to a series of one-frame still pictures from a continuously running movie film. As a result of this discontinuous sampling of a continuous event, we often get the zig-zag patterns of properties with time, which abound in the toxicological journals, complete with the standard errors extending from each point, which often do not even have the decency of overlapping each other. But we must always remember that unless we have variability in our measurements, we have no idea of the uncertainties in our system.

Raw data used to be a very simple concept: they were the numbers actually indicated by a measuring device regardless of their being obtained by summing up the weights on a balance, read from the scale on a buret, determined on an instrument dial, or actually measured on a recorder chart. The analyst had full control and responsibility over the production of the data at every step. He prepared his own reagents, calibrated his weights and volumetric glassware, and standardized the output of his instruments. As efficiency experts and cost accountants penetrated laboratory management, some of these technical responsibilities were delegated to less costly sources: Prepared reagents are purchased from a laboratory supply house or prepared by a central local unit; glassware is washed, stored, and distributed by a specialized organization; responsibility for calibration is assigned to the manufacturer of the equipment; and proper functioning of instruments is assumed to be built-in by the instrument designers and computer operators. This shift in functions is not necessarily bad. It did relieve the analytical chemist of numerous minor, but important, chores which were distractions from higher level responsibilities. But when these functions were placed elsewhere, proper management required that the performance of these professional responsibilities be appropriately monitored to ensure suitable operation. Thus, the production of data shifted from a straight line function, entirely under the direct supervision of the professional scientist, to a maze-type operation characterized by the intermingling of the critical paths of a "PERT" chart, managed by a laboratory director. The demands for efficiency, coupled with the fact that many of our modern measurements cannot be obtained in any other way than by mechanically or electronically controlled automatons, result in machines which measure the samples, execute the manipulations, determine the response, perform the calculations, and present the final answer in whatever form or units are desired. The final value may be copied from a dial, recorded on tape, drawn on a chart, or not presented at all, to be stored in a computer for coordination with past and future values, presenting the entire sequence as the result of the experiment. These final results from machines are raw data just as much as the direct measurements were. Whether or not the final results emanate directly from our manual observations or from our automated instruments is really not asking the right question. The proper question we should constantly be asking is: "Are these data right?" The operational question is: "How do we know that these data are right?"

I intend to discuss that question in this paper. It is a subject which is rarely dealt with in the scientific literature because our journals are not set up to handle this type of discussion. We politely assume that any measurement a scientist makes is correct. In our peer review system, the reviewer assumes that the data reported are correct unless he finds an internal inconsistency which the investigator failed to detect. This is the first line of defense in any investigation--consistency. Very often it is the only line of defense because new information is being developed for which there is no external guide. If there are any guideposts, most likely they are shifting guideposts because many investigations determine how things change with time. The situation is much like being set down in a dense forest and being told to find your way out, with your only direction indicator an occasional glimpse of the sun, whose position shifts with time.

My purpose is to provide a few general guideposts which may be helpful in determining the reliability of chemical and physical data. Most of the problems and concepts discussed here have been developed as a result of the necessity for the review of data produced from the chemical analysis of samples examined for compliance with the Federal Food, Drug, and Cosmetic Act or data submitted to the Food and Drug Administration (FDA) in support of a request for approval of a regulated product. I cannot be of much help in evaluating biological data but you might find the recent publication on quality assurance from the American Public Health Association (1) useful in this respect. It contains chapters describing concepts which have been developed to assure the validity of the product from laboratories involved in such fields as anatomic pathology, clinical chemistry, clinical microbiology, clinical toxicology, cytology, hematology, immunology, and virology. However, you do not need any reference to know that something is wrong when the computer printout shows the results of examination of the uterus of a male rat, the testes of a female mouse, and diet consumption of 9999.9 pounds by a 250 pound rat! Such information appeared in the raw data supporting a recent submission to FDA.

Consistency

The primary guidepost in all data collection activities is consistency. A series of measurements will always fall into one of three categories: They will go up; they will go down; or they will remain constant. This is not as trivial an observation as it may seem. I mean to point out that measurements usually follow a pattern and experiments are usually designed to determine that pattern. If the measurements seem to go up and down without a pattern, that in itself is a pattern. You are observing random variability, which must be factored out to discover the underlying trend. In fact, the reviewer should really begin to worry about the quality of the observations when there is no reasonable

variability component. Less than usual variability suggests that
some averaging has been going on. You can average out quite a few
wild results if they are in opposite directions and get a fairly
decent mean. The statisticians heard of averaging a long time ago
and named it "regression to the mean" for the ability to hide poor
data by taking enough of it.

The data should also be consistent with corresponding information that may exist in the literature or from laboratory experience; if not, an explanation is called for. The whole should equal
the sum of its parts and amounts of products should be chemically
equivalent to the amounts of reactants. Experiments should be
designed to incorporate as many self-checking features as possible, as for example, accounting for all components. However, if
one of the figures is obtained by difference, the self-checking
feature is lost.

There is also a negative aspect to consistency. Data which
are too consistent are also suspect. Variability patterns are
usually quite reproducible from experiment to experiment. Less
than usual variability does not always mean better and more careful experimentation. To an auditor, it may suggest the application of mental telepathy or what, in my student days, was known as
graphite chemistry.

Variability of Measurements

Beyond such simple concepts as consistency of the data and
its additive properties, we must understand the concept of
measurement in analytical chemistry because many of the measurements that the toxicologist makes are chemical in nature. This he
has had to do in self defense because rarely has he had a chemist
at his beck and call. Until the last decade or so, the chemist
largely ignored the area of the analytical chemistry of residues
and metabolites. This is no longer the case. Analytical chemists
in the short space of a few decades have given us some marvelous
tools in the form of the powerful resolutions of chromatography,
the superb sensibility of various kinds of spectroscopy and
polarography, and the exquisite specificity of mass spectrometry.
But despite their power, we must always question the reliability
of the information they are giving us.

There are many causes or sources of variability. Some are
very general and occur in practically all chemical measurements.
Others are specific to the individual methods and thus are difficult to handle in a general way. Therefore, we will concentrate on
the general aspects which must be considered in all analytical
operations. One of the most important is sampling and handling of
the samples and a second is what to do with the final analytical
results. These points are not usually covered in most textbooks
since they are really outside of the analytical operations.
Sampling and handling of the sample is the beginning of the
sequence. The final disposition of the analytical results--how do

you interpret the data--is the end of the sequence. Each of these subjects could support an extensive lecture on its own. All we can do is to point out now that both aspects are important and their neglect can lead to just as much trouble as poor analytical work.

Sampling

Stated simply, the job of the analytical chemist is to report what is in the container that he is given. What the analyst tells you only applies to what he works on. If the toxicologist gives the chemist only half a liver, it will be the toxicologist's job to extrapolate to the whole liver, not the chemist's. The chemist should not take the responsibility for extrapolating the results of analyses to the whole organ, complete tissue, entire animal, or to all animals. The designer of the experiment should have taken into consideration the purpose of the work and built into the material sampled the ability to extrapolate to the desired level of complexity. Therefore, one of the first things that should be looked at is the design of the experiment, to ensure that the proper material was selected for analysis.

Not only must the proper material be selected for analysis but it must be handled properly to avoid contamination and alteration. Plasticizers frequently appear in analyses from contact of the sample with plastic containers or protective films. Metallic contaminants appear from contact with metal instruments, metal and plastic foils and liners, spatulas, and grinders. Adventitious compounds can appear from the most unexpected places. Paper, for example, may contain numerous coating additives; fatty acids and their derivatives appear as coatings on plastic films and aluminum foil; silicones are used to coat glass. Therefore, if samples are in contact with common protective films and containers, they could pick up something which may interfere with your trace analyses. It is good analytical practice to supply to the chemist, portions of all materials which the samples may have contacted. These materials would be examined as potential sources of unidentified materials appearing in recordings or printouts.

Conducting blanks through the entire procedure is an absolute necessity in trace analysis to account for minute amounts of the analyte and interferences in the reagents, absorbents, solvents, water, and other materials which contact the sample and its derivatives during the analysis. Materials which are ordinarily considered inert in most chemical operations (e.g., solvents, filter paper, drying agents such as sodium sulfate) may contribute relatively large quantities of interfering materials as we go lower in the concentration scale.

The preparation, sampling, and analysis of animal feeds deserve special attention. The practicalities of distributing uniformly parts per thousand, parts per million, and even parts per billion of a test material into a heterogeneous feed mixture probably require the talents of a chemical engineer. We have described overcoming the difficulties in the preparation of an analytical sample involved in a feeding study for trace quantities of metals (2). Scaling up this mixing procedure a hundred or a thousand fold undoubtedly requires considerable experimentation and operational control. A summary of the feed mixing procedure for the large scale toxicological study conducted at the National Center for Toxicological Research is given by Oller et al. (3).

We can summarize the importance of sampling by pointing out that in the case of the analysis of peanuts for the mold metabolite aflatoxin, at the parts per billion level, 90% of the total variability is derived from sampling the commodity and preparing the laboratory sample, and only 10% is derived from the analytical operations. Based on the validating collaborative study, the interlaboratory coefficient of variation (CV) of the method of analysis alone in this case is about 30% at the 10 ppb level.

The Reliability of Analytical Methods

The role of analytical methods in modern toxicology and its importance in "risk assessment" can be summarized by a quotation from a recent report to the Environmental Protection Agency (EPA) on pentachlorophenol (PCP) contaminants (4):

"A key problem to overcome in order to make an adequate evaluation of the relative hazard of PCP and its contaminants is the lack of ready availability of suitably sensitive and specific analytical methods. Although progress has been made in developing appropriate analytical capability, routine analysis has been hampered by the unavailability of suitable analytical standards for some of the isomers. In fact, the availability of appropriately specific analytical methods may be the rate limiting factor in assessing the hazard of dioxins and related chemicals. Thus, when there are several isomers with widely differing toxicities, as in the case with hexachlorodibenzo-p-dioxins, analyses of the isomers as a group only permit assessment of hazard based upon the most toxic isomer. This approach may, indeed, lead to overestimates of hazard, but, in the absence of more definitive analyses of specific toxic chemical species, it is necessary to treat contamination data on a toxicologically worst-case basis."

Analytical methods have two types of characteristics--scientific and practical. The scientific characteristics determine the reliability of the analytical data; the practical characteristics determine the utility of the method. The scientific attributes of a method include such things as accuracy, precision, specificity, and limit of reliable measurement; the practical attributes include cost of performance, time required, and level of training needed. For research purposes, the practical aspects are of secondary consideration; for regulatory operations of compliance and surveillance, practicality is of great importance. Little enforcement is possible using a method which turns out one analytical value per day!

Specificity

The fundamental property of all analytical methods is specificity--the tests which are being applied must measure what they purport to measure. For example, many tests which measure chloride also measure bromide and iodide; therefore such tests are for halides. They are useful for chloride determinations because of the absence of the other halides in many materials. Colloquially we speak of methods for chlorides, but strictly speaking such tests are for halides. When we did not have anything better, we measured the organochlorine pesticide residues by extracting the pesticide with an organic solvent and determining total chloride (really total halide). Initially, we called it DDT, but as more related pesticides were introduced, it had to be called organochlorine pesticides, then organochlorine compounds, and now we would have to call it organic solvent-soluble organohalide material. We now know that a lot of what we assumed was DDT or related organochlorine pesticides, even by the early gas chromatographic methods, were in all probability PCBs (polychlorinated biphenyls). Schechter (5) had warned us about this point many years ago with his example of the "pre-DDT era" soil sample that had been kept in a sealed container and had never been exposed to organochlorine pesticides. The gas chromatogram of the multiresidue method exhibited a series of peaks, a number of which had retention times at or close to those of known pesticides. Schechter concludes, "Data reported without application of suitable confirmatory techniques may not only be worthless, but what is worse, incorrect information may be seriously misleading and may be unrectifiable."

We now have much better tools for assessing specificity than we had at the beginning of the pesticide age. Gas and thin layer chromatography can usually detect the presence of mixtures. They do not work so well the other way--proving the identity of a pure compound. For this you have to apply the instruments which work on the whole molecule, or appreciable or critical fractions of the molecule, such as infrared spectroscopy, nuclear magnetic resonance, or best of all, mass spectrometry. But there are always footnotes or reservations to the best of techniques. In this case, for unequivocal identification, apply the techniques only to pure samples; only a small amount is needed, but it must be pure!

The required specificity will depend upon the purpose of the analytical results. The main need for specific identification of analytes lies with the toxicologists. They indicate that many similar compounds have significantly different toxicities. Some examples include the four closely related aflatoxins (B$_1$, B$_2$, G$_1$, G$_2$) whose relative acute toxicities in the duckling cover a range of 10 to 1 (6). In the family of polynuclear hydrocarbons, some are reported as carcinogenic and some are not (7). The most recent and complex example is that of the polychlorinated dibenzo-p-dioxins (CDDs), present as contaminants in 2,4,5-trichlorophen-oxyacetic acid (2,4,5-T) and related ester herbicides. There are 75 possible isomers of CDDs, from monochlorinated to octachlor-inated; there are 22 possible isomers of the tetrachloro compound. The fully chlorinated octachlorinated compound is relatively inert biologically but the symmetrical 2,3,7,8-tetrachlorodibenzo-p-dioxin (2378TCDD) has been characterized as the most potent small molecule toxin known (8). As yet, the toxicologists have not been able to set a limit of toxicological insignificance as a target for method development for this compound. They have merely indicated that the chemist should go as low as he can, certainly into the parts per trillion (ppt) region.

The conflicting requirements of measuring in the low ppt region, and at the same time being sure that the most toxicolog-ically potent isomer is the one which is being measured, presents an interesting dilemma: To obtain specificity for 2378TCDD requires an extensive cleanup from DDE and PCBs, using several adsorption columns and high pressure liquid chromatographic steps, and selective capillary gas chromatography (for isomer separa-tion). With a perfect cleanup, any detector, from a mass spectrometer to an electron capture gas chromatographic detector, may be used to verify the presence of the tetrachloro compound. Specificity also may be obtained through the use of a less rigorous cleanup by relying upon a very expensive high resolution mass spectrometer to measure the exact peak locations of the tetra-chloro compounds to 10 parts in a million. At both extremes, as fewer ions (due to the compound of interest) are monitored or as the cleanup is shortened, a lower level of detectability is achieved, but always at the expense of specificity. Furthermore, as the procedure becomes longer, losses become greater, accuracy and precision deteriorate, and the operation becomes less prac-tical.

Other problems, not necessarily affecting different pro-cedures to the same extent, include lengthy cleanups and gas chromatographic separations, unavailability of isomeric TCDD standards, and impurities in isotopic internal standards. The choice of different signal-to-noise ratios by different labora-tories affects the detection and measurement limits.

Another aspect of the analysis for TCDDs is that the purpose of the work determines the degree of specificity that must be built into an analytical procedure. If you are a regulatory agency scientist who must sustain the burden of proof against potential questions from skeptical scientists and even more skeptical lawyers, you will include every possible point of assistance, even sacrificing a low limit of determination. If you are embarked on

a surveillance program, to determine the extent of TCDD contamination in the environment or in the food supply, or a
metabolism study, you need only satisfy the scientific questioning
of your associates and supervisors.   If you are engaged in
research, following a specific protocol, where there exists
collateral information on presence and absence of the test
material and a dose-response curve to fall back on for the test of
consistency, a minimum amount of characterization is sufficient.

In practice, then, achieving absolute specificity is often
not possible and sometimes not necessary. Absolute specificity in
trace analysis can usually be achieved only at the expense of other
attributes of the procedure.

Accuracy

The accuracy of an analytical result is measured by the difference between the measured value and the true or assigned value.
In most residue or contaminant work, we do not know the true value
of the constituent we are measuring. We therefore have to fall
back on the artificial situation of using the method of additions
to approximate the original content of our analyte, or the far more
difficult task of estimating the true value by more definitive
methods.   But we always seem to be thwarted in our efforts to
obtain reasonable values for our analytes.

I will use as an example a case which you would think would be
a relatively straightforward analytical problem--the determination of the stable inorganic element chromium, which has an
important role in the metabolism of carbohydrates. Figure 1 and
Table 1 show the various published values for the chromium
concentration in human blood or plasma since 1948 as reported by
Mertz in 1975 (9), supplemented by some later values. I have drawn
what appears to be a rough trend line of values generally
decreasing since the early 1960s, which required the use of four
cycle log paper. There is general agreement now that the actual
chromium content of blood is closer to 1 ppb than to 1 ppm, yet
every one of the almost two dozen contributors to Figure 1, using
six different types of methods, was sufficiently convinced of the
soundness of his work to provide a refereed paper offering his
"true" value as developed by the most modern, sensitive, and
reliable procedure and instrumentation available at the time of
presentation.

Trying to discover a pattern among the methods does not seem
to lead anywhere. The spectrophotometric (colorimetric) methods
used initially, which usually have numerous steps, seem to give
high, but not the highest, values. Emission spectrometric methods
appear to cluster in the middle of the scale. Atomic absorption
methods, some of which have extensive preliminary cleanup steps,
have a downward trend, particularly after the introduction of the
graphite furnace. Two of the most recent values were obtained by
neutron activation with chemical separation in one case (0.16 ppb)

TABLE 1. REPORTED CHROMIUM CONCENTRATIONS IN BLOOD

Reference	Year	Method	Concentration ug/L (ppb)
Grushko, Ya. M. (Biokhimiya 13, 124-126; (CA 42, 8302i)	1948	ES	35
Urone, P. F. & Anders, H. K. (Anal. Chem. 22, 1317-1321)	1950	Sp	50
Monacelli, R., et al. (Clin. Chim. Acta 1, 577-582)	1956	ES	180
Miller, D. O. & Yoe, J. H. (Clin. Chim. Acta 4, 378-383)	1959	Sp	30
Paixao, L. M. & Yoe, J. H. (Clin. Chim. Acta 4, 507-514)	1959	ES	24
Herring, W. B., et al. (Am. J. Clin. Nutr. 8, 846-854)	1960	ES	27
Volod'ko, L. V. & Pristupa, Ch. V. (Vestsi Akad. Navuk B. SSR, No. 1, 107-109; CA 57, 11702a)	1962	ES	200
Schroeder, H. A., et al. (J. Chronic Dis. 15, 941-964)	1962	Sp	390
Wolstenholme, W. A. (Nature 203, 1284-1285)	1964	SSMS	1000
Glinsmann, W. H., et al. (Science 152, 1243-1245)	1966	AA	27
Feldman, F. J., et al. (Anal. Chim. Acta 38, 489-497)	1967	AA	29
Niedermeier, W. & Griggs, J. H. (J. Chronic Dis. 23, 527-535)	1971	ES	28
Hambridge, K. M. (Anal. Chem. 43, 103-107)	1971	ES	13

TABLE 1. (continued)

Reference	Year	Method	Concentration ug/L (ppb)
Cary, E. E. & Allaway, W. H. (J. Agric. Food Chem. 19, 1159-1161)	1971	AA	7
Wolff, W. R., et al. (Anal. Chem. 44, 616-618)	1972	GC/MS	10
Davidson, I.W.F. & Secrest, W.L. (Anal. Chem. 44, 1808-1812)	1972	AA	5.1
Davidson, I.W.F. & Burt, R.L. (Am. J. Obstet. Gynecol. 116, 601-608)	1973	AA	4.7
Pekarek, R. S., et al. (Anal. Biochem. 59, 283-292)	1974	AA	1.6
Li, R. T. & Hercules, D. M. (Anal. Chem. 46, 916-920)	1974	Chlm	146
Versieck, J. et al. (Clin. Chem. 24, 303-308)	1978	NA	0.16
Ward, N. I., et al. (Anal. Chim. Acta 110, 9-19)	1979	NA / AA	20 / 20

ES	= Emission spectroscopy
Sp	= Spectrophotometric (diphenylcarbazide)
SSMS	= Spark source mass spectrometry
AA	= Atomic absorption
GC	= Gas chromatography
MS	= Mass spectrometry
Chlm	= Chemiluminescence
NA	= Neutron activation

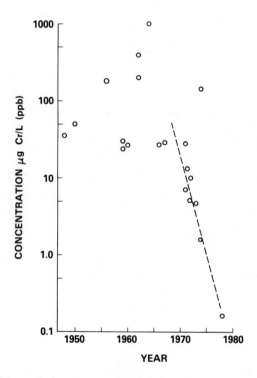

Figure 1. Reported chromium concentration in blood as a function of date of publication

and by standard additions in the other (20 ppb), which differ by
two orders of magnitude.

It is interesting that many of these papers provide linear
calibration curves and recovery of added chromium which approximate 100%. Excellent recoveries were reported despite several
breaches in good analytical practice, such as working at concentrations considerably above the level of interest, operating with
a performance blank that produces a background response that is an
appreciable fraction of the measurement response, and the use of
samples suspected of being contaminated.   Standard Reference
Materials (SRM) of the National Bureau of Standards (NBS) certified for chromium were not available until the middle 1970s. The
original materials, orchard leaves and spinach, contain several
parts per million of chromium, well outside our range of interest.
A brewer's yeast certified specifically for chromium content at
2.12±0.05 ug/g (SRM 1569) also became available in 1976.   Bovine
liver SRM certified at 0.088 ug chromium/g is now also available.

Some of the listed procedures claim high and even unique
specificity, low level of detectability, and extreme rapidity.
Often these claims are made with no mention of the relative magnitude of the accompanying blank and with no evidence of appreciation of the problem of contamination. Frequently comparisons
were made among methods but no mention is made of the starting
point--whether it was the original matrix or a common, prepared
solution. Since in almost all cases where concurrent methods were
used, the various methods gave similar results, it may be assumed
that they shared a common basis for contamination, if it existed.

Thus we see that although six basically different methods
have been used for the determination of chromium in a common, presumably stable and fairly constant biological substrate, blood, we
do not know its chromium content. We cannot assume that the lowest
value is the most correct since there may have been losses; we can
be quite confident that the high values were subject to some contamination. Yet every method was validated by spiking with known
amounts of chromium, and even with labeled ^{51}Cr in some cases, with
"excellent results."   In many cases, blanks were mentioned as
accounted for. If we cannot decide on the concentration of an
inorganic element at trace levels in blood, how can we do better
with more complex and less stable organic molecules in this and
other tissues?

The chromium example is not unique.   We have several other
interesting examples in the area of trace analysis of biological
materials. Most of them are from trace element analysis since this
specialty has been an active area of methods research for at least
half a century, and there are available a number of SRMs from the
NBS for use as reference points.

The remarkable influence of methods of analysis on estimates
of arsenic intake is shown by an evaluation of the data given by
Jelinek and Corneliussen (10) summarizing the arsenic content of
FDA's "total diet" composites during the reporting periods of 1967

through 1975, supplemented by later, as yet unpublished, values
through 1978. The average calculated annual daily intake of
arsenic (as As_2O_3) is shown in Figure 2. Substantial discon-
tinuities occur between 1970 and 1971 and between 1975 and 1976.
In 1970 the program was consolidated in a single laboratory and the
molybdenum blue method of analysis was replaced by the silver
diethyldithiocarbamate procedure, with a resulting lowering of the
blank and operation at a lower limit of reliable measurement. Thus
much of the apparent decrease in the arsenic content of the diet
(as As_2O_3) from an average of 80 ug/day during the 1967-1970 period
to 15 ug/day during 1971-1974 may be an analytical artifact that
does not at all reflect a drastic decrease in the arsenic intake
during this period. The 1975-1976 discontinuity coincides with a
further method change from the silver diethyl-dithiocarbamate
colorimetric procedure to the hydride-atomic absorption procedure. This change brought the total diet values back up to those
originally given by the molybdenum blue method. This interpretation is reinforced by the fact that an identical artifact is
noted in the Canadian total diet program, but a year earlier. The
level of arsenic (as As) found during the first quarter of 1969,
using a modified Guitzeit method, would contribute to the diet not
more than 95 ug/day. Subsequently the method was changed to the
silver diethyldithiocarbamate procedure. The maximum levels or
arsenic (as As) in the total diets dropped to not more than 30
ug/person/day in 1970, 30 in 1971, and 35 in 1972-1973 (11). The
Canadian program was discontinued before any further method change
was introduced.

Precision

Precision is the estimate of variability of measurements. It
is often confused with, or used interchangeably (and incorrectly)
with, accuracy. Accuracy reflects systematic error; precision
reflects random error. The concept is really more complex since
the systematic error term also is subject to random variability,
but for our purpose we can treat the two attributes of analytical
methods as separate characteristics.

Precision is a term which must be handled with care because
there are many different precisions. Any time there is a source of
variability, there is a precision associated with it. It is
usually expressed as a standard deviation at a certain level of
analyte. It can be associated with sampling as a random variability within a single material or as an among samples random
variability of a number of related materials. The most common
analytical precision terms are repeatability, which is the term
associated with a single operator (within-laboratory) and reproducibility, which is the term associated with different operators in different laboratories (between-laboratory). For research
work, repeatability is most often reported; for regulatory work,
the variability between laboratories is the most important. The

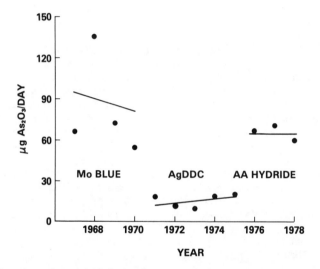

Figure 2.   Annual average daily intake of arsenic (as As_2O_3) in the U.S. total diet as a function of the method of analysis: Mo blue = molybdenum blue method; AgDDC = silver diethyldithiocarbamate method; AA hydride = arsine evolution, atomic absorption determination.

term most often reported in toxicological papers is standard
error, which is a standard deviation of a mean within a laboratory.
Its popularity probably lies in the fact that it results in the
smallest value of all the precisions mentioned. It does not
reflect the variability of individual measurements; rather, it
reflects the variability of means. In comparing precisions one
must be sure that the same types of terms are being compared;
otherwise interpretations are distorted. One of the most important statements of precision is the 95% prediction interval for a
single future assay at a specific concentration that encompasses
all usual analytical variables including different laboratories.
A minimum of 30 data points is needed for a reasonable estimate of
this term.

The next question is what precisions are reasonably expected
in trace analysis. At first glance this would appear to be a very
difficult question to answer when you consider the complicated
environment that analytical chemists and toxicologists must deal
with--mineral and vegetable; solids, liquids, and gases; single
substances and complex mixtures; pure materials to trace organics;
and small molecules to complicated polymeric mixtures. Superim-
pose upon composition variables the variety of techniques at our
disposal--spectrophotometry from infrared to X-rays; chromatography in all of its variations--gas, liquid, and solid; electro-
chemistry and mass spectrometry in all of their modifications; and
the neglected gravimetric and volumetric procedures. Yet we have
found that the results of our total analytical measurement
variability can be summarized, in an oversimplified fashion to be
sure, by plotting the determined mean CV expressed as powers of 2,
against the concentration measured, expressed in powers of 10, as
shown in Figure 3. The sources of the data are the interlaboratory
collaborative studies conducted under the auspices of the Association of Official Analytical Chemists (AOAC) over the past 100
years. The collaborative study technique subjects a clearly
defined individual method to a test by at least a half dozen
laboratories on a series of blind samples. The analytical results
are examined for bias, and for inter- and intra-laboratory variability to determine if the methods are suitable for use in
enforcing laws and regulations by agencies such as the FDA, the
Food Safety and Quality Service of the U. S. Department of
Agriculture (USDA), and the EPA.

The data supporting this relationship have been reviewed in
detail for pharmaceutical preparations at concentration levels of
approximately 0.1 to 100% (12), for pesticide residues at about 1
ppm (13), and for aflatoxins at about 10 ppb (14). We have
recently reviewed the collaboratively studied methods for sulfona-
mides in feeds at about 100 ppm (0.01%), which shows a CV of about
4%, and various drugs as tissue residues at about 1 ppm with a CV
of about 16%. We have also spot checked individual studies of
major nutrients at the 0.1-10% levels, minor nutrients and drugs
at the 10-100 ppm levels, and trace elements by atomic absorption

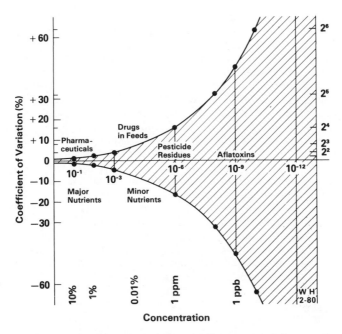

*Figure 3. Variation of the interlaboratory coefficient of variation (relative stand-
ard deviation × 100) with concentration*

and polarographic techniques at the ppm and below levels. They too fall approximately in the region bracketed by the curves of Figure 3. There are no comparable experimental points below 10^{-10} (0.1 ppb) but continuation of the exponential relationship is expected. Some partial studies have been made of methods for dioxins, partial in the sense that either the cleanup or the mass spectrometry has been studied collaboratively but not both together as consecutive steps in a single procedure. The data thus far suggest a CV of about the anticipated 100% at 10 ppt. Even radioimmunoassays appear to correspond to the precision curve. Hunter and McKenzie (15) report what appears to be a final average between-laboratory CV of approximately 30% in the United Kingdom national quality control scheme for the examination of serum growth hormone at the 5-100 ppb level by radioimmunoassay.

It should be remembered that this curve is merely a summary of the available interlaboratory data, covering methodological aspects only. External influences such as sampling and contamination are not involved. The data points are averages of a number of studies of similar analytes over ranges which may cover several orders of magnitude. Any single study may deviate in concentration by an order of magnitude or so. But in general, these values taken from the curve may be interpreted as indicative of satisfactory performance of an analytical method by different laboratories. Methods giving larger variability than those indicated by the curve can stand improvement; those methods giving values inside the curve probably are as good as can be expected.

The data used for Figure 3 are given in Table 2, and are based upon about 50 independent collaborative studies, using five types of determinative systems. On the basis of these data, the interlaboratory precision as a function of concentration appears to be independent of the nature of the analyte or of the analytical technique that was used for the measurement, a rather unexpected conclusion. Note particularly the interesting data from collaborative studies on analysis of metals at decreasing concentration. The methods used in these studies have not been accepted by the AOAC for use at these low levels. These same studies also reveal an interesting relationship between the within-laboratory and between-laboratory variability. The component essentially due to analysts (within-laboratory) is usually one-half to two-thirds that of the total variability (the sum of the within- and between-laboratory error). Ratios of within-laboratory to total variability below 0.5 indicate a very personal method; an analyst can check himself very well but he cannot check other analysts in other laboratories. A high ratio indicates either considerable interaction among laboratories or individual analyst replications so poor that they swamp out the between-laboratory component. This ratio of 0.50-0.67 also appears to be typical of methods utilized in clinical chemistry (16).

There are some independent confirmatory pieces of evidence supporting these values. Quality control studies of pesticide

TABLE 2. INTERLABORATORY COEFFICIENT OF VARIATION AS A FUNCTION
OF CONCENTRATION

Approximate concentration		Analyte (substrate)	Determinative Methods	Approximate Coefficient of variation
Range Units	Mean $(100\% = 10^0)$			$\%$
0.25-20 $\%$	1×10^{-1}	salt (foods)	Potentiometric	$\sqrt{2}$ = 1.4
0.1-60 $\%$	1×10^{-2}	drugs (formulations)	Chromatographic) separations,) spectrophoto- metric,) automated, manual)	2 = 2
0.1-0.05	2×10^{-4}	sulfonamides (feeds)	Spectrophotometric	2^2 = 4
0.37-17 ppm	1×10^{-6}	pesticides (foods, feeds)	Gas chromatographic	2^4 = 16
	1×10^{-6}	trace elements (foods)	Atomic absorption	2^4 = 16
2-200 ppb	1×10^{-8}	aflatoxins B_1, B_2, G_1, G_2 (foods, feeds)	Thin layer chromatography	2^5 = 32
1-100 ppb	1×10^{-8}	pesticide residues (total diet)	Gas chromatographic	2^5 = 32
0.05-5 ppb	1×10^{-9}	aflatoxin M (fluid milk)	Thin layer chromatographic	$2^{5.5}$ = 45
0.5 $\times 10^{-6}$		copper	Atomic absorption	22
0.15 $\times 10^{-6}$		zinc	Atomic absorption	54
0.05 $\times 10^{-6}$		lead	Voltametric	80
0.005 $\times 10^{-6}$		cadmium	Voltametric	220

residue determinations in blood by EPA contractors showed that
their CVs decreased with experience, but only down to an asymp-
totic value approximating the 16% found in the collaborative
studies on foods. Similarly the quality control monitoring of
laboratories determining aflatoxin by the Food Safety and Quality
Service of the USDA gives a value which corresponds to the 32% CV
given for aflatoxins at the 10 ppb level.

It cannot be overemphasized that these values are for data
from many laboratories in blind studies. They are useful for
interpreting the results of analysis of unknown samples, as analyzed by a number of laboratories. They obviously do not correspond to the values for the repeatability (single laboratory)
reported in the literature for standard solutions, recoveries of
added analytes, and comparisons with other methods. Rather, the
values in Figure 3 reflect the expected precision on real blind
samples analyzed under somewhat ideal conditions. Analysis under
practical conditions would be expected to be somewhat poorer;
analysis in a single laboratory by a single analyst would be
expected to be considerably better. On balance, then, Figure 3
approximates what should be expected of methods operated at the
indicated levels.

Limit of Reliable Measurement

The final property of methods which we will consider here is
the limit of reliable measurement. This is the quantitative
aspect of the common limit of detection--the smallest amount or
concentration of a substance which provides a measurable response
by a specified method. Although the limit of detection is a widely
used term, particularly by advertisers of scientific instruments,
it and related terms are not well defined, accepted, or understood. In fact, this characteristic, although intuitively simple,
may not be a stable attribute of analytical methods, but more a
function of external influences such as laboratory environment or
electronic fluctuations.

The limit of detection proved to be quite useless and in fact
rather misleading when applied to the problem of determining
dietary or environmental exposure to contaminants. In survey programs, such as FDA's total diet pesticide intake studies, the diet
of a specific population is analyzed to determine the consumption
of specified components and changes with time. Many of the samples
in such surveys are negative for the analyte of concern, and a
significant proportion are near or at the limit of detection.
Considerable uncertainty exists as to what value should be
assigned, for calculation purposes, to amounts which are detectable, but at a level for which the analyst is unable to assign a
definite quantitative value. In most cases, there are a few
foods, such as animal fats containing organochlorine pesticides,
that usually make such a large contribution to the total dietary
intake of a pesticide that the contribution of trace amounts of the

pesticide in other categories is insignificant. There can be more
generally distributed analytes at "trace" levels that in total may
be toxicologically significant, as in the case of dietary lead
intake (17), where 0.1 ppm is considered the limit of reliable
measurement. The calculated daily dietary lead intake was 57,
159, or 233 ug, depending upon which of the following value
assignments was made: zero for both zero and trace amounts; zero
for zero amount and 0.09 ppm for trace; or 0.05 ppm for zero and
0.09 ppm for trace.

Too often the term "sensitivity" is misapplied to the concept
of limits of detection or determination. Sensitivity is the slope
of the response curve--the change in response per unit measured--
as in almost all other branches of measurement. The concept of
least measurable amount is better described as determinability, or
limit of reliable measurement, and least detectable amount as
detectability, or limit of detection.

Determinability as a property of analytical methods became
important with the passage of the Pesticide Chemicals Amendment to
the Food, Drug, and Cosmetic Act in 1954, which introduced the
concept of "zero residue" to analytical chemistry. The amendment
required that the tolerance for a pesticide residue in food that
has not been shown to be safe should be set at a level no higher
than zero. The Delaney clause of the Food Additive Amendment of
1958 prohibited the acceptance as a regulated food additive of any
substance which was shown to be a carcinogen. The "zero tolerance"
and "no carcinogen" ideals were introduced at a time when a
fraction of a part per million was considered as the limit of
detection. The invention of gas chromatography about this time
revolutionized trace analysis and pushed the limit of detection
for pesticide and drug residues toward the parts per billion
level. Chemists and administrators began to realize that the
terms "zero," "no," and "none" were not absolute entities but
rather were functions of the method employed and the confidence
required. The recognition of this fact resulted in a further
revision of the food additive section of the Act in 1962 which
permitted feeding carcinogenic drugs to animals providing "no
residue of the additive will be found by methods of examination
prescribed or approved by the Secretary..."

But the question remains as to what constitutes "no residue."
Currie (18) examined the corresponding problem of detection limits
in radiochemical procedures and was frustrated by the differences
in terminology and definitions which resulted in a range of three
orders of magnitude for detection limits calculated for the same
system. Figure 4, taken from his paper, shows the situation with
respect to a specific radioactivity process. The horizontal lines
indicate three specific levels: L_C, "decision limit," the level a
signal must exceed to permit a decision as to whether or not the
result of an analysis indicates detection; L_D, "detection limit,"
the level above which an analytical procedure can be relied upon to
lead to detection; and L_Q, "determination limit," the level above

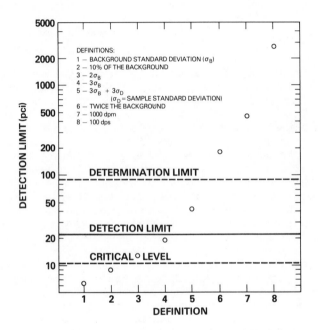

Figure 4. Ordered detection limits—alternative literature definitions and proposed alternatives (18)

which an analytical procedure will be sufficiently precise to
yield a satisfactory quantitative estimate. Currie considered
"sufficiently precise" as the point where the (presumably intra-
laboratory) relative standard deviation was 10%. Figure 3, our
precision curve, shows that this variability is ordinarily reached
at the ppm level, where the intralaboratory CV of 10% is
approximately equivalent to the interlaboratory CV of 15% of the
curve. This presents a real problem with regard to the reliability
of determination limits which necessarily have to be conducted at
lower levels--aflatoxins at parts per billion and dioxins at parts
per trillion.

In his earlier paper, Currie considered only the random error
component. Later, Currie and DeVoe (19) considered the effect of
systematic errors (bias) on detection limits (and by implication
determination limits). At these levels, random error introduces a
sizable component into the presumably stable bias component.
Therefore, in order to detect a systematic error of magnitude
comparable to the standard deviation, one needs at least 15 observations. If the systematic error is not constant, these authors
point out that it becomes impossible to generate meaningful uncertainty bounds for experimental data.

We can begin to analyze the stability of methods at the parts
per trillion level by examining the results from the EPA partial
collaborative study on dioxins (20). In this study, samples of
beef fat and of human milk were extracted; the extracts were
cleaned up in a single laboratory; the cleaned up extracts and
equivalent standards (as unknowns) were supplied to five participants for quantitation by mass spectrometry. Only two of the
laboratories examined all samples. Three laboratories used single
ion monitoring (m/e = 322); two used double ion monitoring (m/e =
320, 322) and the average of the quantitative results was used as
the value found, although both values were reported.

Because of the unbalanced design, the use of different
laboratories for the isolation and determination, and the small
numbers of laboratories involved with each type of sample, the
data cannot be examined by conventional means and consequently
cannot easily be compared with the interlaboratory variability of
methodology for other contaminants.

However, the report shows that the methods are completely
unreliable with respect to negative and lowest level samples. The
number of samples of each type examined and the percent of negative
(no added dioxin) samples reported positive (false positives) are
given in Table 3.

Most of the false negative reports (reporting zero when
dioxin was added) occurred at levels of 9 ppt and below. The only
false negatives at levels above 9 ppt, oddly enough, occurred in
the standard series (no interference). No false negatives were
reported in the beef fat and human milk series above 9 ppt.
Therefore, examination of the data by inspection results in an
estimate of about 10 ppt for the limit of reliable measurement in

TABLE 3. FALSE POSITIVE DIOXIN VALUES REPORTED IN EPA
          RECOVERY STUDIES (20) AT 9 PPT AND BELOW

	No. of labs	No. of samples examined	% False positives
Standards	3	16*	19
Beef fat	4	26*	42
Human milk	3	12*	92

*Where double ion monitoring was used, the value from
each ion was considered as a separate sample. Ignoring
the second ion value (to place all laboratories on a
comparable basis) would not change the % false positives
significantly. Considering only the 322 values (instead
of both 320 (when used) and 322) would give 17%, 43%, and
90% false positives for the three types of samples.
Similarly, the use of two types of methods by one laboratory on beef fat was ignored.

this study. How reliable the measurement at this limit is requires considerably more data than are available. However, a rough estimate of the interlaboratory precision indicates a CV of approximately 100% at 10 ppt (10^{-11}) which can be considered as lying close to our prediction of 90% ($2^{6.5}$) from the precision curve. A large uncertainty is introduced because the extracts were prepared and cleaned up in a single laboratory and examined in different laboratories. If each laboratory had performed its own analytical operations as well as the mass spectrometry, the overall variability probably would have been larger.

Although most procedures for determining limit of detection, determination, or reliable measurement are based upon the calibration curve, this approach does not appear to be practical, based on the limited experience of the TCDD study. The slopes and intercepts at zero concentration of the calibration curves for standards and of the recovery curves for the beef and milk fats vary considerably from laboratory to laboratory with a range of the intercept of the recovery curve from -1.5 to +14 (i.e., 14 ppt must be added to obtain a 0 ppt TCDD found!) and a range of slopes from 0.37 to 1.36, as shown in Figure 5. In the report (20), regression curves and associated limits were also calculated for a single laboratory. Although the confidence interval of the curve (all values except (presumably) zero) was fairly tight (i.e., at 50 ppt, the interval was 12 ppt), the corresponding prediction interval for a single observation was about 50 ppt (100%).

Conclusion

Since important decisions affecting the health and welfare of humanity must be made on the basis of analytical results, considerable effort must be directed toward assuring greater confidence in the reliability of the output of analytical laboratories. The Commission of the European Communities, after performing a study to determine the comparability of chemical analyses for drinking water quality, concluded that analytical quality control must be required as a routine component of analytical work. They state (21), "Only the combination of intralaboratory controls of precision and accuracy complemented by interlaboratory intercomparison tests can lead to a significant evaluation and improvement of analytical results."

The most difficult part of the procedure of producing reliable analytical values will be obtaining a recognition by analysts of the necessity for quality control as an inherent accompaniment of analytical work. If analysts do not utilize this technique voluntarily, outside auditors will insist that such data accompany all regulatory submissions, as part of compliance with good laboratory practice regulations.

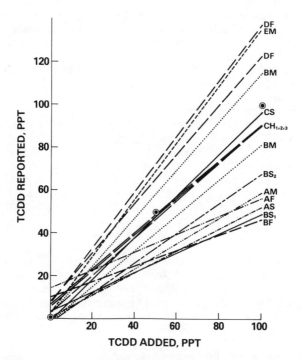

Figure 5.   Regression lines of five individual laboratories (A–E) examining stand-ards (S), and extracts of beef fat (F) and milk fat (M) for TCDD as random un-knowns (EPA data (20))

LITERATURE CITED

1. Inhorn, Stanley L. Ed. "Quality Assurance Practices for
    Health Laboratories"; 1978, American Public Health
    Association: 1015 Eighteenth Street, NW, Washington, DC
    20036.

2. Boyer, Kenneth, W.; Capar, Stephen G.; Fortification
    variability in rat diets fortified with arsenic, cadmium,
    and lead. J. Toxicol. Environ. Health 1977, 3, 745-753.

3. Oller, William L.; Gough, Bobby; Littlefield, Neal A.
    Chemical surveillance and quality assurance for prepara-
    tion of dosed (2-AAF) animal feed (ED_{01} study). J.
    Environ. Pathol. Toxicol. 1980, 3, 203-210.

4. U. S. Environmental Protection Agency, Environmental
    Health Advisory Committee, Science Advisory Board
    (December 29, 1978) Report of the Ad Hoc Study Group on
    Pentachlorophenol Contaminants. EPA/SAB 78/001.
    Washington, DC 20460, P. 4.

5. Schechter, Milton S.; The need for confirmation. Pestic.
    Monit. J. 1968, 2(1), 1.

6. Carnaghan, R. B. A.; Hartley, R., D.; O'Kelly, J.
    Toxicity and fluorescence properties of the aflatoxins.
    Nature 1963, 200, 1101.

7. Dipple, Anthony. Polynuclear Aromatic Carcinogens. In
    "Chemical Carcinogens." Charles E. Searle, Ed. ACS
    Monograph 173. 1976, American Chemical Society,
    Washington, DC.

8. Huff, J. E.; Wassom, J. S. Health hazards from chemical
    impurities: Chlorinated dibenzodioxins and chlorinated
    dibenzofurans. Int. J. Environ. Stud. 1974, 6 , 13-17.

9. Mertz, W. Trace-element nutrition in health and disease:
    Contributions and problems of analysis. Clin. Chem.
    1975, 21, 468-475.

10. Jelinek, C. F.; Corneliussen, P. E. Levels of arsenic in
    the United States food supply. Environ. Health Perspect.
    1977, 19, 83-87.

11. Smith, D. C.; Pesticide residues in the total diet in
    Canada. Pestic. Sci. 1971, 2, 92-95; Smith, D. C. Pesti-
    cide residues in the total diet in Canada II. Pestic.
    Sci. 1972, 3, 207-210; Smith, D. C.; Leduc, R.;

Charbonneau, C. Pesticide residues in the total diet in
Canada, III-1971. Pestic. Sci. 1973, 4, 211-214; Smith,
D. C.; Leduc, R.; Tremblay, L. Pesticide residues in the
total diet of Canada IV. 1972 and 1973. Pestic. Sci.
1975, 6, 75-82.

12. Horwitz, W.; The variability of AOAC methods of analysis
    as used in analytical pharmaceutical chemistry. J. Assoc.
    Off. Anal. Chem. 1977, 60, 1355-1363.

13. Burke, J. A.; The interlaboratory study in pesticide
    residue analysis. Advances in pesticide sciences,
    Part 3. Biochemistry of pests and mode of action of
    pesticides. Pesticide Degradation. Pesticide Residues.
    Formulation Chemistry. Edited by H. Geissbühler.
    Pergamon Press, Oxford and New York, 1979, pp. 633-642.

14. Schuller, P.; Horwitz, W.; Stoloff L. A review of sampl-
    ing plans and collaboratively studied methods of analysis
    for aflatoxins. J. Assoc. Off. Anal. Chem. 1976, 59,
    1315-1343.

15. Hunter, W. M.; McKenzie, I. Quality control of radio-
    immunoassays for proteins: The first two and half years
    of a national scheme for serum growth hormone measure-
    ments. Ann. Clin. Biochem. 1979, 16, 131-146.

16. Steele, Bernard W.; Schauble, Muriel K.; Becktel, Jack
    M.; Bearman, Jacob E.; Evaluation of clinical chemistry
    laboratory performance in twenty Veterans Administration
    hospitals. Am. J. Clin. Pathol. 1977, 67, 594-602.

17. Kolbye, A. C. Jr.; Mahaffey, K. R.; Fiorino, J. A.;
    Corneliussen, P. C.; Jelinek, C. F. Food exposures to
    lead. Environ. Health Perspect. 1974, 65-74.

18. Currie, Lloyd A. Limits for qualitative detection and
    quantitative determination. Anal. Chem. 1968, 40, 586-
    593.

19. Currie, L. A.; DeVoe, J. R. Systematic error in chemical
    analysis. In "Validation of the Measurement Process"
    (1977) James R. DeVoe, Ed.; ACS Symposium Series 63.
    American Chemical Society, Washington, DC 20036.

20. Robert G. Heath; Interlaboratory method validation study
    for dioxins. An interim report. Human effects
    monitoring branch, OPP, OTS, EPA, January 5, 1979.

21. Commission of the European Communities, A study to deter-
    mine the comparability of chemical analyses for drinking
    water quality within the European communities.
    Luxembourg EUR 5542e, August 1976.

RECEIVED February 6, 1981.

Problems and Pitfalls in Analytical Studies in Toxicology

J. D. McKINNEY, P. W. ALBRO, R. H. COX,
J. R. HASS, and D. B. WALTERS

Laboratory of Environmental Chemistry, National Institute of Environmental
Health Sciences, P.O. Box 12233, Research Triangle Park, NC 27709

The major interrelated divisions of environmental health
sciences necessary to define a human health hazard are epidemiology, toxicology and chemistry. If one or more of these areas
is lacking or incomplete in studies of environmental agents such
as pesticides, then health hazard potential can not be fully
assessed. Chemistry is essential in these relationships in order
to provide timely and effective solutions to environmental health
problems. Environmental Health Chemistry has been presented (1)
as a new subdiscipline which emphasizes the chemistry needed to
establish these relationships and permit assessment of the potential human health hazards associated with chemical contamination
of our environment. Fundamental to these various programs is
assurance that good, reliable analytical data is generated
especially in those cases which most directly affect the public's
health and well being. How one generates good, reliable analytical data is the primary focus of the papers in this session. Our
position, which is similar to that taken by others (2), is one
of adopting good analytical practices as an alternative to
standard methods, licensing and certification. Our position is
presented in brief in the following outline of criteria for
analytical protocols.

Good Analytical Practices

General Considerations. It is often more difficult to judge
the validity of an analytical study than to perform the analyses.
Since the difficulty most commonly arises because of failure to
specify the full details of procedures in advance or report them
adequately at the conclusion of a study, the present outline
describes the types of information considered essential for both
preparing protocols for prior approval and submitting final
analytical results in report form. Planning and reporting
categories for the final processing steps, the actual determination procedures, will be the subject of future documents. The
present guidelines pertain to all of those steps preliminary to

determination; that is, from sample collection to presentation of
the final test preparation. The emphasis is upon validation of
procedures, and upon reporting in sufficient detail the procedures
that can in fact be evaluated subsequent to the submission of
analytical results.

Sample Selection. Assuming that the samples to be analyzed
are to be selected from a larger population or populations, the
manner in which samples will be selected must be described. If
selection is intended to be random, enough details must be given
to enable one to distinguish random from haphazard. Since some
statistical analysis may eventually be applied to the analytical
results, one must be able to determine whether the sample groups
are fully independent, or the observations are naturally paired.
Finally, evidence for considering the samples as representative
of the larger populations must be described.

Sample Collection and Storage. This step in an analytical
study offers many opportunities for loss of integrity of samples,
and must be described in full detail. Precisely what tools will
be used to acquire the samples? Disposable scalpels, for example,
are coated with an oil that can contaminate tissues. Metal tools
obviously should not be used to take samples for trace metal
determination. Good judgement can not be assumed; details must
be provided.
      The sample containers must be fully described. What material
is used? Aluminum foil is coated with drawing oil. Bottle cap
liners are a common source of contamination. How are the containers cleaned?
      How will samples be stored prior to analysis? At what
temperatures and for how long? Will they be exposed to light?
Air? Details must be given.

Sample Workup. Most trace level analyses start with some
sort of extraction step. This should be described in detail
("the samples were extracted with chloroform" is not enough).
How long does each step take? For how long are extraction
mixtures stirred or shaken? How many cycles of a Sohxlet extractor?
Enough information must be given to permit someone to duplicate
the procedure. Is the sample finely divided prior to or during
the extraction? What is the solvent to sample ratio?
      Vague terms such as "warm solvent", "extracted exhaustively",
or the like must be avoided. If temperature, time, etc. are
important, they should be precisely specified. Sources of solvents and reagents, and special means used to purify or dry them
must be given. Terms such as "reagent grade" provide no assurance
of freedom from relevant, interfering impurities.

Cleanup Procedures. Most trace level analyses involve some
sort of cleanup of the crude extract. Details of these procedures

must include sources and means of activation of chromatographic media, sources and some discussion of purity of solvents used, and sufficient description of the procedures themselves to permit their duplication.

Validation Studies. Prior to the analysis of unknown samples, all procedures must be validated in terms of recovery, reproducibility, sensitivity, freedom from interference, and accuracy.

Recovery Studies. Since these studies will generally involve the use of "spikes", the first requirement is for the spiking or fortification procedure itself to be validated. Either it must be shown that the "spiked" chemicals equilibrate with the corresponding endogenous ones, or it must be empirically demonstrated that the recovery of exogenous "spike" is the same as the recovery of endogenous compound, over the full range of concentration levels to be sought in the analysis of unknowns. The dependence of both percentage recovery and its standard deviation on concentration must be determined and reported. The actual fortification procedure must be described in sufficient detail to permit duplication.

Limits of Detection. Since the clean-up and recovery achieved in the overall sample workup effects the detection limit of the measurement technique, the actual limit of detection is, relative to the amount of sample available for analysis, always less than the limit of the measurement technique. Thus, if a measurement technique can respond to one microgram of material, but the recovery in the preliminary sample workup is 50%, then the true effective detection limit is two micrograms, and it is this latter value that must be reported. Additionally, the limit of detection of the measurement technique must be taken as not the smallest amount causing a response, but rather the smallest amount to which the analytical criteria employed for qualitative identification can be applied.

Criteria for Qualitative Identification. The criteria to be applied for qualitative identification must be described. Ideally, they should include at least one criterion unique to the compound of interest; failing this, they must include a combination of criteria, which combination is unique to the compound sought. In addition, the criteria by which interfering substances may be recognized must be described.

Criteria for Quantitative Determination. Not all assays provide linear calibration curves. Calibration curves must be constructed from the analysis of spiked samples, to visualize any nonlinearity in recovery. If interpolation is utilized, the interpolation procedure must be fully described; extrapolation should not be used. Standard curves must span the complete range of values found in the set of unknown samples. The samples used for spiking should ideally contain no detectable endogenous compounds responding to the assay, but if zero background can not be achieved, the endogenous (background) level in the spiked samples must be accurately determined and compensated for.

Blanks. "Procedure blanks", samples of the same matrix material that will subsequently be analyzed in the study but known to lack detectable levels of the compounds of interest, must be put through the entire procedure including storage for the same (mean) length of time in the same environment as the test samples. Extraction and cleanup procedures applied to these "blanks" must involve the same amounts of solvents and chromatographic media as will be used on the real samples. In no other way can the likelihood of interferences (false positives) be convincingly assessed.

This paper identifies some unique problems and pitfalls in providing analytical support for toxicological research consistent with our position on good analytical practices. Major areas relevant to toxicological research receiving some attention include environmental analysis, mechanism elucidation, testing programs, toxicity prediction, chemical epidemiology, and safety monitoring.

Environmental Analysis

Assessment of the human health hazard potential of environmental chemicals requires study of the environmental transformation products. The products can include a variety of metabolites generated in plants, soils and animal tissue as well as nonbiological products such as those derived from hydrolysis or photolysis. Therefore, environmental analysis is of basic importance in determining the substances reaching the human environmental interface. The commercial uses of polychlorinated biphenyl mixtures (PCBs) as insulating and dielectric fluids is an example of this problem. The commercial mixture consists of a complex mixture of chlorobiphenyls varying in degrees and positions of chlorine substitution along with a few part-per-million (ppm) chlorinated dibenzofuran contaminants. We now know that the PCBs can be transformed chemically during use (3), biologically in cultures and soil (5), and animal tissues (4) and under physical influence such as ultraviolet light (5) and incineration conditions (6) (Table 1). Residues in humans (7) largely consist of those PCB isomers with high chlorine content which are generally more resistant to chemical and biological transformation. The exact nature of these residues is still being investigated in an attempt to assess the real human health hazard associated with the PCBs (1).

Table I
Classification and Occurrence of Compounds
Associated with the PCB Problem

Compounds	In Commercial Mixtures	Product of Use or Transportation
PCBs and Related Derivatives:		
Non-ortho substituted with high meta-para substitution (Type I)	X	
Ortho substituted Vicinal unsubstituted carbon atoms absent (Type II)	X	
Vicinal unsubstituted carbon atoms present (Type III)	X	
Biphenyl Dimers and Trimers		X[a]
Oxygenated Chlorinated Aromatic Compounds:		
Chlorinated Dibenzofurans	X	X[a]
Chlorinated Diphenylethers		X[a]
Hydroxychlorinated Biphenyls and Related Oxygen Containing Compounds		X[b]

[a] This is based on preliminary data (4) from analysis of the used fluids as well as analysis of model chemical reactions under simulated use conditions.

[b] Biological Transformation.

It has been only recently (8) that the chlorinated dibenzofurans have been tentatively identified along with PCBs in a higher animal. A relatively new technique using negative chemical ionization mass spectrometry provides a rapid and sensitive qualitative screen for these compounds. This technique has had limited application thus far. However, it is not clear whether or not these furans are the original contaminants of the PCB mixtures and/or transformation products of PCBs or other compounds in the environment. Confirmation is needed to rule out possible interfering compounds such as chlorinated diphenyl ethers. The important point to make is that the toxicity of the released products can be considerably different from the modified product which ultimately reaches man or any other higher monogastric animal.

Mechanism Elucidation

One of the more extensively studied areas in drug metabolism
has been the oxidative metabolism of alkenes and polycyclic
aromatic hydrocarbons (PAHs). The evidence is convincing that
this proceeds by formation of epoxides and arene oxides in a
reaction mediated by the cytochrome P-450 dependent monoxygenase
system. The fate of these reactive oxides in biological systems
and the relationships to biological endpoints and toxicity are
receiving considerable attention (9). Improved and more sophisticated methodology for elucidating these mechanisms of action are
continually being developed. Two examples will be described, one
considered to be a major detoxication pathway and the other a
major activating pathway, illustrating some of the problems and
some of the newer analytical approaches to overcoming these
problems.

Conjugation of oxides with glutathione (GSH) catalyzed by
glutathione transferases is a major detoxication process for
removal of these reactive molecules (10). However, studies of
this metabolic process often do not fully recognize the potential
complexity of the metabolic profile. For example, in our studies
(11) of the metabolism of styrene oxide, it has been demonstrated
that both positional and diastereoisomeric metabolite conjugates
of GSH and mercapturic acids are formed. The total characterization of this process has required the synthesis of styrene oxide
conjugates of GSH, cysteinylglycine, cysteine and N-acetylcysteine
(mercapturic acid), both positional and diastereoisomeric using
optically active isomers of styrene oxide. Characterization and
measurement of these isomers in mixtures further required the
development of methodology in ^{13}C nuclear magnetic resonance
(NMR) spectroscopy and high pressure liquid chromatography (HPLC).
The GSH conjugates have also been prepared for α-methylstyrene
oxide, trans-β-methylstyrene oxide, 1,2,3,4-tetrahydronaphthalene-
1,2-oxide, phenanthrene-9,10-oxide, pyrene-4,5-oxide and benzo[a]
pyrene-4,5-oxide (BP-4,5-oxide). Using standards available from
our synthetic studies, we have compared the enzymatic and chemical
conjugation of glutathione with epoxides and found some interesting differences in the regiospecificity and stereospecificity of
the conjugation reaction. For example, the chemical conjugation
of GSH with benzo[a]pyrene-4,5-oxide produces equal amounts of
the 4- and 5-thioether isomers as a mixture of diastereoisomers.
The enzymatic conjugation using rat liver cytosol produces a
mixture of the 4- and 5-thioether conjugates. However, the
diastereoisomers of the conjugates are not formed in equal amounts.
With a purified enzyme from the Little River Skate, an equal
mixture of the 4- and 5-thioether conjugates are produced. But
in this case, only one of the diastereoisomers of each positional
isomer is produced. Product analysis of the GSH conjugates
obtained from ^{13}C-labeled BP-4,5-oxide (4,5-^{13}C) established some
definite stereochemical requirements for the catalytic step (12).

The data demonstrate that the skate liver enzyme has high regioselectivity and stereospecificity for each BP-4,5-oxide enantio-
mer.

The mechanistic implications of these results concerning the
regioselectivity and stereospecificity of enzymatic conjugation
are being evaluated further. The importance and potent biological
activity of certain glutathione conjugates is being increasingly
recognized (13).

Most chemical carcinogens require metabolic activation to
highly reactive electrophilic intermediates to be carcinogenic.
Such intermediates can bind covalently to cellular constituents
such as RNA, DNA, and proteins. Therefore, one approach to the
study of chemical carcinogenesis is to determine the nature and
degree of covalent binding processes in biological systems. The
role of arene oxides as reactive metabolic intermediates has
been investigated extensively, and the subject of reactive metabolites is addressed in a separate paper in this conference.
Evidence is increasing to support the formation of arene oxides
during the metabolism of certain PCBs (14). In addition, chemical
synthesis of certain PCB arene oxides has been completed and
these oxides show the expected chemical properties consistent
with their potential for rearrangement and covalent binding to
various nucleophiles.

However, reports of PCB binding to biopolymers in vivo and
in vitro generally do not differentiate between bound and simply
adsorbed residues. In order to prove that a chemical binds
covalently to a biopolymer, it is necessary to isolate and
characterize the modified polymers and monomers. Simple failures
to extract PCBs from tissues with organic solvents or physical
methods of fractionation alone, do not constitute evidence of
covalent binding. Such studies are greatly facilitated by the
use of radioisotopes since the amount of covalent binding is
usually quite small.

Recent work (15) in mouse liver (in vivo) with a slowly
metabolized PCB (2,4,5,2',4',5'-hexachlorobiphenyl) and a rapidly
metabolized PCB (2,3,6,2',3',6'-hexachlorobiphenyl) along with
the appropriate controls has clearly demonstrated increased
binding of the more rapidly metabolized isomer to biopolymers.
This was determined through isolation and characterization of PCB
bound biopolymers and monomers. The greatest binding was observed
in RNA followed by protein and DNA, respectively, and binding
occurs in tissues other than liver as well. This binding is
likely to be covalent and the result of metabolic activation, but
proof of this awaits further chemical characterization of the
isolated materials. Field desorption mass spectrometry (MS) and
other specialized MS techniques should be useful in characterizing
such adducts (16).

The analytical capability now exists to determine the nature
and extent of covalently bound chemicals in tissues. Further
work should provide a data base upon which to draw some general-

izations concerning likely targets for attack by reactive metabolites. However, the quantitative aspects of the problem have
received little attention so far. In elucidating the mechanism
of chemical carcinogenesis, one must determine the nature and
degree of "effective binding" to biopolymers, i.e. what specific
perturbations of the macromolecules are sufficient to induce
neoplastic transformation. Progress in this area is likely to be
dependent upon the development of good in vitro models for studying
neoplastic transformation in cells (17).

Structure-Activity and Toxicity Prediction

    The toxic propensity of a molecule resides in the chemical
makeup of the molecule. The availability of multiple measures of
physical/chemical, toxicological and pharmacokinetic properties
allows us to understand the toxicity of a molecule. One can
measure or calculate molecular properties and compare the magnitude of these properties with the magnitudes of observed biological responses of animals exposed to the molecule. It is essential
to have reliable and accurate biological measures of toxicity for
correlation purposes. We have used the exquisite sensitivity
(LD_{50}) of the guinea pig to the toxic effects of halogenated
aromatic hydrocarbons as a biological response for comparison
with some measured molecular properties considered important in a
specific receptor interaction (18). Measured molecular features
include size, shape, symmetry and polarizability. A variety of
techniques have been used to make these measurements including X-
ray crystallography, nuclear magnetic resonance and mass spectroscopy and gas chromatography.
    A number of isomers and homologs in the halogenated dibenzo-
p-dioxin, dibenzofuran, biphenyl and naphthalene classes have
been tested in the guinea pig. Based on these and other relevant
studies, some generalizations about structural properties important
in their toxicity can be made. The critical halogenation pattern
takes the approximate form of a 3 X 10 Å box for chlorines and
can be somewhat smaller and triangular for bromines. Planarity or
coplanarity of rings is necessary only to effect juxtaposition of
four lateral chlorines. This imparts a certain degree of symmetry
to the molecule, but symmetry per se is not a requirement.
Studies (19) of various isomers and homologs in the biphenyl
series suggest that the underlying factor responsible for binding
is net molecular polarizability which has a preferred distribution
along the 10 A receptor distance in the molecule. We are currently investigating ways to measure this property directly.
Some variation in these properties through molecular conforma-
tional preferences in the biphenyls is thought to explain the
apparent "mixed induction" activity seen for certain PCB isomers.
    Pharmacokinetic studies are obviously important for deter-
mining the fate and disposition of chemicals in biological systems.
The cost effectiveness of such work is greatly increased by the

use of radiolabeled compounds to facilitate the monitoring of
absorption, distribution and excretion of the parent compound
and its metabolites. A few examples will be described that
illustrate some of the problems in using both radioactive and
stable labels as biological tracers.

   The availability of labeled materials with the desired
properties can be a problem. Interest in carrying out pharmacokinetic studies on the toxic environmental contaminant 2,3,7,8-
tetrachlorodibenzofuran (2,3,7,8-TCDF) engendered a need for
synthesis of the compound ^{14}C labeled at a high specific activity
(greater than 50 mCi/mmol). After considerable synthetic work
(20), the labeled 2,3,7,8-TCDF was obtained in low yield but at
high purity via Pschorr cyclization of o-phenoxyaniline-U^{14}C,
chlorination of the resultant dibenzofuran and separation of the
tetrachloro isomers by high pressure liquid chromatography.
Completely anomalous results were obtained when more convenient
synthetic routes were tried that had been previously reported for
"cold" material. These results were interpreted in terms of the
intervention of "hot" free radical intermediates. The effects
observed in the promotion of reactive free radical formation are
perhaps not widely known. Similar effects could possibly obtain
in the metabolic degradation of the ^{14}C-2,3,7,8-TCDF affording
again reactive radical intermediates which could alter the
"normal" course of metabolism.

   There is at least one report (21) describing the significantly different gas chromatographic behavior of ^{14}C-labeled
2,3,7,8-tetrachlorodibenzo-p-dioxin (2,3,7,8-TCDD), specific
activity greater than 150 mCi/mmole. This result suggests that
at sufficiently high specific activities there can be an effect
on physical/chemical properties. Whether the highly labeled ^{14}C-
2,3,7,8-TCDD behaves in biological systems as unlabeled TCDD can
not be answered definitively, but it is important to be aware of
this potential problem.

   The use of stable isotopes is becoming increasingly popular
in biological work (22). Progress has been made in using ^{13}C-
labeled compounds to facilitate metabolite isolation and identification by MS techniques. Other work has been done using ^{13}C-
labels to help elucidate biomechanism. Several ^{13}C-labeled
benzo[a]pyrene metabolites were prepared with specific incorporation of ^{13}C-labels. These labeled compounds proved to be useful
in the assignment of ^{13}C-NMR spectra to the BP metabolites. The
(±) benzo[a]pyrene-4,5-oxide-4,5 ^{13}C also proved to be useful in
determining the positional and diastereoisomeric glutathione
conjugates as a result of its reaction with glutathione transferase in biological systems (12). The ^{13}C enriched compound
significantly simplified the analytical problem by enabling
measurements to be made on much smaller amounts of metabolites.
The double label at C-4 and C-5 in the BP-oxide was intended to
facilitate the determination of regiospecificity of its reactions,
but a single label at either C-4 or C-5 would have been just as

useful for this purpose and would have further improved the
analytical problem since the reduction in signal by $^{13}C-^{13}C$
coupling would have been eliminated. The availability of the
synthetic conjugates as standard reference compounds was essential
in this work.

Chemical Epidemiology

The importance of chemical epidemiology in assessing the
health hazard potential of environmental chemicals is obvious.
The variety of demands that are made on the analytical chemist in
support of epidemiology studies has recently been discussed (23)
and major factors identified which impact on the chemist.
Two areas of concern currently receiving attention by the
epidemiologist will be described. The first is the problem of
PCBs in mother's milk and the potential transfer of these chemicals to the baby during breast feeding; the second is the potential problem of dioxin (2,3,7,8-TCDD) residues in humans (milk
and fat) associated with exposure to 2,4,5-T and related herbicides and their precursors.
A study (24) is in progress that will attempt to correlate
health effects in the developing infant with levels of PCBs and
related compounds measured in mother's milk. Figure 1 illustrates
the specific problem of analyzing for PCB in human milk by electron capture-gas chromatography. Although several peaks appear
to coincide with the Aroclor standard, the pattern in milk is
clearly different from the standard pattern. One really can not
compare these peaks since each one likely contains more than one
compound with varying relative detector response. Therefore,
there is no true standard for quantitation purposes and absolute
quantitation by this method is not possible. The best that can
be hoped for is a reproducibly quantitative method that can be
used for relative comparisons. In developing such reproducible
methods, our experience has shown that extensive method validation
is required for each sample matrix of interest. Literature
methods have been of little value in facilitating the validation
work. Reproducibility of methods is clearly a function of both
method technology and operational techniques.
However, the analytical problems of PCBs in human tissues is
further complicated by the uniqueness of the pattern found in the
general population (7). Residues in humans appear to be largely
the ortho-substituted type PCBs with high chlorine content which
are not readily metabolized and eliminated from the body. These
isomers are also not particularly toxic on a short term basis.
Therefore, it is the PCBs which can not be measured by an extraction method that may be of greater biological consequence. As a
part of this epidemiological study, a method relying on neutron
activation analysis has been developed (25) that will allow
determination of total organic chlorine residues in body fluids
and tissues which includes both bound and unbound materials.

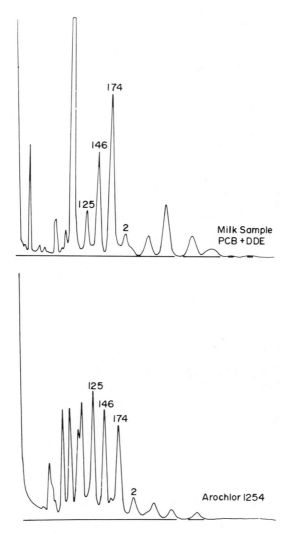

Figure 1. PCB pattern in human milk compared with Arochlor 1254 Standard. Peak off-scale is DDE. GC parameters: 3% OV-1, 220°C, 6 ft × 4 mm column.

The analytical problems associated with dioxin residues in humans are similar in many ways but there are some key differences which should be pointed out (26) (Table 2). Absolute quantitation is achievable since standard reference compounds are available or can be made available. However, because of the high toxicity of these compounds, the desired detection limit is in the low part-per-trillion (ppt) range instead of the ppb to ppm range required for PCBs. Dioxins are also not generally considered to have ubiquitous distribution in the environment although this has recently been challenged by the "chemistry of fire" proposal of Dow Chemical Company (27). As a result of the low concentrations being sought and the non-ubiquity of these residues, a considerably more complex and sophisticated analytical method is required to achieve the desired sensitivity and specificity needed. The method presently in use generally consist of some form of low or high resolution chromatography and high or low resolution mass spectrometry.

Table 2
Problems in Development of TCDD Analysis in Human Fat

1) Sample size usually small (fat biopsy - 0.5 g or less).

2) Low part-per-trillion (ppt) sensitivity desired because of exquisite toxicity of the compound (1 ppt on 0.5 g sample requires 0.5 picogram in total sample).

3) The analysis must not only be highly sensitive but highly specific for the 2,3,7,8-tetra isomer.

4) Requires synthetic work to provide analytical standards and highly compatible separation science and specific measurement capabilities under high resolution conditions (must eliminate interferences).

5) Requires stringent analytical protocols and safe handling procedures and facilities to maintain sample integrity and avoid contamination of facilities and exposure of personnel.

Although existing methods for TCDD have already been put into use for supporting limited epidemiological studies, they can not be defended as unequivocal for determination of 2,3,7,8-TCDD. The methods available lack complete validation in the appropriate sample matrices and sizes. This in turn is due to the lack of sufficient analytical standards including other tetra dioxins, internal standards and possible interfering compounds. These standards are needed to validate all aspects of the method including spiking, extraction, cleanup, and measurement for the sample type, sizes and concentrations of interest. Such extensive method validation and elaboration are needed for unequiv-ocal determinations of 2,3,7,8-TCDD residues in human samples.

Of particular concern is the problem of potentially ubiquitous PCB metabolites that are exact mass equivalents of dioxins
and can lead to false positives (28). Such tetrachlorobenzoqui-
none metabolites are known to occur in the metabolism of certain
hexachlorobiphenyl isomers. At least one metabolite of this type
can be drawn which shows a remarkable resemblance to the dioxin
molecule and at the same time can possibly depolarize itself
through electronic interactions between the two ring systems.

Our present technique involves high resolution mass spectrometry using an instrument with reversed geometry capable of
doing mass analyzed ion kinetic energy mass spectrometry (MIKES)
experiments. Sample introduction is by capillary gas chromatography. The information generated is the elemental composition
of the molecular ion, the elemental composition of the most
unique fragment ion (M-COCl), the intensity ratio of these two
ions, and the retention time of the presumptive TCDD.

The availability of radioimmunoassay (RIA) procedures for
environmental agents holds some promise in minimizing the need
for the more sophisticated and expensive instrumental methods of
analysis by eliminating "negative" samples and for routine monitoring of exposure in environments known to be contaminated by
certain classes of compounds. There are a number of fundamental
problems involved in development of such RIA procedures and in
their use (Table 3). Double-antibody RIA's have been developed
(29) for quantitating a number of chlorinated aromatic hydrocarbons of current concern from environmental samples including
animal tissues. These chlorinated hydrocarbons include members
of the dibenzo-p-dioxin, dibenzofuran, and biphenyl classes of
compounds. The use of RIA procedures for trace residue analysis
is discussed further in another paper in this conference.

These examples hopefully illustrate the complexity and
considerations necessary for developing and applying analytical
methodology to support regulatory decisions made in the interest
of protecting human health.

Table 3
Problems in Development of Radioimmunoassay for TCDD

1) Choice of Dibenzodioxin Derivative

2) Choice of Coupling Method

3) Choice of Carrier Protein

4) Characterization of Antigen

5) Immunization Schedule

6) Solubilization of TCDD

7) Selection of Antisera

8) Partial Purification of Antibodies to Improve Specificity.

9) Selection of Appropriate ^{125}I-labeled Reagent.

10) Prevention of Nonspecific Adsorption.

Toxicity Testing

The National Toxicology Program (NTP) was formed in 1978 to coordinate toxicological testing efforts of the Department of Health and Human Services (previously DHEW) and amounted to $42 million for FY-79. Components of the National Toxicology Program include in vitro and in vivo bioassay testing programs at the National Institute of Environmental Health Sciences (NIEHS), the National Cancer Institute (NCI), the National Center for Toxicological Research (NCTR) and the National Institute for Occupational Safety and Health (NIOSH) (30).

Sound analytical chemistry support is essential for the performance of these tasks. Chemistry assistance for the NTP in vitro genetic toxicology and certain in vivo programs is currently provided by contract laboratories. Chemistry problems are complicated by the sheer numbers of chemical requiring testing including storage, distribution, analysis, disposal, computer inventory, etc. Analytical capabilities include:

1. Chemical assay to determine the purity of the principal chemical component(s) of commercial chemicals and major impurities (% level) which may be present.

2. Stability and solubility determinations performed as necessary to ascertain fate and distribution of a chemical under specific conditions (time duration, temperature, light, solution media, etc.) as required for the particular bioassay experiment.

3. Comprehensive analysis to determine all possible chemical components with identification and quantitation of trace impurities at the residue level.

The first two capabilities are straight forward; however, an explanation of the need for comprehensive analysis is useful. The first step in the tier system of genetic toxicology blind bioassay testing is the Salmonella, microbial (Ames) test. Because of the low cost of this bioassay test ($500-1,000/chemical) relative to the often unpredictable and always more costly demands of chemical analysis, most chemicals are bioassayed by Salmonella before they are chemically analyzed. Plans call for 300 chemicals to be tested in the microbial system in FY-80 with a eventual goal of 1,000 bioassays of new chemicals a year by 1983. Test chemicals producing positive, ambiguous and selected negative bioassay results are then subjected to comprehensive chemical analysis. An important aspect is that the separation technique used in the chemical analysis must insure:

1.   Composition of biologically active compounds is not altered.

2.   Maximum conservation of all components.

3.   Maximum separation of chemical groups.

4.   Minimum introduction of impurities.

After separation, the resultant fractions are retested and the positive fractions are further fractionated or subjected to complete component identification and quantitation using sophisticated chromatographic and spectroscopic techniques, GC/MS (EI and CI), HPLC, NMR, IR, etc. Each component of the final active fraction is then bioassayed separately to determine the chemical(s) responsible for the original observed biological activity. It may also become necessary to synthesize the isolated active chemical for confirmation purposes.

The in vivo NTP efforts include the NCI lifetime rodent bioassay for carcinogenesis. Approximately 75-100 new chemicals are started on test each year at a total bioassay cost of about $500,000 per chemical. Chemical support for the bioassay testing, as shown in Figure 2, is also provided by contract laboratories and includes:

1.   Chemical purity and stability analysis of bulk chemicals before testing and development of protocols for reanalysis of bulk chemicals.

2.   Development of protocols for assay and stability determinations of chemical/vehicle mixes and dosage analysis.

3.   Development of procedures for analysis and stability of reprocured as well as residual bulk chemicals.

4.   Development of procedures and protocols for a chemical/ vehicle quality control analysis program. Such programs are designed to insure that reliable procedures are used by the bioassay and chemistry laboratories for the analysis of bulk chemicals and chemicals in the dosage/feed mixtures.

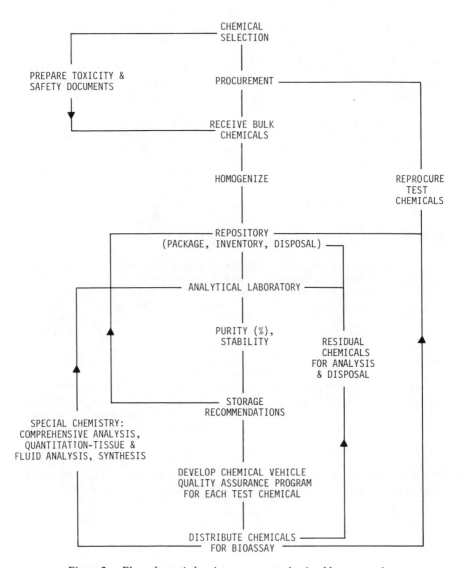

Figure 2.   Flow chart of chemistry support to in vivo bioassay testing

5. Special analyses as needed for:

a. Identification and quantitation of impurities in bulk chemicals.

b. Analysis of vehicles for toxic components.

c. Analysis of chemicals in body fluids and tissues.

d. Development of chemical monitoring techniques to be used for safety and pollution concerns.

Analysis to Insure Safe Working Environment

Chemistry operations to support the NTP in vitro efforts are performed in a specially designed Hazardous Materials Laboratory. The Laboratory (31,32,33) is under negative atmospheric pressure and all air and water effluents are filtered through particulate, charcoal and other appropriate media before being discharged. The design of the Laboratory is based on the principles of containment and the effective use of engineering controls rather than reliance on personnel protection (34,35). Overreliance on personnel protection can lead to a false sense of security resulting in an overt exposure to hazardous chemicals. Only through an understanding of the substances' chemical, physical and toxicological properties can adequate facilities be designed and monitored for the safe use of these chemicals.

Inherent with the use of containment facilities is a routine monitoring program which should include laboratory air, treated waste water and suitable surface areas. Because of the wide range of compound types and classes used in the Hazardous Materials Laboratory, a general monitoring procedure is necessary.

Laboratory air is routinely monitored quarterly by the NIOSH charcoal tube sampling procedure. Laboratory air is drawn through the tube for an 8 hour period and the charcoal adsorbant is extracted with carbon disulfide or other suitable solvents. The extract is analyzed by gas chromatography using both flame ionization and electron capture detectors. Chromatograms from each sample are compared to those of blank samples collected prior to initiation of Hazardous Materials Laboratory operations. Standard analytical techniques (HPLC, GC/MS, etc.) are used, as required, for identification, confirmation and quantitation.

Similarly, surface samples are collected quarterly. Cotton swabs saturated with acetone are used to collect samples from six 100 cm^2 surface areas in the Laboratory. The swabs are extracted with acetone and analyzed by methods analogous to the charcoal extract above.

Samples from the waste water effluent purification system are collected and analyzed quarterly. The samples are extracted according to the EPA's Priority Pollutant Protocol and analyzed analogously to the above method.

Coupled with these facilities and analytical procedures are programs for routine weekly decontamination; waste disposal by incineration and burial; personnel protection and safety training.

Chemical analysis is used in the in vivo program to insure that a safe working environment exists in which to do toxicological research. In addition to all the problems faced by the in vitro support laboratory the in vivo facility is particularly concerned about weighing and preparation of feed mixtures containing large quantities (several kg) of hazardous chemical. A similar negative pressure type mixing and feed preparation area is used for this work. Several papers have appeared describing potential hazards of such operations (36,37,38,39). Only through effective chemical monitoring and periodic use of marker compounds such as fluorescein can the extent of the problem be realized and assurance for safe working environments be given.

The NIEHS has two high hazard containment laboratories; one for chemical research (34) and the other for biological research involving hazardous chemicals (35). Programs are currently being developed for monitoring these rooms. The Chemical Containment Laboratory is designed for three purposes:

(1) organic synthesis of hazardous materials;

(2) analysis by GC and HPLC of hazardous reaction mixtures and products; and

(3) routine weighing of mg quantities of hazardous compounds.

Monitoring procedures and frequencies for air, water and surfaces are simplified since the room is used almost exclusively for aromatic halide compounds such as 2,3,7,8-tetrachlorodibenzo-p-dioxin (TCDD) and similar compounds. Samples are prepared from air sampling through a suitable sorbate (polyurethane foam, PUF) followed by extraction or chromatography with an appropriate solvent(s) and analysis by GC/MS. Water samples are extracted and concentrated before analysis using a Kuderna-Danish receiver. Surface wipe samples are generally extracted overnight on a Soxhlet apparatus using a solvent chosen on the basis of compound classification; methanol, toluene or methylene chloride.

The High Hazard Laboratory for Life Scientists involves a wide assortment of compounds. For this reason, a broader general method for chemical monitoring was needed which could be followed up when necessary with a specific technique. Such a scheme is currently being developed and involves monitoring of surface samples by the researcher using an inexpensive spectrophotofluorimeter. If a reading is recorded above the previously determined background level of fluorescence the sample is submitted to the Laboratory of Environmental Chemistry for identification and confirmation. The only requirement for this technique is a short training period for all users of the High Hazard Laboratory; submission of a small amount (mg) of all research chemicals used in the facility as reference material and back-up by more sophisticated in-house capabilities. The benefits are speed, self-monitoring, low cost and minimal loss of research time.

In conclusion, analytical chemistry is an underlying factor in essentially all aspects of toxicological work. It is evident from these examples that the sophistication of the analytical methods available for use can to a large extent determine the complexity of the toxicological problem that can be approached and solved. The best analytical approach is designed to meet the specific needs and emphasis of the toxicological research and is consistent with good analytical practices.

Literature Cited

1.   McKinney, J.D., Environmental Health Chemistry. Definitions and Interrelationships. A Case in Point. Environmental Health Chemistry. Chemistry of Environmental Agents as Potential Human Hazards, J.D. McKinney (ed.), Ann Arbor Science Publishing Co., Ann Arbor, Michigan, 1980, Chapter 1.

2.   Amore, F., Good Analytical Practices as an Alternative to Standard Methods, Licensing and Certification. In Quality Assurance of Environmental Measurements, Information Transfer, Inc., Silver Spring, MD, 1979, pp. 183-190.

3.   Albro, P. and C. Parker, General Approach to the Fractionation and Class Determination of Complex Mixtures of Chlorinated Aromatic Compounds, J. Chromatog., In Press, 1980.

4.   Hutzinger, O., S. Safe and V. Zitko, Metabolism of Chlorobiphenyl. In The Chemistry of PCB's, Hutzinger, Safe and Zitko (eds.), CRC Press, Cleveland, Ohio, 1974, Chapter 7 for general review.

5.   Hutzinger, O., S. Safe and V. Zitko, Photodegradation of Chlorobiphenyls, In The Chemistry of PCB's, Hutzinger, Safe and Zitko (eds.), CRC Press, Cleveland, Ohio, 1974, Chapter 6.

6.   Buser, H.R. and C. Rappe, Formation of Polychlorinated Dibenzofurans (PCDFs) from the Pyrolysis of Individual PCB Isomers, Chemosphere, $\underline{8}$(3), 115-124 (1979).

7.   Jensen, S. and G. Sundstrom, Structure and Levels of Most Chlorobiphenyls in Two Technical PCB Products and in Human Adipose Tissue, Ambio $\underline{3}$, 70-76 (1974).

8.   Kuehl, D.W., R.C. Dougherty, Y. Tondeur, D.L. Stalling and C. Rappe, Negative Chemical Ionization Studies of Polychlorodibenzodioxins and Dibenzofurans in Environmental Samples, In Environmental Health Chemistry. The Chemistry of Environmental Agents As Potential Human Hazards, J.D. McKinney (ed.), Ann Arbor Science Publishing Co., Ann Arbor, Michigan, 1980, Chapter 12.

9.   Miller, J.A., Carcinogenesis by Chemicals: An Overview-G.H.A. Clones Memorial Lecture, Cancer Res., $\underline{30}$, 559 (1970).

10.   Jerina, D.M. and J.R. Bend, Glutathione S-Transferases. In Biological Reactive Intermediates, D.J. Jallow, J.J. Koesis, R. Snyder and H. Vainio (eds.), Plenum Press, New York, 1977, pp. 207-236.

11.   Cox, R.H., O. Hernandez, B. Yagen, B. Smith, J.D.
McKinney, and J.R. Bend, ^{13}C NMR Studies of the Structure and
Stereochemistry of Products Derived from the Conjugation of
Glutathione with Alkene and Arene Oxides. In Environmental
Health Chemistry. The Chemistry of Environmental Agents as
Potential Human Hazards, J.D. McKinney (ed.), Ann Arbor Science
Publishing Co., Ann Arbor, Michigan, 1980, Chapter 20.
12.   Hernandez, O., M. Walker, R. Cox, G.L. Foureman, B.
Smith, and J.R. Bend, Regioselectivity and Stereospecificity in
the Enzymatic Conjugation of Glutathione with (+) Benzo(a)pyrene
4,5-oxide, Biochem. Biophys. Res. Comm. in press (1980).
13.   Corey, E.J., D.A. Clark, G. Goto, A. Marfat, C. Mioskow-
ski, B. Samuelsson and S. Hammerstrom, Stereospecific Total
Synthesis of a "Slow Reacting Substance" of Anaphylaxis, Leuk-
striene C-1, J. Am. Chem. Soc. 102(4), 1436-1439 (1980).
14.   Forgue, S.T., B.D. Preston, W.A. Hargraves, I.L. Reich,
and J.R. Allen, Direct Evidence that an Arene Oxide is a Metabolic
Intermediate of 2,2',5,5'-Tetrachlorobiphenyl, Biophys. Res.
Commun. 91(2), 475-483 (1979) and references therein.
15.   Morales, N.M. and H.B. Matthews, In vivo Binding of
2,3,6,2',3',6'-Hexachlorobiphenyl and 2,4,5,2',4',5'-Hexachloro-
biphenyl to Mouse Liver Macromolecules, Chem.-Biol. Interact.,
27, 99-110 (1979).
16.   Harvan, D., J.R. Hass, and M. Lieberman, Interaction of
N-acetyl-N-acetoxyaminofluorene with Synthetic DNA. Chem.-Biol.
Interact., 17, 203-210 (1977).
17.   Barrett, J.C., Neoplastic Transformation Induced by a
Direct Perturbation of DNA, Nature, 274, 229 (1978).
18.   McKinney, J.D., P. Singh, L. Levy, and M. Walker, High
Toxicity and Cocarcinogenic Potential of Certain Halogenated
Aromatic Hydrocarbons. In Safe Handling of Chemical Carcinogens,
Mutagens and Teratogens: Chemist's Viewpoint. Vol. II, D.
Walters (ed.), Ann Arbor Science Publishing Co., Ann Arbor,
Michigan, 1980, Chapter 22.
19.   McKinney, J.D. and P. Singh, Structure-Activity Rela-
tionships in Halogenated Biphenyls: Molecular Polarizability in
the Most Toxic Halogenated Biphenyls, Chem.-Biol. Interact.,
submitted, 1980.
20.   Gray, A.P., W.J. McClellan and V.M. Dipinto, Carbon-14-
Labeled 2,3,7,8-and 1,2,7,8-Tetrachlorodibenzofuran, J. Labelled
Compounds and Radiopharmaceuticals, in press, 1980.
21.   Oswald, E.O., P.W. Albro, and J.D. McKinney, Utilization
of Gas Liquid Chromatography Coupled with Chemical Ionization and
Electron Impact Mass Spectrometry for the Analysis and Characterization of Potentially Hazardous Environmental Agents and their
Metabolites. J. Chromatog. 98, 363 (1974).
22.   Tolbert, B.M. and N.A. Matwiyoff, Medical and Biological
Applications of Stable Carbon, Nitrogen and Oxygen Isotopes,
Radiopharm. Label. Compounds, Proc. Symp. New Develop. Radiopharm.
Label. Compounds, Vienna, IAEA, 2, 241-253 (1973).

23.   Rogan, W., Analytical Chemistry Needs for Environmental Epidemiology. In Environmental Health Chemistry. Chemistry of Environmental Agents as Potential Human Hazards. J.D. McKinney (ed.), Ann Arbor Science Publishing Co., Ann Arbor, Michigan, 1980, Chapter 6.

24.   Rogan, W.J. and B. Gladen, The Breast Milk and Formula Project - Current Findings. Environ. Hlth. Perspect., in press (abst.).

25.   Fawkes, J., D.B. Walters and J.D. McKinney, Neutron Activation Analysis of Organically Bound Chlorine. Precautions to Minimize Extraneous Halide Contamination in Collection, Storage, and Analysis of Human Milk and Formula, 178th ACS National Meeting, Washington, DC, September, 1979, abst. No. 83.

26.   McKinney, J.D., Analysis of TCDD in Environmental Samples. Proceedings of the Conference on Chlorinated Phenoxy Acids and Their Dioxins. In Ecol. Bull. (Stockholm), C. Ramel (ed.), 53-66 (1978).

27.   Report by Rebecca L. Rawls in Chemical Engineering News, February 12, 1979, Dow Finds support, doubt for dioxin ideas.

28.   McKinney, J.D., Toxicology of Selected Symmetrical Hexachlorobiphenyls. Correlating Biological Effects with Chemical Structure. Proceedings of the National Conference on PCB's, Chicago, Illinois, November, 1975. In EPA Publication-560/6-75-004, pp. 73-76.

29.   Luster, M.I., P.W. Albro, K. Chae, S.K. Chaudhary, and J.D. McKinney, Development of Radioimmunoassays for Chlorinated Aromatic Hydrocarbons. In Environmental Health Chemistry. Chemistry of Environmental Agents as Potential Human Hazards, J.D. McKinney (ed.), Ann Arbor Science Publishing Co., Ann Arbor, Michigan, 1980, Chapter 14.

30.   National Toxicology Program, Annual Plan for Fiscal Year 1980 (NTP-79-8), November, 1979.

31.   Walters, D.B., L.H. Keith and J.M. Harless, Chemical Selection and Handling Aspects of the National Toxicology Program, In Environmental Health Chemistry: The Chemistry of Environmental Agents As Potential Human Hazards, J.D. McKinney (ed.), Ann Arbor Science Publishing Co., Ann Arbor, Michigan, 1981 (in Press).

32.   Keith, L.H., J.M. Harless and D.B. Walters, Analysis and Storage of Hazardous Environmental Chemicals for Toxicological Testing, In Environmental Health Chemistry: The Chemistry of Environmental Agents as Potential Human Hazards, J.D. McKinney (ed.), Ann Arbor Science Publishing Co., Ann Arbor, Michigan, 1981 (in press).

33.   Harless, J.M., K.E. Baxter, L.H. Keith, D.B. Walters, Design and Operation of a Hazardous Materials Laboratory, In Safe Handling of Chemical Carcinogens, Mutagens, Teratogens and Other Hazardous Materials, Vol. I, D.B. Walters (ed.), Ann Arbor Science Publishing Co., Ann Arbor, Michigan, 1980, Chapter 4.

34. Walters, D.B. J.D. McKinney, A. Norstrom and D. DeWitt,
Control of Potential Carcinogenic, Mutagenic and Toxic Chemicals
via a Protocol Review Concept and Chemistry Containment Laboratory,
In Safe Handling of Chemical Carcinogens, Mutagens, Teratogens
and Other Hazardous Materials, Vol. I, D.B. Walters (ed.), Ann
Arbor Science Publishing Co., Ann Arbor, Michigan, 1980, Chapter
1.
35. Hunt, C.L., Jr., D.B. Walters and E. Zeiger, Approaches
for Safe Handling Procedures and Design of a High Hazard Laboratory for Life Scientists, In Safe Handling of Chemical Carcinogens,
Mutagens, Teratogens and Other Hazardous Materials, Vol. I, D.B.
Walters (ed.), Ann Arbor Science Publishing Co., Ann Arbor,
Michigan, 1980, Chapter 2.
36. Sansone, E.B. and M.W. Slein, Application of the Microbiological Safety Experience to Work with Chemical Carcinogens,
Am. Ind. Hyg. Assoc. J., 37, 711-720 (1976).
37. Sansone, E.B. A.M. Losikoff, R.A. Pendleton, Sources
and Dissemination of Contamination in Material Handling Operations, Am. Ind. Hyg. Assoc. J., 38, 433-442 (1977).
38. Sansone, E.B. A.M. Losikoff, Contamination from Feeding
Volatile Test Chemicals, Toxicol. Appl. Pharmacol. 46, 703-708
(1978).
39. Sansone, E.B. and A.M. Losikoff, Chemical Contamination
Resulting from the Transfer of Solid and Liquid Materials in
Hoods, Am. Ind. Hyg. Assoc. J., 38, 489-491 (1977).

RECEIVED February 2, 1981.

Analytical Aspects: A Summary

WILLIAM HORWITZ

Food and Drug Administration, HFF-101, 200 C Street, NW,
Washington, DC 20204

Modern toxicology is a multidiscipline approach to providing information on materials to which consumers are exposed. The primary test for safety is still a long-term animal study which must be monitored at every stage by the techniques of analytical chemistry.

Dr. Cairns described a unique and probably never to be repeated experiment involving almost 25,000 mice handled over a 33-month period, requiring the services of analytical chemistry from beginning to end, including:

1. Identity, purity, properties, and stability of the test substance;
2. Handling and storage of the test substance;
3. Analysis of the feed and other essential bioassay supplies for essential and deleterious ingredients;
4. Homogeneity, stability, and proper concentration of the test substance in the dosage form;
5. Safety surveillance of personnel and work areas;
6. Safe disposal of the test chemical and contaminated experimental material.

For monitoring the test, environmental, and experimental systems, both Dr. Fishbein and Dr. McKinney described some of the powerful tools which can be applied to explore, interpret, and understand situations affecting our health and safety. These tools are applied to nitrosamines and dioxins, which are families of toxic chemicals isolated, purified, and characterized by chemical and physical techniques operating at levels of parts per billion and below. (Remember that 1 part per billion is one second in 33 years; 1 teaspoonful of vermouth in a 40,000 gallon tank of gin.)

Yet when we operate at such exquisitely low levels, as well as in all of our scientific work, we are constantly confronted by the fact of variability. Dr. Tiede described statistical tools which have been found useful to describe and summarize this variability. But the purpose of statistics is to manage data; statistics cannot eliminate variability. Statistics can

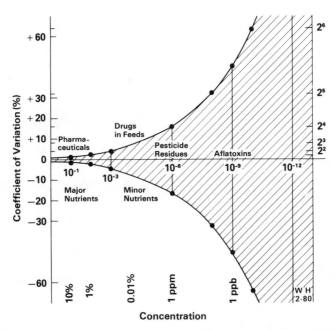

Figure 1. Variation of the interlaboratory coefficient of variation (relative standard deviation × 100) withconcentration

help us sort out important variables from unimportant ones. One of the speakers in another session provided a useful insight into the uses of statistics. He said, "If you have to use numbers to answer questions regarding your statistics, you haven't adequately interpreted your data " (1).

In my analytical chemistry paper I tried to give you some practical information about the variability of analytical systems at the trace levels where toxicologists and residue chemists must operate. The lower you go, the more variable will be the results. A copy of my curve relating precision to concentration is repeated below. Eventually, and probably before the parts per trillion level, at the present state of the art (picograms), the results become so bad that the false positives and false negatives will determine the limit of detection and determination.

Nevertheless, despite the high variability of our analytical chemistry values at trace levels, they are infinitely better and more stable than the results of our biological tests. Therefore both Dr. Cairns and Dr. McKinney aspire to be able to predict biological properties from chemical structure. This is an aspiration we hope can be accomplished. But the basic data for such a deduction will have to be reliable and accurate biological measures of toxicity for correlation purposes.

These papers have shown that the analytical chemist has served the toxicologist well in identifying compounds and determining their amounts. However, do not tempt him to push his art and science too far or your reward may be the receipt of faulty data without even recognizing this fact. As Dr. Cairns described trace analysis in one of his papers, "The analytical chemist has his feet firmly planted in midair."

Literature Cited

1. Oller, William L.; Gough, Bobby; Littlefield, Neal A. Chemical surveillance and quality assurance for preparation of dosed (2-AAF) animal feed (ED$_{01}$ study). J. Environ. Pathol. Toxicol. 1980, 3, 203-210.

RECEIVED February 20, 1981.

REGULATORY ASPECTS

Regulatory Aspects: An Introduction

PHILIP C. KEARNEY

Chief of Pesticide Degradation Laboratory, U.S. Department of Agriculture, Beltsville, MD 20705

Toxicology is playing an increasingly important role in the pesticide registration process. As the various mammalian toxicological tests for carcinogenicity, mutagenicity, and teratogenicity become more sophisticated and complex, their interpretation vis-a-vis adverse effects on man also becomes increasingly complex and, therefore, subject to considerable debate. These data are used now as a major component in risk-benefit decisionmaking for the registration of new pesticides and the reevaluation of registered pesticides under the Rebuttable Presumption Against Registration process. The debate involves industry, represented by the manufacturers; government, represented by the regulatory agencies; and the public sector, represented largely by environmental and consumer groups. Major issues in this wide-ranging debate include questions of the need for additional toxicological testing and ensuing guidelines, regulations and rules, the quality and validity of data submitted to the regulatory agencies, the use of laboratory animals as human surrogates for evaluating safety, and the development of a reasonable formula for evaluating both risks and benefits. Each group may perceive the need for and interpretation of toxicological data in judging human safety from a different perspective. The viewpoint of each of these groups and the issues are presented in this symposium.

The Congress of the United States, in the 1972 amendments to the Federal Insecticide, Fungicide, and Rodenticide Act (FIFRA), granted the Environmental Protection Agency (EPA) broad powers to regulate pesticides. The toxicological requirements under EPA are defined in a series of expanding guidelines. On an international basis, requirements for toxicological data vary considerably among the nations that make extensive use of pesticides for crop production or public health programs. In Western Europe, the current registration process is less stringent than that in the United States. There is emerging among the various countries in Western Europe an effort to harmonize the needs for toxicological data in pesticide registration. In contrast to the regulations evolved in North America, represented by the United

States and Canada, and the policies and needs of Western Europe, the Peoples' Republic of China is only now in the process of developing registration standards and determining the role that toxicology will play in these standards. This symposium presents the differences and similarities between the United States, Canada, Western Europe, and the Peoples' Republic of China as to toxicology and its role in pesticide registration.

RECEIVED March 10, 1981.

Risk Benefit Analysis:
Role in Regulation of Pesticide Registration

ROBERT A. NEAL

Center in Toxicology, Department of Biochemistry, Vanderbilt University
School of Medicine, Nashville, TN 37232

The Environmental Protection Agency regulates the use of
pesticides in such a way that they are permitted for use when
the beneficial effects are deemed to outweigh the risks that may
occur to man and his environment on use of those pesticides.
Thus, in regulating pesticides, the EPA resorts to benefit and
risk analysis as pertains to specific pesticides and specific
uses of pesticides.

Risk-benefit analysis as related to pesticides will be discussed from the following points of view; (1) how does one
determine risk, (2) what are the potential adverse health effects in man from exposure to pesticides, (it must be recognized
that adverse effects of pesticides on wildlife and non-target
organisms are also an important part of risk-benefit analysis.
However, because of time limitations we will restrict our consideration to adverse health effects in man), (3) principles and
problems concerning the estimation of risk to man from exposure
to pesticides and (4) effects of pesticides that are considered
to be beneficial.

Determination of Potential Risk

There are a number of procedures that may be used in determining potential adverse health risk in man from exposure to
pesticides. These include epidemiology, which can be applied to
pesticides currently in use in an attempt to determine if any
adverse health effects in man are evident from the registered
uses of the pesticide. Animal bioassays, perhaps the most
important methodology used in determining potential adverse
health risk to man from exposure to pesticides, can be performed

0097-6156/81/0160-0469$05.00/0

on new pesticide products which are intended for use. The
animal bioassays can also be used to assess potential adverse
health effects of pesticides currently in use for which the data
base is considered by EPA to be inadequate. Biochemical or
morphological changes which occur in nonmammalian species or in
bacterial or mammalian cells on exposure to pesticides can also
be used to assess the potential adverse health effects of pesticides. The major current use of these latter tests is the determination of the mutagenic properties of pesticides or contaminants of these pesticides.

Potential Adverse Health Effects

There are a number of potential adverse health effects
which may occur in man on exposure to pesticides. These include
acute toxicity or toxicity which occurs in a very short period
of time after exposure to the pesticide, usually less than 24
hours; organ and organ system toxicity, that is, damage to organ
systems such as kidneys or liver; damage to the nervous system;
to the blood forming system (the hematopoetic system); toxicity
to the reproductive system of males or females of the species;
teratogenesis or birth defects; immune system toxicity; carcinogenicity; and mutations of both somatic and germ cells. In examining the potential toxicity of a pesticide intended for use
or reexamining pesticides currently in use these are the major
toxicity endpoints which are considered. As noted above these
toxicity endpoints are examined for using epidemiology, rodent
bioassays and examination for biochemical or morphological
changes in nonmammalian species or in bacterial or mammalian
cells in culture.

Principles and Problems in Risk Assessment

A commonly used principle in risk assessment is the "no
observed effect level" which is defined as the dosage of the
compound at which no adverse health effect is detected either in
epidemiological studies in man or in rodent bioassays. The "no
observed effect level" may apply to both a variety of toxic effects or to a specific toxic effect. The "no observed effect
level" is an important principle in estimating and controlling
risk to man from exposure to toxic chemicals. A second important principle in risk assessment are the so-called safety
factors. These factors are applied to the "no observed effect
level" for a particular toxic effect of a chemical to provide an
additional margin of safety for humans exposed to the chemical.
The numerical value of the safety factors range from approximately 10 for inhibition of enzymes such as acetylcholinesterase
to 100 for organ system effects. Higher safety factors are
sometimes applied for toxic effects on reproduction, teratogenesis etc. Another principle used in risk assessment is the

"maximum tolerated dose". This is the highest dose which is administered in a rodent bioassay. It is a dose which causes some toxicological or pharmacological effects in the experimental animal. However, these toxicological-pharmacological effects should, ideally, not interfere with the determination of the ability of the compound to cause the toxic effect in question. The administration of the maximum tolerated dose is a controversial practice. On the one hand it is argued that to determine the potential for the occurrence of a toxic effect in a large human population using a small number of animals, a dose which is orders of magnitude higher than the expected human exposure must be administered. On the other hand, it is argued that the maximum tolerated dose overloads normal metabolic pathways and exceeds the capacity for excretion of a chemical and, therefore, is not physiological. Thus, data obtained using the "maximum tolerated dose" have little or no validity in predicting toxic effects in man. In spite of this argument, the "maximum tolerated dose" is still an established part of testing for toxicity endpoints and will likely continue to be used until additional data is brought to bear that indicates that some of the larger doses being administered in rodent bioassays are invalid in predicting the potential adverse health effects in man from exposure to the pesticide. Another area of controversy in risk assessment is the existence or lack of existence of toxicity thresholds for a chemical for irreversible effects such as carcinogenicity. I think most toxicologists will agree there is a threshold for any biological effect of a chemical, including carcinogenesis. In other words there is a level of exposure below which normal repair mechanisms, metabolic mechanisms for inactivation and barriers to penetration of a chemical to a target site which would not allow the compound to exert its toxic effect. However, for irreversible effects such as cancer, the existence of such a threshold is difficult to demonstrate experimentally. Therefore, the discussion is likely to continue until experimental data demonstrating the existence of a threshold is obtained. In the meantime regulatory agencies will continue to regulate on the basis that there is no threshold for irreversible effects such as carcinogenicity and mutagenicity. Another area of considerable uncertainty is the estimation of the risk to man from exposure to chemicals which have been shown to be mutagenic. It is clear that exposure of man to mutagenic chemicals needs to be controlled. However, it is not clear how one goes about determining the risk to man implied by exposure to mutagenic chemicals. Much more work needs to be done into trying to determine what is an acceptable level of exposure to a mutagenic chemical. Finally, another area of controversy is the practice of estimating the number of tumors that may occur in a human population on exposure to a certain level of a chemical as calculated using data on cancer incidence obtained by exposure of experimental animals to various levels of the chemical. Some

argue that the inaccuracy of this quantitative risk extrapolation practice is so great that it should not be used. Others,
including myself, believe that quantitative risk extrapolation
is a useful exercise that tends to account for the potency of
oncogenic chemicals and is an indispensible tool to the regulator in making judgments about chemicals which cause irreversible
effects.

Beneficial Effects of Pesticides

I think the effects of a pesticide which are generally
considered beneficial are, for example, its effect on the cost
of agricultural products. A new pesticide may decrease the cost
of production of a food or fiber product. Therefore it should
be considered beneficial for that reason. Likewise, the elimination of the use of an old pesticide may increase the cost of
the food or fiber. Use of pesticides may provide increased re-
creational opportunities. Elimination or control of human
disease organism would of course be considered a benefit. Likewise, the elimination or control of unwanted animals or plants
would also be considered a beneficial effect of pesticides.

Summary

With these various data elements in hand, the regulator is
in a position to make judgments about the desirability of
allowing continued marketing of a pesticidal product that has
the potential to cause adverse health effect in man.
Of most importance in this decision process is the toxic
potency of the chemical, the degree of human exposure and the
reversibility (i.e. organ damage) or irreversibility (i.e.
cancer, teratogenesis) of the toxic effect. The options available to the regulator are a ban on the use of the chemical,
restriction of some uses and methods of application of the
pesticide, or relatively unrestricted use. Which of these
options is chosen will depend on the consideration of the scientific assessment of the risk, the economic consideration of the
benefits and, unfortunately too often, the political climate at
the time the decision is made and the political persuasion of
the person making the decision. Of these various elements, the
toxicology data relative to the potential toxicity of the compound in question is of greater importance. However, the experience and judgment of the person evaluating the data is as important as the data itself. Toxicology is not yet an exact
science; nor is it likely to become one in the foreseeable future.
Thus, the ability of an experienced and objective toxicologist to
examine competently derived data and arrive at an assessment of
the potential risk to man from exposure to the chemical is the
key element in the process of risk-benefit analysis.

RECEIVED March 18, 1981.

An Environmentalist's View of Toxicology and Pesticide Regulation

JACQUELINE M. WARREN

Environmental Defense Fund, 475 Park Avenue South, New York, NY 10016

The Environmental Defense Fund has been actively involved in the pesticide regulatory process for more than 12 years. Beginning in 1967, EDF lawyers and scientists have participated in administrative and judicial proceedings to suspend or cancel DDT, mirex, aldrin, dieldrin, chlordane, heptachlor and, currently, the herbicide 2,4,5-T. Along with pesticide manufacturers and user groups, EDF comments extensively on the development of the pesticide registration guidelines under §3 of FIFRA. EDF has also submitted comments and data in proceedings relating to a wide variety of regulatory and administrative actions proposed by the EPA.

EDF's concerns about pesticide use have focused primarily upon involuntary human exposure to hazardous compounds, especially through the food chain, and upon detrimental effects of pesticide use on non-target species. Since pesticides are poisons by definition, and are specifically designed to kill, EDF has sought comprehensive testing and assessment of the health and environmental effects of pesticides before they are widely used. This preventive or "test-first" approach was incorporated into FIFRA in 1972, when Congress gave EPA the authority to develop hazard evaluation guidelines for the registration of pesticides. The basic regulatory standard, which requires that pesticides be shown to pose no unreasonable adverse effects on the environment, places the burden of establishing that a pesticide may be safely used directly upon the registrant. Since all pesticides are likely to be hazardous unless properly used, an applicant for registration must not only provide the EPA with comprehensive evidence of the potential adverse health and environmental effects of a particular compound, but must also demonstrate that the risks outweigh the benefits when the pesticide is properly used. The burden that FIFRA imposes upon registrants is a difficult one to meet, but the cost to society of a "use-first-test-later" approach has been firmly rejected by Congress and the courts.

0097–6156/81/0160–0473$05.00/0

In the face of extensive evidence of widespread contamination of the environment, the food chain, and human tissues with
pesticide residues, EPA has acted to remove a small number of
persistent carcinogenic or otherwise highly toxic pesticides from
the marketplace. The nature and extent of the exposure resulting
from the use of some of those compounds was well illustrated in
the proceedings for suspension and cancellation of aldrin,
dieldrin, chlordane and heptachlor. For example, the FDA's
annual Market Basket Surveys showed consistently high and increasing residues of dieldrin and heptachlor epoxide in meat, fish,
poultry and dairy products. In addition, the EPA's Human
Monitoring Survey, which analyzes human adipose tissue samples
from surgical procedures and autopsies, from 1970 to 1972 found
measurable dieldrin residues in 96.5, 99.5 and 98.2% of the
samples tested. Similarly, between 1971 and 1974 the Survey
found heptachlor epoxide and oxychlordane residues, both metabolites of heptachlor and chlordane, in more than 90% of the
tissue samples analyzed (Table I). The average residues of
heptachlor epoxide and oxychlordane ranged from 0.08-0.09 ppm and
10-12 ppm, respectively.

Table I
Percentages of Positive Samples Found from 1971-1974

Year	Heptachlor Epoxide	Oxychlordane
1971	96.2	93.3
1972	90.3	92.3
1973	97.7	98.3
1974	96.3	98.6

Faced with evidence of the carcinogenicity in rodents of the
pesticides in question, EPA cancelled their registrations for
most uses. Those actions have been affirmed in a series of court
decisions upholding the Agency's preventive approach to the
regulation of carcinogens. The basic principle is that "if
regulation were withheld until the danger was demonstrated conclusively, untold injury to public health could result." EDF v.
EPA, 598 F. 2d 62 (D.C. Cir. 1978).

The same stringent registration standards applied to the
chlorinated hydrocarbon pesticides mentioned above, and that all
new pesticides must meet, are slowly being applied to 1500
existing active pesticide ingredients, the various formulations
of which comprise the approximately 45,000 pesticide products
currently registered. Over the next several years, all of these
registrations must be reviewed by EPA and decisions made either
to continue or to cancel them. The Agency's review process
involves evaluation of the toxicology data supporting the registrations as well as consideration of the benefits provided by use
of the pesticides.

The unreasonable adverse effects standard set forth in FIFRA requires the EPA to conduct a risk-benefit balancing in which the economic, social, and environmental costs and benefits of the use of each pesticide are taken into account. To perform the statutorily mandated balancing, EPA must have the most complete knowledge possible about the potential risks posed by use of a pesticide, as well as the relative costs, availability and effectiveness of substitute compounds. The importance of accurate toxicological information for the performance of the risk-benefit balancing cannot be over-emphasized. For this reason, EPA has not only specified what data must be provided to support a pesticide registration, but has also recently proposed Good Laboratory Practice Guidelines for Toxicology Testing[1] in an effort to assure the integrity of the data.

There are good reasons supporting EPA's decision to propose Good Laboratory Practice Guidelines. Beginning in 1975, there have been numerous disclosures of "irregularities", both deliberate and inadvertent, in data submitted to FDA to support various regulated food and drug products, and to EPA to support pesticide registrations and food tolerances. Such guidelines are long overdue. Indeed, several Congressional hearings and subsequent regulatory agency investigations document drastic shortcomings in the integrity of many toxicity studies and reports submitted to FDA and EPA. "Unacceptable laboratory practices" have been found at several contract laboratories, and at drug and pesticide manufacturing plants, that raise very serious questions about the validity of the data generated there. According to the EPA,

> "The unacceptable practices noted included selective reporting and underreporting of test results, lack of adherence to specified protocols, inadequate qualification and supervision of personnel, poor animal care procedures, poor record-keeping techniques and the general failure of sponsors to monitor studies"[2].

In addition to these deficiencies, the data have also often been so poorly tabulated and summarized that a conclusion as to carcinogenicity could not be made. As a result of the problems that were created by these and other methodological shortcomings, such as an inadequate number of animals, failure to report findings on all tissues studied, and lack of data on statistical differences in effects on treated as compared with control animals, a significant portion of the data supporting pesticide tolerances and registrations has been discredited as unreliable.

In testimony before a Senate committee in 1976, EPA Deputy Administrator John Quarles admitted that his Agency's investigation showed that "serious problems" might exist with the toxicology data supporting pesticide registrations. One example of such "problems" was the deliberate withholding of valid results

because a laboratory might be so dependent upon a pesticide pro-
ducer for contract work that its independent scientific judgment
could be imparied by the close economic relationship. Quarles
also suggested that a laboratory might intentionally misrepresent
results at the request of the manufacturer. A vivid illustration
of such a situation was the 1978 federal grand jury indictment in
Chicago of six corporate officers of Velsicol Chemical Corporation for allegedly withholding studies indicating the carcinogenicity of heptachlor and chlordane from the Environmental
Protection Agency.

Other laboratories' procedures have also been found to be so
questionable by a joint EPA-FDA audit program that some of the
audited facilities have been referred to the Department of Jus-
tice for possible prosecution. Thus, public confidence in Agency
regulatory decisions based upon toxicology data generated and
submitted by drug manufacturers and pesticide registrants, often
the same companies, has been seriously undermined. The publication and implementation of Good Laboratory Practice requirements
by EPA and FDA should help to alleviate some of the public's
misgivings about the validity of data generated by the pesticide
and drug industry, but it is clear that rigorous enforcement of
these requirements will be necessary if the regulatory process is
to function effectively to protect public health and the
environment.

Perhaps the major area of disagreement between environmentalists and the pesticide industry is the question of how much
evidence of risk is necessary to justify regulatory action. It
is not merely a question of balancing risks and benefits, but of
defining risks in the first place. Much of the public debate
surrounding the regulation of carcinogenic pesticides, and of all
environmental carcinogens, for that matter, has focused upon the
histopathological definition of what is a carcinogen, the extrapolation of animal carcinogenicity evidence to humans, and the
quantification of cancer risk in humans. The science involved in
these determinations is so imprecise and so subject to varying
interpretations that definitive conclusions cannot be drawn. The
debate is further complicated by the background of mistrust and
suspicion about the quality of industry-supplied toxicity data,
that resulted from the aforementioned disclosures of
improprieties in laboratory practice.

The task of balancing risks and benefits is extremely difficult when there is no agreement about what is on either side of
the equation. The evidence of potential hazard to humans is very
often hypothetical or suggestive rather than conclusive. Often
the regulatory agencies are dealing with reasonable medical
theory or epidemiological evidence suggesting, but not proving,
associations between exposure to particular chemicals and
increased incidence of disease. At the same time, the benefits
of continued use of the compound in question, be it a pesticide,
drug or industrial chemical, are usually claimed by the affected

industry and users to be very extensive. Government officials charged with protecting public health and the environment must therefore make decisions which will have measurable economic impacts in order to prevent potential harm, the magnitude of which cannot be readily ascertained. In such situations, the manufacturer and users of a pesticide, for example, argue for continued use, while environmentalists, consumers and public health officials strongly urge a preventive approach.

Based on past experience, Agency decisions to take protective action will generally be upheld by the courts. In a long line of cases liberally construing the Agencies' authority to act to protect public health, the courts have recognized that the social cost of a wrong decision is far less where only the benefit of use of a product has been foregone; irreparable injury to health or the environment cannot be so readily recouped.

These are some of the concerns that have directed EDF's involvement in the pesticide regulatory process. EPA faces an immense task in reviewing the toxicological support for tens of thousands of pesticide registrations, which consists of more than a million studies. Only time and conscientious effort will enable the agency to complete the job. To ensure that past abuses will not occur in the future, however, the generation of toxicological data should be guided by careful compliance with the Good Laboratory Practice standards currently being developed by EPA for the testing of pesticides.

Scientists have a responsibility to their profession and to future generations to facilitate the objective and complete evaluation of the potential health and environmental hazards of pesticides. We are all living today with the consequences of the failures of the past. It is time to put into practice the old adage that "an ounce of prevention is worth a pound of cure."

Literature Cited

1. Guidelines for Registering Pesticides in the United States; Proposed Good Laboratory Practice Guidelines for Toxicology Testing, 45 Fed. Reg. 26373 (April 8, 1980).

2. Good Laboratory Practices Standards for Health Effects, 44 Fed. Reg. 27364 (May 9, 1979). [EPA Standards for Testing under §4 of the Toxic Substances Control Act].

RECEIVED March 31, 1981.

Industrial View of Toxicology and Pesticide Regulation

J. L. EMMERSON

Lilly Research Laboratories, Eli Lilly and Company, Greenfield, IN 46140

Those of us who are engaged in the development and testing
of new pesticides have, over the past ten years, been witness
to a radical change in the regulation of these products. The
authority for a more active role of government was granted by
the Congress in the 1972 amendments to the Federal Insecticide,
Fungicide, and Rodenticide Act (FIFRA). It is through the issuance of regulations, however, that we learn how this authority
is to be exercised. The EPA, being the responsible agency, has,
during the interim, proposed and re-proposed rules, regulations,
and guidelines which encompass virtually every aspect of the
testing and use of pesticide products. Although the regulations
are not fully in place and are yet evolving, their scope is
evident and the underlying philosophy unmistakable. In the
formation of the regulations, which are detailed and prescrip-
tive, the originators apparently presume that the public
interest will be best served if the registration process can be
standardized and if latitude in the design, conduct and reporting of experimental work can be minimized. In the preface to
EPA regulations which issued on July 3, 1975, the following
purposes were cited to show why the Congress chose to amend the
existing law. (Table I) That purposes one, two, and four have
been realized as a result of the amendment is not questioned; I
have serious misgivings, however, that the remaining two
purposes, numbers three and five, will ever be brought about.
The whole process of pesticide registration has become
infinitely more complex, and with the increased complexity has
come delay.

To understand why we have come to the present state of
affairs, it is necessary to provide some historical background
and to discuss some of the factors that have influenced EPA
policy during the 1970 decade. It is my intention then to
review some past and present consequences of the changing
regulatory scene. I am a toxicologist, one whose job it is to

0097–6156/81/0160–0479$05.00/0

assess the relative toxicity of new chemical substances, pharmaceuticals, feed additives, and agricultural chemicals, in an industrial laboratory. And as one who has been engaged in safety evaluation work for almost 20 years, I have had an opportunity to view the regulatory process over a span of time that precedes and encompasses the present era. It is from this perspective that I review the status of pesticide regulation.

TABLE I

CONGRESSIONAL PURPOSE IN ENACTING
1972 AMENDMENT TO FIFRA

1. Strengthening regulatory controls on the uses and users of pesticides;

2. Speeding up procedures for barring pesticides found to be undesirable;

3. Streamlining procedures for making valuable new measures, procedures, and materials broadly available;

4. Strengthening enforcement procedures to protect against misuse of these biologically effective materials;

5. Creating an administrative and legal framework under which continued research can produce more knowledge about better ways to use existing pesticides as well as developing alternative materials and methods of pest control.

From the summary of the House Committee on Agriculture (Federal Register 40, No. 129, p. 28242, 1975).

Scope of the Regulatory Task

I am convinced that at the inception no one, neither the Congress, the EPA, nor the regulated industries, comprehended the magnitude of the regulatory task faced by the EPA as a result of the 1972 amendments. The new requirements which are shown in Table II appear on first analysis to be logical and desirable. It is only when one begins to reduce to specific cases the general principles that one can perceive the difficulties. As shown in Table III there were estimated to be at the time of the initial regulatory proposals 1400 individual chemicals and 33,000 formulations of these agents. Most have multiple uses; each use must be regulated. Represented in each of the 15 classes of pesticides shown are tens and hundreds of individual compounds with the most diverse chemical, physical and biological properties imaginable.

TABLE II

NEW REQUIREMENTS POSED BY 1972 AMENDMENTS TO FIFRA

1. Intrastate products became subject to federal regulation.

2. In conditions of use, pesticides must not have an unreasonable adverse effect on the environment.

3. Pesticides shall be classified for restricted use if in general use unreasonable adverse effects on the environment may occur.

4. Regulations shall be issued for the registration and classification of pesticides (new) as well as the registration and re-classification of pesticides registered prior to October 21, 1974.

TABLE III

SCOPE OF THE REGULATORY TASK

Estimated number of registered pesticides (October 1974):

    1. Individual pesticides: 1400
    2. Formulated products: 33,000

Classes of Pesticides Covered:

    1. Amphibian and reptile poisons or repellents
    2. Antimicrobial agents
    3. Attractants
    4. Bird poisons or repellents
    5. Defoliants
    6. Desiccants
    7. Fish poisons or repellents
    8. Fungicides
    9. Herbicides
   10. Insecticides
   11. Invertebrate animal poisons or repellents
   12. Mammal poisons or repellents
   13. Nematocides
   14. Plant regulators
   15. Rodenticides

Let us examine for a moment the implications of one of the new requirements. Pesticides shall be classified for restricted use if in general use unreasonable adverse effects on the environment may occur. A number of formidable questions readily come to mind: What constitutes an unreasonable adverse effect? What tests are appropriate for the detection of an adverse environmental effect? - tests on which organisms? - at what levels of exposure? - for how long? One cannot adequately judge the potential adverse effects of a chemical unless the fate of the compound in the environment is known. Is it photolyzed by sunlight? - degraded by soil bacteria? - taken up by plants? - consumed by animals? Is it translocated, volatilized or bound to soil? Is it persistent? One question leads to another ad infinitum. It was clear that some criteria would have to be established and in the interim those criteria have appeared as regulations.

Status of Regulations (Table IV)

Just prior to the deadline set by Congress, the EPA published proposed registration, reregistration and classification procedures. After a period for comment and revision, these procedures were published in final form on July 3, 1975. Broad, general requirements were given for the types of tests to be done for the determination of product hazard to humans as well as non-target organisms (environmental effects). At the same time as the general regulations were issued, however, testing guidelines were also published. The guidelines for data requirements, which were first proposed on June 25, 1975, were reissued as proposals in July and August of 1978. While one who is not acquainted with pesticide research cannot easily grasp the ramifications of these proposals, a listing of the major section headings can give an appreciation of the scope and the areas affected (Table V). The July proposal covered product specifications, studies on environmental fate, and toxicity studies in birds and aquatic species, as well as a detailed description of toxicity studies to be done in laboratory animals. In the proposal of these "guidelines," the EPA departed from conventionally accepted domestic and international regulatory practice in two ways:

1. The guidelines specified in considerable detail the elements of experimental design; and
2. The guidelines were published as proposed rules which would, if formally adopted, become regulations with the force of law.

TABLE IV

STATUS OF MAJOR EPA REGULATIONS ON PESTICIDES

Subject	Date of Publication	Status
Proposed Registration, Reregistration and Classification Procedures	Oct. 16, 1974	Proposed Rules
Guidelines for Registering Pesticides in the United States	Jun. 25, 1975	Proposed Rules
Regulations for the Enforcement of the Federal Insecticide, Fungicide, and Rodenticide Act	Jul. 3, 1975	Final Rules and Regulations
Proposed Guidelines for Registering Pesticides in the United States	Jul. 10, 1978	Proposed Rules
Proposed Guidelines for Registering Pesticides in the United States; Hazard Evaluation: Humans and Domestic Animals	Aug. 22, 1978	Proposed Rules
Enforcement Policy Regarding Failures to Report Information Under Section 6(a)(2) of the Federal Insecticide, Fungicide and Rodenticide Act	Jul. 12, 1979	Final Rules and Regulations
Guidelines for Registering Pesticides in the United States; Proposed Good Laboratory Practice Guidelines for Toxicology Testing	Apr. 18, 1980	Proposed Rules

During the 1970's, the guidelines have slowly evolved to their present state where they have been re-proposed in 1978 for additional comment. Unofficial draft copies had been circulating for years prior to publication; before the first proposal in 1975, there were said to have been seven consecutive drafts. While many changes have been made in response to comments received, much to the dismay of many scientists in

testing laboratories, the latest proposals for safety evaluation
testing are more structured than the earlier versions. This
has occurred even though most experienced toxicologists have
continually advocated the use of general and unofficial testing
guidelines, a method of operation which the FDA has developed
and followed successfully for many years in the application of
similar data to the evaluation of the safety of drugs and food
additives. While the latter approach provides flexibility and
permits the exercise of scientific judgment in experimental
design, it is clear that the EPA has chosen a different
regulatory posture, one in which one receives something akin to
a recipe for registration. That the EPA has chosen this
approach may be attributed to four factors:

1. Several early EPA decisions were contested in
    court. Internal support for a structured
    approach was strengthened in the belief that the
    Agency's position would be more defensible.
2. Some manufacturers, in fact, sought prescribed
    rules in an attempt to eliminate uncertainties in
    the registration process.
3. The Agency has not been successful in attracting
    and retaining experienced toxicologists to review
    data. Neither have scientific personnel with
    experience in safety evaluation been sought out
    for high-ranking administrative positions. The
    present approach seems to be promoted on the
    belief that safety evaluation studies are so
    routine that they can be codified and given
    sufficiently detailed instructions, the need for
    scientific expertise can be minimized.
4. The promulgation of guidelines is an attempt to
    cope with a vast and changing market, which stems
    from the need to reduce all facets to writing so
    that the research and developmental process will
    hold still for viewing and can thereby be
    controlled.

TABLE V

PESTICIDE GUIDELINES: TOXICOLOGY STUDY REQUIREMENTS

Acute Testing

        Acute Oral Toxicity Study
        Acute Dermal Toxicity Study
        Acute Inhalation Toxicity Study
        Primary Eye Irritation Study
        Primary Dermal Irritation Study
        Dermal Sensitization Study
        Acute Delayed Neurotoxicity Study

TABLE V (CONTD.)

Subchronic Testing

      Subchronic Oral Dosing Studies
      Subchronic 21-Day Dermal Toxicity Study
      Subchronic 90-Day Dermal Toxicity Study
      Subchronic Inhalation Toxicity Study
      Subchronic Neurotoxicity Studies

Chronic Testing

      Chronic Feeding Study
      Oncogenicity Studies
      Teratogenicity Studies
      Reproduction Study

Mutagenicity Testing

      Test Standards for Detecting Gene Mutations
      Test Standards for Detecting Heritable Chromosomal
        Mutations
      Test Standards for Detecting Effects on DNA Repair or
        Recombination

Special Testing

      General Metabolism Study

Special Requirements

      Domestic Animal Safety Testing

Avian and Mammalian Testing

      Avian Single Dose Oral LD_{50}
      Avian Dietary LC_{50}
      Mammalian Acute Toxicity
      Avian Reproduction
      Simulated and Actual Field Testing for Mammals and Birds

Aquatic Organism Testing

      Fish Acute LC_{50}
      Acute Toxicity to Aquatic Invertebrates
      Acute Toxicity to Estuarine and Marine Organisms
      Embryo Larvae and Life-Cycle Studies of Fish and Aquatic
        Invertebrates
      Aquatic Organism Toxicity and Residue Studies

What were the fundamental changes wrought by the 1975
regulations?

1. The negligible residue concept was abandoned. Previously,
   it was possible to obtain the registration of a product for
   which there was a residue on a food crop if that residue
   did not exceed 1/2000th of the no-effect dose in 90-day
   animal studies. In the new rules, this procedure, which
   did not provide the degree of certainty desired, was
   rejected as a means of determining what was a toxicologically insignificant residue. Given the sensitivity of
   modern analytical techniques and the resourcefulness of the
   residue chemists, one seldom encounters a pesticide which,
   in use, produces no residue. To obtain a tolerance for a
   residue, one must establish a no-effect dose in 18-24 month
   rodent studies. The net effect of this change is to
   require long-term studies in laboratory animals for
   virtually every pesticide.

2. Studies on non-target organisms (environmental studies)
   were required. Toxicity studies on wildfowl, fish and
   other vertebrate and invertebrate organisms had, in 1975,
   been conducted in a few laboratories, but the methods were
   rudimentary and experimental. Nevertheless, laboratory
   tests were proposed without the scientific basis that would
   permit the formulation of sound procedures. The proposal
   did stimulate research in this area; the methods are still
   in a state of flux and the general implications of adverse
   findings poorly understood. It is ironic to note in the
   proposed guidelines that the more uncertainty that
   surrounds a given test, the more apt the guideline is to
   specify with great exactitude what must be done.

3. In long-term studies in rodents, no distinction would be
   made between benign and malignant tumors. This approach
   was adopted following court rulings on aldrin and dieldrin
   in which the Court of Appeals upheld the position of the
   Administrator of EPA in his contention that for purposes of
   hazard evaluation the two findings should be considered
   synonymous. With this ruling the regulatory task was
   simplified, but was good science served? Among those who
   are professionally trained as pathologists, this position
   does not have general support. In essence, the diagnosis
   is made immaterial.

4. An assessment of applicator or user hazard would be
   required. The law, which specified that all pesticides had
   to be classified for restricted or for general use, prompted
   a review of the toxic properties of each chemical, not only
   for exposure through crop residues but also for acute dermal
   or inhalation exposure to the user. Depending upon the
   degree of toxicity shown in acute or single exposure
   studies, materials were to be classified in one of four
   categories with accompanying label and use restrictions.

Present Consequences

1. The registrations for the use of a number of pesticides
    have been cancelled (Table VI). Although proceedings
    against some of these agents occurred prior to the formal
    promulgation of pesticide regulations, the rules, which
    were being proposed, were used to guide policy, even as is
    done today. The pesticides shown were either banned or
    their use severely restricted because the properties of
    these chemicals (it was decided) did not permit their use
    without "an unreasonable adverse effect on the environment."
2. The registration process has inexorably slowed. In the
    re-proposed toxicology guidelines, in addition to the major
    new test requirements, applicants are now directed to submit
    much of their original or raw data which was, in times past,
    only presented in summary. We have estimated that in one
    long-term rat study, including the tabulation and summariza-
    tion of data, individual data points may exceed 500,000.
    Not only does the preparation of such a report require a
    tremendous effort, all this data must be reviewed. One can
    safely predict an increasing review time as studies done
    and reports prepared under the guidelines issue.
3. Publication of the EPA guidelines for safety evaluation
    presaged a spate of guideline writing activity that
    continues unabated today. Governmental agencies (domestic
    and foreign), interagency committees, joint international
    groups, ad hoc committees from professional societies,
    trade organizations - everyone, it seems, feels compelled
    to formulate guidelines as to how toxicology should be
    done. Needless to say, each group is promoting its own set
    as authoritative and that each proposal, just to ensure
    attention, has introduced a variation that guarantees that
    the recommended procedure will not be compatible with any
    other. If one wishes to comment authoritatively (that is,
    with appropriate references), it is necessary to assign
    several full-time, experienced toxicologists to do nothing
    but review guidelines and prepare responses.

TABLE VI

SUSPENDED AND CANCELLED PESTICIDES

Aldrin	DDT	Lindane
Chloranil	Dieldrin	Mirex
Chlordane	Heptaclor	Toxaphene
DBCP	Kepone	

Partial list of pesticides whose uses are banned or
restricted. (Taken from a May 1978 EPA listing of
suspended and cancelled pesticides.)

Prospects for the Future

1. New pesticides that reach the marketplace will be selective
    in activity and few in number. Many that ultimately
    receive approval will be available only under restricted
    use. Two factors prompt these conclusions: a) the
    complexity of the registration process; and b) the chemical
    and physical properties of a pesticide that confer useful
    activity (biological activity with a degree of persistence)
    will, in multiple safety testing, ensure that an effect in
    some non-target organism is detected. The judgment as to
    whether that effect is not "adverse" is a difficult
    decision, one that is sure to be argued, reviewed and
    re-reviewed before a responsible agency manager affixes his
    name to the decision.
2. The increased reliance on existing products will ultimately
    diminish the usefulness of the chemical tools we have.
    Experience has shown that many target organisms become more
    resistant as treatment is repeated from year to year and
    generation to generation; new strains emerge as a result of
    selection and adaptation.
3. The cancellation of old products and the slowed availability
    of new agents will, however, serve to create opportunities.
    To the laboratory that is able to discover that miraculous
    agent that is able to effect its activity without harm to
    the myriad of life-forms, microbial, plant and animal, that
    are found in the world, ample rewards exist.
4. The costs of research and development as a result of
    regulatory requirements will diminish the number of
    laboratories competing. Only those corporations with the
    size and capital needed to sustain the long and expensive
    developmental process will be able to persevere.

Recommendations

I wish to end this review with some recommendations. We
who are engaged in research on pesticides stand or fall with
the success or failure of the EPA. By success I mean the
expeditious review and approval of those agents that are
effective and can be used safely. Any means to further that
goal serves not only the interests of the EPA, as well as those
engaged in research and development in private industry, but
also the public who will ultimately benefit.

In my review of the status of pesticide regulation in the
United States, I have come to a disquieting conclusion. I do
not believe that the EPA or any organization, no matter how
well organized and managed, is equal to the task mandated by
the Congress in the 1972 amendments to FIFRA. That the EPA has
in their perception of their charge tended to expand the scope
of their activities has made the goal of efficient pesticide

regulation that much more unachievable. For the task to become
manageable, a change in philosophy or policy as well as a
reduction in the scope of the program would be necessary;
neither appears to be likely. The comments that I offer,
therefore, are made not with the hope of stirring some radical
change, but with the desire that some alternatives to the
present mode of operation be considered.

1. The Use of Certified Summaries. The preparation and review
    of toxicological data would be greatly simplified if the
    EPA would permit the use of certified summaries. Instead
    of the voluminous reports now required, the applicant would
    have the option of supplying in summary form all pertinent
    data certified as to its accuracy. In the summary reports,
    the applicant would prepare a complete description of the
    work done as would be expected for a scientific publication.
    All adverse effects would be identified and relevant data
    supplied. If in the Agency review, individual data were
    needed for reference, these data could be quickly supplied.
    A policy which permits the use of certified summaries would
    complement and be consistent with the purposes of the EPA
    laboratory audit program. The latter program and the Good
    Laboratory Practice guidelines were instituted to assure
    that work of acceptable quality was done and that pertinent
    data were accurately reported.
2. A Clarification of the Objectives of the EPA GLP Program.
    Every toxicology laboratory that does safety evaluation
    work in support of a new drug or a pesticide product is
    subject to inspection by FDA or EPA personnel. Good
    Laboratory Practice regulations, which authorize this
    activity, have been published by the FDA and proposed by
    the EPA. The inspectors, who arrive on short notice,
    undertake an exhaustive comparison of raw data, point by
    point, with that contained in internal documents and in the
    study report. Errors are tabulated and inspection reports
    issued. In response to the GLP inspection program many
    laboratories are engaged in what I regard as a quest for
    zero defects. This is done, in part, out of fear that the
    reputation of the laboratory will be damaged by the
    mindless reporting of errors. This fear is fostered by the
    approach of the inspectors whose business it is to find
    "errors" and who simply tabulate errors, typographical,
    transcriptional, major and minor, all apparently being
    given equal weight. In this context a minor error that has
    no bearing whatsoever on the scientific validity of the
    study is assigned a disproportionate regard, and extra-
    ordinary and expensive measures are being employed to guard
    against that error. In safety evaluation work, as in all
    human endeavors, errors will be made. Major errors that
    affect the conclusions of the study are prominent and are

easily detected. Minor errors can be minimized but their
complete elimination is only achieved slowly and at great
expense. The whole subject needs to be addressed and a
forthright policy put forth that assures that the proper
distinctions will be made, that perspective will be
maintained, and that measures will be taken to ensure
against the inadertent release of unevaluated inspection
reports.

3. The Modification of Current Proprosed Toxicology
    Guidelines. The EPA has a representative on each of
    several groups that are proposing guidelines, e.g., the
    Interagency Regulatory Liaison Group. If the guidelines
    prepared by the joint agency groups could be adopted, the
    benefits would be immediate and protracted. The adoption
    of a uniform set of toxicology guidelines by United States
    agencies would serve as a very strong impetus for foreign
    governments to subscribe to the same conventional
    procedures.

4. The Acquisition of a Stable, Authoritative Body of
    Scientific Personnel. It may be inferred from an
    examination of EPA regulations and guidelines including the
    comments in the preambles to these documents that the
    motivating force is a belief that a) every problem in the
    administration of various pesticide programs can be handled
    by innovative regulation and b) detailed and prescriptive
    regulations provide the only truly objective and impartial
    procedure. While this philosophy in theory is seen to have
    merit, in practice it offers only delay and frustration.
    To be successful in the Agency approach would require that
    every contingency be anticipated. This is clearly only
    possible in broad and general terms. No provision is made
    for the ultimate questions, those that are inevitable, are
    the most perplexing and are those which require scientific
    judgment.

    - Is an observed effect an adverse effect?
    - How large a safety factor is appropriate?
    - Were the studies done properly designed in light of
      the chemical properties, the biological activity
      and the eventual use?
    - To approve an experimental use permit, for an
      herbicide for corn, for an insecticide for
      coniferous trees, for a soil sterilant, what
      studies should be required?

While these questions can be postulated in general terms,
each can only be answered for a specific pesticide for a
specific use. The needs are different. The imposition of
additional guidelines with the vain hope that the problems
will yield to regulation will finally only encumber. The
solution lies in the presence of scientific expertise.

It has been my experience to find Agency scientific
personnel continually in motion – coming, leaving, moving,
shifting jobs. One seldom talks to the same scientist in
the course of one study and never throughout a research
program. A concerted effort should be made to hire
experienced, nationally recognized experts in toxicology,
seasoned scientists who could speak with authority. The
second and inseparable requisite is that authority be
given. It is not possible to attract scientists of the
caliber I mention unless provision is made for them to
exercise their authority. The unacceptably slow pace of
the present review process is directly related to the
absence of experienced scientific personnel with decision-
making authority.

Competition for talent in the area of safety evaluation work is intense, the demand for experienced scientists
being largely a function of increased regulatory pressures.
Those of us in industry and in professional societies of
toxicology have a responsibility to work with government
agencies to find ways to help them recruit scientific
personnel. I regard success in this effort as a necessary
prerequisite for any real progress in the attempt to
regulate the introduction and use of pesticides.

RECEIVED March 6, 1981.

Human Risk Assessment from Animal Data

ROBERT A. SQUIRE

Division of Comparative Medicine, Johns Hopkins School of Medicine,
Baltimore, MD 21205

A forum entitled "Animal Tests and Human Cancer" was
reported in Chemical and Engineering News, June 27, 1977, with an
eye-catching cover photo of a rat. The forum was prompted by the
proposed ban on saccharin following the Canadian study which
showed the induction of bladder cancer in rats. For the first
time, it appeared the public was acutely aware of the impact of
governmental regulatory legislation upon their personal lives.
The possibility of this ban has stimulated widespread awareness
and uncertainty about regulatory policies, and raised many questions about the state of the art in animal testing.

There are three fundamental methods for estimating potential
human risk as the result of exposure to toxic substances, whether
the exposure be from food, drugs, air, water or the workplace.
These are: (1) epidemiological studies, (2) animal tests, and
(3) short-term or in vitro analyses such as studies of DNA damage
or mutagenesis. Of the three methods, the greatest confidence is
placed upon epidemiological studies in humans. Unfortunately,
however, epidemiology is limited in its sensitivity and its application to toxicity assessment. The greatest value has derived
from recognition of occupational hazards where there is high
exposure to well defined human populations, or from studies like
those on cigarette smoking, where exposure may be clearly
defined. However, in the case of more ubiquitous, ill-defined
and low level exposure to toxic substances, observations in
humans often lack the sensitivity to discern possible toxic
effects. We are left then with the next alternative, the use of
other mammalian species as human surrogates. Intact mammalian
systems are considered most relevant to human risk because no
other methods can simulate the complex biological systems which
allow us to survive and even thrive in an environment replete
with natural and man-made chemical poisons as well as harmful
physical and biological agents. There is an efficient homeostatic apparatus in the intact animal system which determines the
safe versus toxic levels of exogenous substances.

0097-6156/81/0160-0493$05.00/0

Table I
Rationale for Use of Animals in Toxicological Testing

1.   Mammals are anatomically, physiologically and
     biochemically similar.

2.   Mammals have similar health and disease manifestations
     and causes.

3.   Mammals respond similarly to exogenous chemical,
     biological and physical agents (differences are
     primarily quantitative rather than qualitative).

The three statements in Table I are simple and axiomatic.
They form the basis for all of the comparative medicine, and the
use of animals in medical research and in toxicological and carcinogenicity testing. They are however, broad generalizations of
biological truths. They do not necessarily apply in every
instance nor do they tell us which species is most similar to
human. It is generally impossible to predict how any given
species or individual will respond to a potentially toxic
substance without detailed metabolic and pharmacokinetic studies.
Despite proclamations to the contrary by some scientists, poli-
ticians, lawyers and others, the extrapolation of animal toxicity
data to human risk assessment, although necessary in the regulatory sense, is often based upon several unproven assumptions. It
is this very uncertainty which allows and encourages diverse and
opposing claims, predictions and warnings from those advocating
one or another viewpoint, since most of the claims, no matter how
extreme, cannot be proven or disproven. Litigation proceedings
and hearings have not, in my view, provided the most efficient or
rational means to resolving these questions.
     It is particularly true with respect to the assessment of
carcinogenic risks that we have had to place virtually complete
reliance upon extrapolation from animal tests. This has resulted
in the restriction of use or removal from the marketplace of
several chemical substances including pesticides - such as DDT,
Aldrin, Dieldrin, Chlordane and Heptachlor. Other pesticides and
chemicals already in use have also been found to be carcinogenic
to one or more animal species in the National Cancer Institute
testing program - now part of the National Toxicology Program.
It is probably safe to assume that such animal testing activities
will increase, at least for the immediate future.
     Since we have learned that finding a carcinogenic response
in a test animal may trigger decisions which have far reaching
impact on our society and economy, it is useful to examine some
of the procedures involved in evaluating toxicological data and
estimating potential human risk. In doing so, it should become
apparent that several areas of uncertainty and potential error

can influence major regulatory decisions and mislead an
uninformed public.

Although the procedures could be discussed in many ways, I
have chosen to use four main categories: (1) validity of animal
data; (2) weight of toxicological evidence; (3) characteristics
of the test substance; and finally, (4) quantitative risk
assessment.

The first step is to insure that the evidence that a substance actually is an animal carcinogen is sufficient and
persuasive.   The degree of evidence for animal carcinogenicity
varies considerably from very weak to overwhelming, but this
aspect is often overlooked when a positive finding is reported.

The next two categories, the weight of toxicological
evidence and the chemical and biological characteristics of the
test substance, can discriminate among the relative potencies or
virulence of potential human carcinogens and hopefully dispell
the impression that all animal carcinogens pose equal threats to
man.   Lastly, I will briefly discuss quantitative risk assessment
which attempts to predict a numerical incidence or range of potential toxic or carcinogenic responses in the human population.

The validity of the animal data addresses not only the
accuracy of the findings but also the relevance of the experimental data for man.   If comparative metabolic or pharmacokinetic
studies reveal a quantitative difference between the test animal
and human responses or routes of exposure, the findings may
totally lack predictive value.   Such studies are rarely performed
because of the limitations imposed by time and funding; thus
there is usually no alternative but to err on the side of pru-
dence and accept positive animal findings.   Unless there is evidence to the contrary, a regulator has no choice but to assume
that test animal data may be predictive of the response among at
least some individuals in the heterogeneous human population.

Dose levels employed particularly in carcinogenesis testing
remain an area of controversy.   The rationale for the maximum
tolerated dose concept is based in part upon the insensitivity of
tests which use small numbers of animals as compared to the large
human population at risk.   This is a valid toxicological premise
for safety testing <u>provided</u> we assume the phenomenological events
in carcinogenesis are dose related in a relatively linear
fashion.   That is, the pharmacokinetics, metabolism, extent of
DNA damage versus repair, etc. are directly proportional to the
dose; that toxic effects observed at high test doses accurately
predict, qualitatively and quantitatively, the effects at actual
low exposure levels.   We have heard and read much about this
issue for some time, and it is the basis for controversy surrounding not only the choice of the maximum tolerated dose but also
the selection of mathematical models when attempting to predict
specific levels of human risk.   In truth, the facts are usually
not known since the necessary experiments are not performed.   The

choice of the high test dose, like that of mathematical models, is based more upon conviction or theory than upon scientific evidence, and yet these are two of the most important factors in extrapolating animal results to human risk.

There are known animal carcinogens, or tumor inducers, if you prefer, which probably would not have been detected if animals had been treated at doses which were not overtly toxic. So, an important question is: Do we really want to know if something can cause cancer in animals at very high doses, even if they are considered excessive or unphysiologic? I think it important to have this information so I do not object to high test doses. But, in using this information to extrapolate to human risk, additional factors should also be considered other than the fact that a substance can induce cancer in test animals.

Next, I cannot fail to stress the importance and the potential errors involved in the pathologic evaluation. In the present political climate a carcinogen may be identified or obscured because there is a statistically significant difference in one or more tumor types in treated animals as compared to control animals. The process, thus, may amount to a numbers game rather than a reliance upon the biomedical judgment which is required. Particularly important is knowledge of the spontaneous diseases in laboratory animals, a specialty field in itself.

Proper pathological evaluation requires a relatively complete and, above all, uniform examination of tissues in treated and control groups in order to determine actual tumor incidences. In past research studies dealing with known, strong carcinogens, pathologic accuracy was less important. Similarly today, if test compound X turns out to be a strong carcinogen, this fact will be readily apparent -- probably as early as the necropsy examinations before any precise tissue counts or histologic examinations are performed. However, when trying to discern weak carcinogens from non-carcinogens, as most chemical testing now attempts to do, the addition or substraction of very few tumors from any one animal test group or another can statistically create a safe substance or a carcinogen!

Realize that aged control mice and rats may have high and variable spontaneous tumor rates, varying in some tissues from 5-40% among different control groups. These differences may be highly significant, and undoubtedly result from the many environmental modifying factors which influence tumor incidences. This emphatically points out the possibility of spurious results in some carcinogenesis tests. The subject of false positive and false negative results in identifying weak animal carcinogens, therefore, requires more recognition and evaluation than it has received.

Another important issue is that the diagnostic terms used by pathologists determine how many of each type of preneoplastic lesions or of benign or malignant tumors is reported. On the

basis of pathogenesis or etiology some different tumor types
should be lumped together for assessment of carcinogenic effects
and others should not be. This, of course, is critical to the
statistical analysis and the final conclusion. These are medical
decisions which must be made by the pathologists on a case-by-
case basis, and if there is controversy surrounding the
diagnosis, or if a significant regulatory decision rests upon the
pathologic classification of certain lesions, then a peer review
with consensus is essential. The scope and impact of such decisions requires that the pathologic interpretations not be left to
a contradictory testimony or a single judgment.

Finally, following the enumeration of pathologic diagnoses,
the choice of the statistical model can, in itself, affect the
conclusion. This is especially true in discerning a negative
from a weak positive effect. Thus, before we even approach the
area of human risk assessment, or extrapolation, the complex test
required to determine whether or not a chemical is an animal
carcinogen, i.e., the basis for the qualitative decision, is
already encumbered by many possible errors of procedure or
judgment.

Table II

Weight of Evidence from Test Animal Data

> Number of species affected
> Number of tissue sites affected
> Latency periods
> Dose-response relationships
> Nature (severity) of lesions induced

Table II represents an important aspect of animal to human
extrapolation. These five points are, to me, the biological
parameters which best determine the potency or virulence of an
animal carcinogen, which is to say the potency of a human carcinogen according to current regulatory policies. To ignore this
type of information, which is often done, and consider all animal
carcinogens as equal threats to man is ludicrous in light of our
knowledge. We know that there is wide variation in species and
tissue susceptibility to carcinogens. However, the more animal
species which are susceptible the more confident we may be that
man is likely to be susceptible rather than unique in his
response. We also know that carcinogenic response is dose and
time related, and that some carcinogens induce more malignant
tumors than others. Thus, unless there is evidence to the contrary, the highest degree of potential human hazard should be
attributed to chemicals which induce primarily malignant tumors,
at multiple sites, in short periods of time, at low doses, and in
both sexes of several species. If the type of induced tumor in
the animals is normally rare, this should also be taken into
consideration since the enhancement of tumors with genetically
determined high spontaneous frequency may also be accomplished by

numerous dietary and other environmental modifiers which are not in themselves carcinogenic.

Conversely the least concern might be attributed to a chemical which, after multiple species tests, is found to only enhance the incidence of common tumors, in one site, in one sex and species, and only following long exposure at high and toxic dose levels.

Table III describes the characteristics of the test substance to be considered. Although in the case of carcinogenesis, we do not know the mechanisms involved, there are inherent biological and chemical properties which can indicate limits to the potential reactivity of chemicals with mammalian cell constituents. These include chemical similarity to other known toxins, binding or adduct formation with cell macromolecules, genotoxicity or activity in short-term tests for carcinogenicity, metabolic and pharmacokinetic data, and other pertinent physiological, pharmacological or biochemical properties.

Table III
Characteristics of Test Substance

Chemical similarity to other known toxins
Binding to DNA, RNA, protein
Genotoxicity or activity in short-term tests for
    carcinogenicity
Metabolic and pharmacokinetic data
Physiological, pharmacological, and biochemical
    properties

Unless we are to ignore all of our heavily financed research on carcinogenesis to date, we must assume that biologically inert substances, and those in which the parent compound or its metabolites do not alter DNA, are not genotoxic and do not induce cell transformation, are not likely to be genetic-type carcinogens. If they do induce tumors, it can hardly be by a one-hit, mutagenic-like event, but rather by non-genetic mechanisms including chronic tissue injury. This type of in vitro and biochemical data, together with the weight of evidence from animal test results can contribute to a rational basis for regulatory judgment.

Finally, there is the area of quantitative risk assessment (Table IV). This subject has recently assumed an importance and prominence which tend to obscure the underlying ignorance involved. Such procedures are attempts to predict the magnitude or incidence of toxicological responses in humans at low levels of exposure based upon responses observed in animals at high levels of exposure. Assumptions, again which are largely theoretical, must be made concerning not only high to low dose extrapolation but also concerning interspecies extrapolation.

Table IV
Quantitative Risk Assessment

1. Sensitivity of test animals versus humans
2. Arbitrary safe factors
3. Biological assumptions and mathematical models
    a. one-hit
    b. multi-hit
    c. multi-stage

The reason we are struggling in this area is that the mechanisms of carcinogenesis remain obscure. We simply do not know what the biological events or risks are at low level exposures to carcinogens where most human exposure occurs, and which is beyond the sensitivity of test animal observations. Thus, for example, thresholds or no-effect levels cannot be proven or disproven, and we do not know which mathematical model is best, or even if any come close to reflecting the actual biological process.

The primary advantage of extrapolation using mathematical models is that it avoids the necessity of debating a no-effect or threshold level, which cannot be scientifically documented. Rather it provides an estimate of risk which can be judged as acceptable or unacceptable and such a decision is a societal rather than a scientific one.

The one-hit, multi-hit or multi-stage mathematical models listed in Table IV reflect the range of current theories surrounding the molecular events in carcinogenesis. The one-hit model presumes that a single mutagenic-like event can initiate the neoplastic process. This implies a linear dose-response relationship at low actual exposure levels and thus usually results in a prediction of the highest incidences of cancer and the lowest acceptable exposure levels of a chemical. The multi-hit and multi-stage models, on the other hand, do not a priori assume a one-hit mechanism at any exposure level.

A critical factor in the assumption of low-dose linearity is the background of spontaneous tumor or disease rates, i.e. an additive effect from exposure to multiple carcinogens. Thus, the carcinogenic responses in the liver or lymphoreticular systems of mice generally give a linear response regardless of the model employed. But mice have extremely high spontaneous rates of liver cancer and lymphoma, and it must be assumed that there is a significant population of initiated or transformed cells in the mouse whether they be virus induced or otherwise. One cannot necessarily assume a similar process in humans, since no cancer incidence approximates those of the liver or lymphoreticular systems in mice. Conversely, however, one perhaps could expect a linear or even a concave response in some humans exposed to a new animal lung carcinogen as the result of the high existing lung cancer rate. The important point here again is that all positive

animal carcinogenesis data do not necessarily indicate equal
human hazard. The many factors I have outlined should be
considered when estimating potential human risk, including the
choice of a mathematical model which best reflects the total
biological and chemical evidence available concerning the
substance in question. And in light of our recent experience
with mathematical predictions as they relate to animal tumor
studies, the existing background of tumors in specific tissues in
humans should perhaps receive greater attention in quantitative
risk assessments and regulatory decisions.

One final point is the sensitivity of test animals versus
humans. Certain committees and individuals have, on the basis of
very few comparative observations, expressed the view that humans
are generally more susceptible than the test animals, since carcinogenic response appears to be directly proportional to total
carcinogenic dose. Inasmuch as humans live longer, potential
exposure, and thus cancer risk, is assumed to be greather than
observed in test animals. In answer to this, note that the
spontaneous cancer and other disease rates in aged or 2 year old
mice and rats are comparable to those in man at 70 years of age.
Assuming that most cancers are, in the broad sense, environmental
diseases in both man and animals, as the evidence strongly
supports, it may be equally or more plausible to assume that the
sensitivity of rodents is greater if carcinogenic response
depends upon total exposure, since they only live a fraction of
the human lifespan. The same inference can be drawn from other
mammalian species. Degenerative diseases and cancer reach high
levels at the end of their natural lifespans. Such responses
therefore seem to depend upon biological processes which are not
time related in the absolute sense.

In summary, I would like to stress several points. Extrapolation from experimental carcinogenesis data to human risk is
essential. Aside from the limited information derived from human
epidemiological studies, it is the only means of regulating carcinogens. However, our present policies may not be the best to
serve the public interest. They thrive on the critical areas of
scientific ignorance in this field, and, unfortunately, there is
much to be gained -- financially, professionally, and politically
-- by exploiting some of the uncertainties which exist. If the
public were fully aware of the uncertainties rather than being
confused by conflicting claims, each sounding as if it were a
proven fact, our cancer education and prevention efforts might be
more effective in the long run.

We cannot continue to propagate the notion that all animal
carcinogens are equally hazardous any more than all other toxins
are equally hazardous. This totally discourages any attempts by
individuals in society to prioritize and discriminate in their
own risk/benefit analyses, and the public should clearly have
this privilege. Much of the misinformation admittedly is the
result of media coverage and sensationalism. But it should be

the responsibility at least of government agencies to correct
this by proper educational efforts. The public canot discriminate between press releases which announce positive carcinogenic
findings unless the relative weights of evidence are also
prominently presented in an understandable manner.

As I have pointed out, there are many types and levels of
animal evidence to be weighed, and when combined with genotoxic,
biochemical and other data -- we see a whole spectrum of evidence
for carcinogenic potential. No rigid system of classification or
of regulation can accommodate these biological variables.

We can and should attempt to rank carcinogens by the nature
and extent of the experimental data in mammalian and in vitro
systems. This ranking based upon a spectrum of biological
evidence, together with the use of mathematical models, when
approprite, can provide a more rational basis for quantitative
risk assessment on a case-by-case basis.

RECEIVED March 12, 1981.

Pesticide Regulation:
Toxicology and Risk Evaluation

EDWIN L. JOHNSON

Deputy Assistant Administrator, Office of Pesticide Programs,
U.S. Environmental Protection Agency, 401 M Street, SW,
Washington, DC 20460

It is a pleasure to be here today to discuss the role of
toxicology in pesticide regulation - one of the Environmental
Protection Agency's most controversial and difficult jobs.
The Federal pesticide regulatory laws pose particular, even
unique challenges within the broad spectrum of public policy
decisions which EPA must take. The science of toxicology
plays a leading role in that decision-making process.

Pesticides are of tremendous value to society in
agricultural and forest production, disease vector control and
other areas. These benefits are fairly plain to see so that
even as individuals, without recourse to sophisticated analytical techniques, we are usually capable of making at least
rudimentary estimates of the benefits of the pesticides we
personally choose to use.

However, in the last decade or so, we as a society have
also become increasingly aware that pesticides can have the
potential for causing significant adverse human health and
environmental effects. This knowledge is particularly disquieting because we also know that we do not have a good
understanding of what some of those potential health effects
really are. Indeed, for certain pesticides whose use contributes substantially to the total environmental burden of pesticides, we have positive test evidence of risks which are as
yet unexamined. Moreover, as individuals, we are usually not
so able to assess potential risks of pesticides to our health
and well-being, as we are the benefits which we see as individuals and agricultural producers in the use of pesticides we
choose to employ. In fact, we are often unable to choose
whether or not we will be exposed to many of the pesticides
now used, and thus cannot choose that degree of risk we wish
to accept.

It falls to EPA, on behalf of the many involved sectors
of the public, to assess the degree of risks posed by the
numerous pesticides available, and to determine what level of
risk, tempered by benefit, society ought to accept. This
level of risk is not necessarily the level which each of us
individually would choose, nor is it perfectly adapted to
every particular use situation, but is rather a guarantee of a
minimum standard of protection combined with the opportunity
to enjoy at least certain benefits of the pesticide as well.

Risk/benefit balancing is the chief tool established in
the basic pesticide law, the Federal Insecticide, Fungicide,
and Rodenticide Act (FIFRA), for reaching regulatory decisions. We may not ban a pesticide simply because it has a
potential for causing harm. What we need to know is how much
risk a pesticide may pose and what are its benefits to society
before taking action for or against use of a pesticide. One
of our first goals, then, is to assure that these objective
building blocks - risks and benefits - of what ultimately must
be a subjective regulatory decision are evaluated on the best
scientific basis attainable. And obviously, toxicology is a
major contributor to the building of the risk side of the
balance which is finally struck.

Described in this manner, pesticide decision making seems
very precise and clear. But we all known that it's not. What
needs to be added into the process is a large element of
uncertainty and a pressing need for timeliness. Uncertainty,
because science itself is often uncertain, despite its mantle
of truth and objectivity. Technology is changing, methods of
evaluating risk are imperfect, theories on the environmental
causes of cancer and other adverse chronic health effects are
still evolving, and much remains unknown. Timeliness, because
inaction for a regulator is action. No decision is a decision
to take the risk of allowing a human effect to occur or continue unchecked. Thus, in exercising its public responsibility
by balancing risks with benefits, EPA is also required to
balance scientific certainty against timeliness of decisions.
We must decide on a weight of contemporaneous evidence, and
not have as an objective, the resolution of each and every
scientific issue related to the decision. When do we have
enough? The National Academy of Sciences has put this problem
well in observing that, "Environmental regulation is not a
detached leisurely process of transferring verified results of
objective scientific research into clearly indicated environmental decisions." (1) Frankly, it often seems more like a
juggling act than a balancing routine.

Keeping this regulatory scene in mind, a process which depends heavily on assessment of potential risk, you can see that the role of toxicology in the process is a critical one. Because Pesticide Programs has been grappling with the problems of gathering data and making regulatory decisions on pesticides for some years now, we have developed procedures for reviewing a pesticide for registration, which we think help to reduce time-certainty conflicts for our regulatory staff and enable us to respond in a consistent and rational manner.

The driving criteria for determining the extent of review a pesticide is to be given are risk related. To perform this assessment, we require of the registrant, as a condition for registration, or continuation of registration, without which a pesticide may not be marketed in the U.S., a wide array of data dealing with:

- chemical composition of the product including an assessment of impurities and other incidental contaminants;

- biological test data on acute effects such as oral LD_{50}, dermal LD_{50}, eye and skin irritation;

- test data on chronic effects which help us judge such hazards as cancer, mutations, impaired reproductive ability, nerve disorders, and

- test data on equivalent risk indices for fish and wildlife which could be exposed to the pesticide.

The results of these studies are compared initially to finite regulatory criteria which place the pesticide into one of several categories. In the most usual case the criteria of risk potential are not exceeded and the mere fact that the pesticide performs its intended function is adequate to determine that benefits exceed risks. Such products are registered with little if any sophisticated assessments of actual hazard, exposure or economic benefit.

Other pesticides demonstrate a risk potential that warrants keeping them from the general public and restricting their use to specially trained applicators or in other ways. The assessment here is somewhat more complex and intensive and involves deciding whether the reduction in risk achieved by the restriction is sufficient to outweigh the additional costs to society imposed by restriction.

And a relative handful of pesticides demonstrate a risk potential, when judged by our criteria, of such magnitude that it is presumed that they ought not to be registered at all - unless an intensive evaluation of the risks and benefits of each use of the pesticide demonstrates that the benefits of such use warrant the acceptance of the risks associated with that use. As many of you are aware, this intensive review has been termed the rebuttable presumption against registration - or RPAR - process. That is, the risk potential of the pesticide is such that the Agency presumes the pesticide should not be permitted for use. However, that presumption on the part of EPA may be rebutted by showing our data are incorrect or invalid, that exposure to humans is inconsequential or that benefits of use warrant taking the risk. Since the RPAR process has been in effect for several years now, and many academic, industry and user groups are familiar with it, I won't go into detail explaining its mechanics.

Risk measures derived from animal test data, coupled in some few cases with human experience, are used for a variety of administrative purposes as well. For example, acute toxicity and skin and eye effects levels are used to classify all pesticide products into four toxicological categories for purposes of warning and caution statements on labeling, to decide on the need for child resistant packaging and a variety of similar determinations. However, the chemicals which trigger an RPAR are those presenting the biggest challenge to the public decision-maker and the bulk of my remarks will be colored by those cases rather than the most simple cases of low potential risk.

Hazard Assessment

Since valid scientific data are needed to trigger a full RPAR review in the first instance, I'd like to discuss briefly the use of data from animal tests to estimate human risks. Certainly the most compelling data are human epidemiological data demonstrating that an adverse effect indeed occurs in humans. However, prudent public administration does not, indeed cannot wait to act until an effect is observed in a human population. As Russ Train once remarked, "We must put chemicals to the test, not people." This does not mean of course that we never have evidence of human effects of a pesticide, but where we do, it must be considered a measure of program failure, not program success, since we should have dealt with the health issues before the effects became widely evidenced in a human population. This is particularly true because humans are exposed to a broad range of stresses, chemical and other, which can result in illness, disease, and

death, and it is only the most significant of these that we
can hope to ferret out by studying general human populations.

Thus we rely first on animal studies designed to show the
relations of pesticide exposure and toxicological effects.
These are then translated to the possible human experience by
adjusting for any known or hypothesized differences in human
sensitivity and the dose a human may reasonably be expected to
get. These animal data come from pesticide manufacturers who
have the primary responsibility for testing under the statute,
from extensive searching of the scientific literature, and
from direct communications with research scientists. Thus,
our risk assessment most often must rely solely on data from
animal tests since reliable human epidemiological data are,
and perhaps almost always will be, unavailable.

When a variety of data are available, EPA attempts to
follow a weight of evidence approach which acknowledges differences in data types - that is human-epidemiology versus
animal bioassay versus short term, in vitro (test tube) tests;
the central tendencies of the data toward suggesting the same
effect; and the scientific adequacy of the studies involved.
Extrapolation from animal data to human risk is, while scientifically supported, still full of uncertainty, but as I indicated earlier, regulators cannot wait for certainty, nor can
the public. In the face of serious threats to public health
we must act on the best assessment of these data available at
the time.

Risk assessment - quantitative risk assessment in
particular - is a very controversial subject. The Agency
faces this controversy head on when considering whether a food
tolerance or acceptable registration can be established for a
pesticide which, for example, may be a possible cancer agent.
Traditional methods which the Agency has used in assessing the
acceptability of an exposure level - the determination of a
No Observed Effects Level used to calculate an Acceptable
Daily Intake Level for food, or an acceptable level of exposure
from other sources, won't work for nonthreshold effects. Thus
the Agency has in such cases made use of quantitative risk
assessment procedures in making decisions for carcinogenic
pesticides.

There are, of course, two extreme views of the validity
of such quantitative risk assessment. On the one hand, it is
held that there is no valid method for extrapolating cancer
data in animals to arrive at risk assessments for humans.
Thus there can be no scientifically accurate weighing of the
potential cancer risks of a chemical against its benefits.
The other view is that valid animal data on a chemical's

carcinogenic potential <u>can</u> be used to estimate human cancer
risk by employing one or more statistical methods to quantify
risks with various degrees of certainty. EPA has chosen to
accept quantitative risk assessment for non-threshold effects
such as cancer; however limited and imperfect they may be,
they do provide the ability to discriminate among potential
levels of risk posed by alternative pesticides and provide a
measure of risk for priority setting on a risk basis.

I would also like to consider briefly the role, or perhaps
more accurately, the potential role of short-term tests in
pesticide regulation. Historically, EPA has focused on pesticide active ingredients in its regulatory activities. As we
have grown more sophisticated in our knowledge of potential
effects and have sought ways to test for them, the range of
short and long term data required for these active ingredients
has increased. Long term, whole animal bioassays have become
a routine element in the required testing for registration.

But while active ingredients have received and will
increasingly receive close scrutiny, pesticidally inert ingredients and contaminants have <u>not</u> typically been subject to
much testing or careful scientific scrutiny. This is so, even
though we know inerts and contaminants may also potentially
pose significant health risks. One solution would simply be
to require the same battery of toxicity tests for inerts and
contaminants as are now required for active ingredients. If
this approach were practically possible, it would certainly
provide us with the best available data for conducting risk
assessments for inerts and contaminants.

However, manufacturers and laboratories will be taxed
close to their limit by the testing requirements for active
ingredients alone. Thus, from the regulator's point of view,
short term bioassays which could at least reliably identify
carcinogens, for instance, would be a godsend for setting
priorities for more extensive inerts testing.

But one of the most crucial questions to be asked in
considering the use of short term bioassays is what is the
probability of false negatives? Or what is the chance that a
chemical found <u>not</u> positive in a short-term assay is indeed
<u>not</u> a carcinogen? It does not seem clear within the scientific
community that the positive correlation between mutagenic
effects in short term tests and chemicals which are carcinogenic in whole animal bioassays is sufficient to permit regulatory reliance on short term tests. Appropriate batteries of
short term tests may reduce the probability of false negatives,
but this apparent shortcoming to regulatory scientists remains
a practical limitation on the usefulness of short term tests
in regulatory decision making.

We are, however, considering the use of microbial
bioassays in our Product Chemistry Guidelines as a means of
rapidly screening pesticides and inert ingredients for detecting the presence of potentially toxic impurities. The battery
of microbial bioassays we will be proposing are tests which
can be performed speedily, at reasonable cost, are reproducible in a variety of laboratories, and are sensitive at the
microgram level. These tests would be used only as a means of
rapid screening, not as a basis for developing quantitative
assessment of human risks. Positive results would lead to
minimizing any such impurities, establishing certified upper
composition limits for them and developing analytical methodology suitable for monitoring commercial products.

   When this part of the Product Chemistry Guidelines is
publicly proposed (sometime this summer) we will be looking
for some strong, constructive reaction to the proposal. In
considering the usefulness of these microbial bioassays it is
important to keep in mind the difference between the needs of
the scientific community and those of the public administrator. While scientists seek certainty, public administrators
must usually make do with information which is indicative, but
not definitive. Thus, despite current limitations on the
usefulness of short term testing, I believe these limitations
can be reduced sufficiently for regulatory purposes and we
will be finding ways to use short term testing, at least as as
a screen or indicator for deciding where our limited resources
should be concentrated.

   As you can see, toxicology, or the characterization of a
pesticide's risk <u>potential</u>, is central to the assessment of
the risk a pesticide poses, but it does <u>not</u> complete that
assessment. Risk is, of course, the product of the ability to
cause harm coupled with human exposure to the compound.

   So, what remains is to assess human <u>exposure</u> to the
pesticide. While I do not want to discuss exposure asessment
extensively today - it is not our topic - I think a brief
discussion would clarify how risk assessments based on toxicity
data are used in completing a full hazard assessment. Exposure assessment is not a new art. It is, however, an area
that requires a lot of work. Most often we do not have monitoring data to precisely describe human exposure, for example.
Models are bing used to predict these exposure patterns but no
standard process for making these predictions has yet been
established.

   Thus, the problem for us is that exposure data are often
sketchy and meager. Often there are <u>no</u> available exposure
data and the Agency must develop reasonable worst case assumptions to assess potential risk.

Chemical analysis of low level residues is often critical to successful assessment of exposure and such work is often conducted at the state of the art levels of detection. Often new methods pushing levels even lower are needed to characterize residues. Generally speaking though, your ability as chemists to detect residues has out-stripped our ability to deal with the "so-what" question of the potential hazard of such exposure.

In assessing risk it is necessary to consider each use of the pesticide, the potential alternatives which may be used to fill a pest control vacuum and their relative risk, the individual and collective risks to users of the pesticide, workers exposed in subsequent agricultural or other activities, consumers of treated food, people who live nearby and those exposed through environmental contamination. Other descriptors of the nature of the risks are also germane:

- who receives the benefits compared to who bears the risks - equity

- voluntary vs. involuntary exposure

- the type of health effect, whether it is fatal or not, whether it is reversable or not; whether it is an effect realized in old age or among the young.

Thus the characterization of risk is a multidimensional and complex concept.

At this stage we have completed a risk assessment - coupled the potential hazard demonstrated by animal, epidemiological or similar information with our best estimates of exposure. In theory, this is purely a scientific analysis. In practice, the uncertainties and data gaps associated with time-constrained regulatory analyses require the infusion of some public policy choices into the process. If we are unsure of the level of effect, we presume, as a matter of public policy, that it is the higher or more significant of a reasonable range of choices. If we are unsure of the actual exposure, then we presume a reasonable worst exposure scenario. Unknowns are resolved so that errors work in favor of reduced public risk - we try to err on the side of safety. This means attempting to minimize the chance of rejecting significant adverse public health outcome, which indeed is the true state of nature, in favor of a possible but less adverse consequence. Thus we arrive at a scientifically derived assessment of risk which is perhaps more qualitative than quantitative, laden with uncertainties and tempered with public policy values introduced to span the lacunae in our knowledge and our data.

I have today discussed only one side of the regulatory
decision-making process - I have not touched on benefits
assessment or the role of scientific and public review. I
will only stress that benefits assessment and outside review
and participation in the regulatory process are equally
important.

In conclusion, there is much room for improvement - for
instance, in the application of statistical decision theory,
improved risk assessment methods and generally better data.
But it is important to remember that the limiting factor may
be one single part of the analytical chain - chemical detection, or exposure, or benefits - or ultimately, it may be the
limited ability of the human mind to attribute social values
to the myriad of risk and benefit descriptors and to calculate
conclusions accounting for all of them. Improvements in the
process of risk assessment must therefore achieve balance both
among the various types of scientific and economic inputs to a
decision <u>and</u> with the ability of the decision maker to use the
inputs. But even then we all do not, nor should we necessarily, have the same set of values to apply to the decision,
assuring interesting and controversial pesticide risk debates
for some time in the future.

Literature Cited

1.   National Academy of Sciences, National Research Council,
Environmental Studies Board, "Perspectives on Technical
Information for Environmental Protection", Vol. I, Washington,
D. C., March 1977, p.20.

RECEIVED March 19, 1981.

Pesticide Regulation in Europe

PETER DUBACH

CIBA–GEIGY Limited, Basel, Switzerland

In the last decade many countries throughout the world
have changed their relevant laws and regulations and, to be
on the safe side, they have made the list of requirements an
exhaustive one. This has happened also in Europe and requirements thus have become rather uniform today. While in daily
practice an applicant still has a hard life to cope with the
numerous country-specific different details of study protocols
and in the preparing, launching and following up of an application for registering a product, I certainly would bore you
if I tried to instruct you here on details. Although I have
thus to stick to rather general terms, I do hope that some
selected remarks on aspects of European Registration Requirements, on European Registration Policies and Procedures, and
on Activities of the European Economic Community will find
your interest. These recapitulative remarks are of course
shaped by my eight years of industry experience in registration
and contain some very personal opinions.

European Registration Requirements

As indicated briefly in the introduction, requirements in
Europe today do not differ basically from those in the U.S. or
elsewhere. Harmonizing efforts and scientific and political
environment protection activities have led to a sort of maximum requirement checklists in nearly all countries. Often the
requirements are listed on a form which has to be filled in
upon application. All in all, these checklists have been
applied so far rather reasonably in Europe, i.e. they are
used by the authorities to check that no testing aspect is
overlooked while accepting valid arguments that in many
specific cases particular tests on the checklist are not
relevant and need not be performed. It is understandable that
the more established in expertise an authority is the more

0097–6156/81/0160–0513$05.00/0
© 1981 American Chemical Society

such flexibility that authority is likely to show as long as
its experts have a real administrative influence. For example
provisional or, in minor use situations, even full registrations
may be granted without chronic toxicity/carcinogenicity studies,
although contained in every checklist, provided the other data
including mutagenicity and residue data are favorable. This
possibility arises also from the fact that in most instances
the seasonal exposure of applicators in Europe lasts only a
few days and that re-entry problems largely are nonexistent.
In respect to mutagenicity many countries request data from
three independent test systems, sometimes more tests are required.

A pronounced harmonizing influence is being exerted by a
Council of Europe publication "Pesticides" (formerly "Agricultural Pesticides"), which gives guidance to authorities
responsible for registration and which soon will go into its
5th edition (1). In the Council of Europe, Europe's first
embracing political institution, which operates by recommendations and agreements only, twenty member states are
represented. Registration experts of a great number of these
member states are responsible for the drafting and up-dating
of this booklet on registration guidance. In about fifty
pages the requirements of safety testing of pesticides are
described, including Chemical Identity and Properties, Toxicology, Residues, Environment Phenomena, Disposal, Classification, and Labeling. The spirit of the authors emerges by
citing a few general, introductory remarks: "It (the booklet)
does not deal with these questions (data requirements) ex-
haustively, nor are the various proposals to be taken as
final. Manufacturers must realize that they can always be asked
to supply additional information about their products. - Although the features of a chemical which may need investigation
can be listed ..., it must be realized that what is needed is
an appropriate ... examination of the substance and not the
mere completion of some predetermined list of tests. - It is
strongly suggested that manufacturers should discuss their
program of investigation with the responsible authorities at
an early stage."

This type of flexibility is reflected in the laws and
regulations of all European countries, which, of course, has
two sides: it can be applied reasonably by experts but it can
also be misused to ask for ever more data by non-experts or
bureaucrats. It is practically impossible for today's manufacturers of new active ingredients to live up to the proposal
to involve the responsible authorities in a specific testing
program, because for economic reasons they have usually to
seek a quick market introduction simultaneously in many

countries. This fact is realized by European authorities and
they are ready to accept that there might be more than one
sound approach to investigate the safety of a chemical.

It is only recently that a few countries in Europe have
become active in issuing rather detailed testing guidelines in
the field of toxicology, environment and residues. These
guidelines are not fully harmonized with other countries,
which is not a problem as long as they are applied with
flexibility. However, it seems unavoidable that the mere
existence of issued guidelines produces a loss of flexibility
and invites bureaucratic, unscientific formalism. This is
observable to some extent even with authorities which in the
past have impressed by their pragmatic approach.

Integrating the data requirements in European countries,
which concentrate alternatively on different registration
aspects, reflecting the influence and interest of individual
officials, I am not in a position to say that they are less
exhaustive than in the USA, as Hahn (2) did when he reviewed
pesticide regulation in Europe in 1972. However, I feel that
certainly the political and legal possibilities and perhaps
the status of the authorities to apply the flexibility of the
regulations in individual cases are still more favorable in
Europe than e.g. presently in the U.S.

In respect of the extent of data requirements in the field
of ecological testing European authorities so far have been
rather reasonable, but I feel that some are about to develop
a tendency to go too far. Are we not going to ask for the
impossible, if we require a pesticide to display effective
toxicity towards target organisms, to stay in the area where
it is applied, to degrade in due time, not to accumulate, and
yet to be harmless to other organisms which often have no real
place in that cultivated area anyway. Maybe we could allow
mankind to become again a little more egocentric and be ready
to accept that a pesticide, which is considered to be safe to
man based on favorable results in today's sophisticated toxicological and biochemical testing scheme, will not be destructive to eco-organisms. It is doubtful whether such a
pesticide which implicitly degrades at a reasonable rate and
stays predominantly in the place where it is applied can
materially or permanently increase the effects on the ecology
produced by nonchemical agricultural practices alone. It is
accepted that such a pesticide, according to its use, indicates testing of acute effects on wildlife such as birds and
fish. However, tests on organisms outside the cultivated
area, e.g. on Daphnia, are not indicated, because adverse
effects could hardly justify the luxury to ban or restrict
essential uses of such a unique compound without questioning

other basic elements of our agricultural systems. Of course,
one can argue that careless applicators could still contaminate
surface waters. Certainly, care must be taken that the present
day's sophisticated product safety evaluation is matched by an
adequate and monitored responsibility in the application of the
products. It would seem that only for chemicals with problems
in mobility and/or degradation, or with aquatic uses, additional
tests with other eco-organisms, e.g. Daphnia, are indicated.
It is hoped that a flexible sequential approach will be
followed in Europe also in testing for effects on eco-organisms.

The danger of requesting checklist data which cannot be
considered in the final evaluation and decision should be more
carefully recognized by some European authorities. This danger
seems to be particularly apparent in areas which offer fascinating scientific playgrounds, e.g. environmental studies and
analytical work, particularly with radiolabeled compounds. I
admit that extremely interesting redundant data are not only
produced on request by authorities but often also freely
offered by uncritical industrial scientists. One may have to
realize that data requiring/producing has reached a stage of
self-propelling dynamics as it offers interesting business
opportunities and helps in job enlargement and job security
in government and in the industry. Accepting that there
exists no absolute safety and that there is a limit given to
the number of data which reasonably need and can be considered
in a safety assessment, the time will come, or is already
here, when adding new tests to the checklist has to imply the
elimination of others. I am positive that the accumulation
of experience in the ranking of tests and in the full use of
data will allow a stop to the proliferation of tests without
loss of elements for adequate regulatory decision making. An
entry into the molecular biology field opened with mutagenicity studies may be instrumental in this endeavor.

European Registration Policies and Procedures

Practically in all European countries the key role in the
administrative handling of applications for registration lies
with offices of the Ministries of Agriculture. They accept the
applications, organize the involvement of experts of the
Ministries of Health and Ministries or Agencies of Environment, and they issue the final registration decisions. These
decisions are often reached in an interministerial special
commission which may include experts from academia, or even
industry.

Noteworthy and unique is the nonstatutory registration
scheme in the U.K. The Pesticide Safety Precautions Scheme,

as it is called, was drawn up in 1957 between the interested
Ministries and Industrial Associations. If agreements are to
stay in such a field they have to have the same enforcement
power in practice as corresponding laws. But the negotiation
and up-dating of agreements may create a climate for more easy
understanding among partners with divergent interests.

Because of the key involvement of the Ministries of
Agriculture in the registration process their experimental
stations are extensively involved in the evaluation of the
biological efficacy and the practical usefulness of the products. In this evaluation aspects of integrated pest control
are being increasingly considered and this proves to be not
the least hurdle to take in registering a new product. Although
efficacy is not a registration aspect in the U.K., a
corresponding approval by official testing is needed for
successful marketing.

While the data actually required in practice vary from
country to country according to specific local interests and
the possibility or readiness for flexibility, the basic
concepts of laws and regulations are homogeneous and not
substantially different from those of the U.S. The generally
prevailing conciseness of European pesticide laws and regulations would indicate that only a rather limited number of
government lawyers have found a full-time engagement in
pesticide regulatory affairs. Private lawyers may not earn an
appreciable amount of money in that domain since suing the
authorities is only rarely done by companies or industry
associations.

Registration officials in the various European countries
have rather close contacts with each other. Although some
aspects may be reviewed less thoroughly than others by individual authorities, their differently oriented primary interests coupled with their bilateral contacts finally result
in a rather thorough evaluation. It is interesting to note
that in a given case requests for further information or
data usually do not focus on the same question but may relate
to as many areas as countries are involved. A standard request
in the application for registration is to state where else
registration has been asked for, has been already granted,
or has been denied. Also the applicant must report when data
or facts become known which may put those submitted in doubt.
European authorities have so far been prudent in not trying
to handle this domain formalistically as this can easily lead
to a voluminous bureaucracy. They consider that finally it is
the manufacturer who has to cope with a severe, rather
consumer-friendly product liability and that he will be
freely ready to involve them in cases of a certain importance.

Fundamental differences exist between European countries in respect of the administrative handling of crop residues: Some authorities have deliberately not introduced a residue tolerance-setting system, arguing that careful evaluation of the residue situation at the time of registration and subsequent periodical monitoring of the practical situation will take care of the problem. Others, mainly the countries which import agricultural produce, operate systematically a tolerance system very much like that of the U.S. The remaining countries operate tolerance systems only for produce to be exported, for certain product categories, or on a case-by-case basis after overall judgment of the product and its use.

In a number of important countries registrations of new products are in principle granted only for a limited time, mostly for three years, with or without an accompanying request for additional data for re-registration after that period. Some countries operate a phased registration scheme giving provisional registrations on a reduced but favorable database to allow early market introduction of new products. As an example, phased registration schemes have allowed CIBA-GEIGY to start selling TILT®, a new fungicide, this year in France and the U.K., but it will be 1984 when it will be available to U.S. farmers. The data which may possibly be developed during a provisional registration phase involve chronic toxicity, soil metabolism, and other environmental studies. As mentioned earlier, final registrations are granted without chronic toxicity studies in minor use situations.

European authorities are in principle sympathetic to industry's plea to protect the property rights of companies with respect to the data developed and submitted during the registration procedure and not to use them in favor of following registrants. They are aware of the large and continuous investment in personnel, space, equipment and time necessary to produce registration data. However, unlike industry, which feels that this plea is well covered by the various constitutional laws in European countries, most authorities stress that corresponding regulations need first to be created, which so far has not been done in Europe. Presently some authorities seem to respect informally, at least partly, property rights on data, others are ready to use data in favor of second registrants only when they can demonstrate that their technical active ingredient has the same specifications as that of the first registrant; and the rest of the authorities are not able or willing to pay attention to that problem.

Although safety and environmental questions have reached and do reach high and sometimes explosive momentum in Europe

too, I feel that officials still can reach registration decisions in a climate without too much political pressure. They
can demonstrate rather severe legislation and are thus presently not forced to make public the detailed arguments on
which they base individual registration decisions.

Activities of the European Economic Community (EEC)

Within the general endeavor of removing barriers to trade
between the nine present member states of the Community, the
EEC Commission has started to become active in harmonizing
laws in the field of chemicals and pesticides. A Council
Directive of the year 1967 on the approximation of the laws of
member states relating to the classification, packaging, and
labeling of dangerous substances basically covers all chemical
and has been extended with six modifications to an equivalent
of the U.S. Toxic Substances Control Act (3). This directive
will involve a notification of authorities on the market
introduction of new chemicals, excluding chemicals regulated
otherwise, such as pesticides. However, the harmonization of
the existing pesticide laws within the Community proves to
be a particularly delicate affair, much in the same manner as
other questions related to agriculture, but particularly so
because national pesticide legislations and registration
authorities in the member states have been well established
since a rather long time.

Based on the above-mentioned directive on classification,
packaging and labeling of chemicals, a special directive (4)
to harmonize these aspects for pesticide formulations as sold
to the user has been adopted by the Council of Ministers of
the member states and will be in force as of January 1, 1981.
At the same time, an additional directive (5) will be in force
prohibiting or restricting the use of certain pesticides in
all member states.

The discrepancies mentioned earlier in the approach
followed by European countries in regulating pesticide
residues in agricultural produce is reflected by the difficulties encountered in harmonizing this domain. So far only
a directive relating to residues in fruit and vegetables has
been adopted. Three others on residues in cereals for human
consumption, on residues in products of animal origin, and
on residues in animal feeds have remained in the proposal
stage for quite a long time.

In 1976 a first attempt to harmonize pesticide
registration procedures was made in the form of a proposal of
a directive which would create the possibility of a type of
EEC-accepted plant protection product with the aim of having

in the future a catalogue of EEC-accepted products besides
the nationally registered ones. As the EEC Commission does
accept that most pesticides have specific importance for local
or regional agricultural needs and that they do have to
correspond to local or regional ecological and environmental
situations, the Commission also accepts that member states
should stay free to regulate pesticides for their own terri-
tories in accordance with their national laws. Therefore, in
this proposed directive on putting into circulation EEC-
accepted plant protection products (6), due flexibility has
been incorporated. The option is left open to seek either, as
up to now, national registrations in the various countries or
to apply for EEC acceptance of a product in a member state of
free choice. That member state would have to notify the other
member states and the Commission on the receipt of such an
application and respect a related directive in draft stage
on uniform principles (i.e. the tests required) for checking
compliance with the requirements for EEC acceptance of a
plant protection product. If the member state grants EEC
acceptance to a product it has to inform the other member
states and the Commission on the elements of that decision
documented with a label copy. Member states and the Commission
may request a full copy of the application submission for
their own review and make known possible reservations within
one year's time. Afterwards no restrictions on the free
circulation of the product may be imposed by member states.
It is required that EEC-accepted plant protection products
contain exclusively active ingredients listed in a
corresponding appendix of the directive.

A Scientific Committee for Pesticides established in
1978 consisting of highly qualified scientists is to be
mentioned, which may be consulted by the Commission on
scientific and technical problems relating to the use and
marketing of pesticides and to their residues.

Concluding Remarks

No fundamental differences is pesticide regulation
between the U.S. and European countries can be noted. However, due attention has to be drawn to some relevant
differences in flexibility of laws, guidelines and requirements, as well as in the political regulatory environment.

Although it still may take a long time until a central
pesticide registration becomes a practical reality in the
European Economic Community, the initiated activities will
certainly speed up the harmonization of regulations and
decision making in the member states.

Officials from European countries, supported by experts from industry, are actively cooperating in the OECD efforts to harmonize the testing of chemicals. The need for agreed tests and for guidelines in Good Laboratory Practice is well recognized. Most European delegates will be active to secure optimal flexibility in emerging guidelines.

The proliferation of data requested for registration, specifically in the field of environmental aspects, may still continue for some time. But it is hoped that finally experience in integrated human and environmental safety testing will make it possible to reach better decisions on a smaller number of relevant key data.

Literature Cited

1. Council of Europe, Ed. "Pesticides"; 4th ed., Strasbourg, France, 1977.
2. Hahn, S.; Proc. 11th Br. Weed Control Conf., 1972, 1028.
3. Off. J. of the Europ. Comm., 1967, No 196, 1.
4. Off. J. of the Europ. Comm., 1978, No L206, 13.
5. Off. J. of the Europ. Comm., 1979, No L33, 36.
6. Off. J. of the Europ. Comm., 1976, No C212, 3.

RECEIVED February 2, 1981.

Pesticide Chemistry and Regulation in the People's Republic of China

ZHENG–MING LI (CHENG–MING LEE)[1]

Research Institute of Elemento–Organic Chemistry, Nankai University, Tianjin, People's Republic of China

Development of Pesticide Chemistry in China

In comparison with the United States, China has similarities in size of territory, and geographical latitude. But China has a population of 1 billion (4.5 times that of the U.S.) and a harvested land of 100 million hectares (about 80% of that of the U.S.). China grows rice which comprises nearly half of the national grain production in the southern provinces, while in the northern provinces wheat, corn, millet and sorghum are produced. Due to nation-wide irrigation construction, popularization of high-yielding seeds, usage of futilizer and pesticides, and intensive cultivation, China has made great strides in providing adequate food for her large population. In 1979, China's grain output reached 324.9 million tons, the output of oil-bearing crops topped 6.43 million tons, and of cotton amounted to 2.2 million tons (1). In these achievements, the research and development of Pesticide Chemistry has contributed its part.

The pest situation in China is somewhat different from that in the United States. Some of the major pests are listed below:

Insects: rice paddy borer (_Tryporuza incertules_ Walker)
plant hopper (_Nilaparvata lugens_ stäl)
leafhopper (_Nephotettix cinciticeps_ Uhler)
wheat armyworm (_Leucania separata_ Walker)
corn borer (_Ostrinia nubilalis_ Hübner)
cotton aphid (_Aphis gossypii_ Glover)
cotton bollworm (_Heliothis armigera_ Hübner)

Fungi: Rice blast (_Piricularia oryzae_ Cavara)
Rice bacterial blight (_Xanthomonas oryzae_ Dowson)
Sheath blight of rice (_Pellicularia sasakii_ (Shirai) S. Ito)

---

[1] Current address: Visiting Scientist, OCSL, AEQI, USDA, Beltsville, MD 20705.

Fusarium wilt of cotton (Fusarium vasinfectum Atk.)
Wheat rust (Puccinia triticina Eriksson, and
Puccinia glumarum (Schmidt) Eriksson et Henning)

Weeds: Nutgrass flatsedge (Cyperus rotundus Linnaeus)
Barnyard grass (Echinochloa crusgalli (L.) Beauvois)
Wild oat (Avena fatua Linnaeus)
Cogangrass (Imperata cylindrica (L.) Beauvois)
Bermuda grass (Cynodon dactylon Persoon)

Due to the requirements of the expanding agriculture in
China, pesticide industry has now grown from virtually nil before
the liberation in 1949 into large-scale production. It was
reported that pesticide production in China in 1977 was probably
500,000 tons (in gross weight) (2). Though many pesticides are
produced in terms of large tonnage, the supply can only meet
about 60% of the present demand.
Following is a partial list of the pesticides produced in
China (structures of pesticides of Chinese origin are shown in
Table I.):

Insecticides

Organic Cl: BHC (benzene hexachloride), lindane, DDT,
toxaphene, tetradifon

Organic P: Ethion, phorate (Thimet), parathion,
methylparathion, demeton, trichlorfon
(Dipterex), dichlorvos (DDVP), dimethoate
(Rogor), Malathion, phosphamidon

Carbamates: carbaryl (Sevin)

Fungicides

EMC (ethyl mercuric chloride), DD mixture, HCB
(hexachlorobenzene), PCNB, Ambam (diammonium ethylene
bis(dithiocarbamate)), Zineb, Urbazid (methylarsine
bis(dimethyldithiocarbamate)), Captan, Captafol,
Asozine methylarsine sulfide, p-aminobenzene sulfonic
acid
401, Jingangmycin, Duo-Jun-Ling

Herbicides and Plant Growth Regulators

PCP, Herbicide no. 1, nitrofen, 2,4-D, dalapon, propanil, NAA, 920 (gibberellins), MCPA, Ethephon

Since the fifties, many research institutions in China have
screened about 10,000 of their synthetic compounds and new

antibiotics in bioassays. Many candidate pesticides have gone
under different tests, and some of the pesticides shown in Table
I have gone into production. These are China's first efforts at
originating her own pesticides to adapt to her peculiar needs.
In recent years, certain research groups in China have also been
active in the fields as insect pheromones (3), pyrethroids and
natural bio-active products. Nevertheless, they are in the stage
of research and development rather than large-scale production.
At the present 90% of all pesticides produced in China are
insecticides, but recent emphasis has been placed on the research
and development of fungicides, herbicides, and plant growth
regulators. China also imports new pesticides from abroad to
supplement some of her special needs.

The Impact of Modern Toxicology

     From Rachel Carson's "Silent Spring" (15) published in 1963
as a turning point, many environmentalists have become much
concerned with the environmental pollution arising from the
widespread use of pesticides. Since then, many countries,
including China, have strengthened their control over the
regulation of pesticides. Today research on pesticide chemistry
not only involves the collaborative efforts of chemists and
biologists, but also that of biochemists, pharmacologists,
toxicologists and environmental scientists.
     It might be beneficial to review some development stages of
pesticide chemistry in China. In the mid-fifties, when highly
active organic-phosphorus insecticides were first introduced to
the market, their large-scale production was encouraged not-
withstanding the fact that many of them are highly toxic substances. Many "highly active, highly toxic" pesticides actually
went into production, and only acute toxicity data was required
at that time. In the sixties, due to the growing awareness of
some serious accidents that occured during the handling of these
highly toxic substances, the preference was shifted gradually to
produce new "highly active, low toxic" pesticides. Besides the
acute toxicity data, subchronic data emerged as an important
problem. Thus some relatively low-toxicity pesticides came
into production in place of some old ones. Starting in the
seventies, due to the rapid advances in the field of toxicology,
several national pesticide conferences were held in China to
discuss guidelines for the future development of pesticides. New
pesticides have been required to be "highly active, highly
selective, low residual, less expensive". Checking residual
effects on nontarget organisms has been greatly emphasized. Now
for every new pesticide comprehensive carcinogenic, mutagenic
and teratogenic data must be submitted before production. An
integrated Pest Management (IPM) policy has been officially
announced and is greatly encouraged. All kinds of cultural and
biological control methods have also been popularized nation-wide

Table I. Pesticides developed in China

Year Discovered	Pesticide Name	Formula	Usage(a)	Institution	Reference
1958	Herbicide no. 1	$(CH_3)_2NCN$ (with $\overset{S}{\overset{\|}{}}$ and phenyl-Cl)	H	Nankai Univ.	(4)
1958	P47(b)	$(C_2H_5O)_2PSCH_2OC_2H_4SC_2H_5$ (with $\overset{S}{\overset{\|}{}}$)	I	Nankai Univ.	(5,6)
1958	401	$C_2H_5SSC_2H_5$ (with $\to O$)	F	Res. Inst. of Organic Chemistry (Shanghai)	(7)
1970	Duo-Jun-Ling	(benzothiazole ring) N=C–NHC–OCH$_3$, with $\overset{O}{\overset{\|}{}}$ and N–H	F	Shen-Yang Chem. Eng. Res. Inst., Shanghai Pesticide Res. Inst.	(8)
1971	Ai-Jian-Su	$\left[CH_2=C\overset{CH_3}{\underset{Cl}{|}}-CH_2N\overset{CH_3}{\underset{CH_3}{<}}\right]^{+} Cl^{-}$	PGR	Nankai Univ.	(9)
1971	Di-Ku-Shuang(c)	(thiadiazole)–NHCH$_2$NH–(thiadiazole)	F	Sichuan Pesticide Res. Inst., Sichuan Res. Inst. of Chem. Eng.	(10)
1973	Jinggangmycin	antibiotics	F	Shanghai Pesticide Res. Inst.	(11)

Year	Name	Structure	Type	Source	Ref.
1974	Sha-Chung-Shuang	$(CH_3)_2N-CH$ with CH_2SSO_3Na / CH_2SSO_3Na	I	Guizhou Chem. Eng Res. Inst.	(12)
1977	771-20[d]	$(C_2H_5O)_2P-S-S-P(OC_2H_5)_2$ (each P bearing a double-bonded S)	F	Nankai Univ.	(13,14)

a. In this column, H = herbicide, I = insecticide, F = fungicides, and PGR = plant growth regulators.

b. Though the acute toxicity of P47 is lower than Demeton, due to higher production cost of P47 its development was stopped.

c. Due to the unsatisfactory results of 2-year chronic toxicological tests, its production was discontinued.

d. Activity against bacterial leaf blight of rice was discovered. It is undergoing large scale field tests and chronic toxicity examination.

(16,17,18). Production, storage and application of mercury-
containing pesticides have been strictly prohibited. Tin and
Arsenic-containing pesticides have been discarded. Organochlorine pesticides have been greatly restricted and are
gradually being phased out. Though some were reported recently
to cause delayed neurotoxicity, other organophosphorus and
carbamate pesticides are still considered to be acceptable due
to the relative ease with which they are degraded biologically
and chemically.

The Regulation of Pesticides

Pesticide regulation became complicated when chronic
toxicological problems stepped into the picture. In 1974 when
the National Pesticide Information Conference was held in China,
Wuhan Medical College presented their own toxicological data
about an imported low-toxicity fungicide dichlozoline (Sclex),
[3-(3,5-dichlorophenyl)-5,5-dimethyl-2,4-oxazolidinedione]. In
their chronic tests, they confirmed that dichlozoline could cause
cataracts and induce malignant tumors in experimental mice.
Subsequently there was the chlordimeform (Fundal) case. At that
time Chinese scientists were trying to use the low-toxicity
chlordimeform (Fundal), [N'-(4-chloro-o-tolyl)-N,N-dimethyl
formamidine] to replace the traditional organochlorine insecticides in combating rice paddy borers. Reports from abroad
that chlordimeform was possibly carcinogenic surprised
everyone and the application of chlordimeform was abruptly
stopped. These two events spurred the Chinese workers to start
their own toxicological research. Later a series of national
conferences were held. In 1976, the National Forum on Pesticide
Toxicology and Residues was held. In 1978, the Ministry of
Chemical Engineering, the Ministry of Agriculture and the
Ministry of Public Health called for a national meeting to
discuss the toxicology and residue problems of pesticides and
finally drafted "Proposed Regulations of Experimental Methods
for Pesticide Toxicology" (19) and "Proposed Regulations for
Pesticide Toxicology and Residues" (20). This was an attempt
for the first time in China to standardize the methodology
involved in the toxicological tests. Soon after another paper
"Proposed Regulations of Toxicity Experiments for Fishes" was
also drafted (21). In April 1980, a decisive step was taken
jointly by the Ministry of Agriculture, the Ministry of Chemical
Engineering, the Ministry of Public Health, and the Environmental
Protection Agency (an organization directly responsible to the
State Council). After consulting with the specialists and
practitioners concerned, the four departments jointly drafted
out the "Regulations for Pesticide Management" (22). At present
it is submitted to State Council for final promulgation. It
will be the legal basis for regulating and monitoring the
research production and application of all pesticides in the
future.

The "Regulations for Pesticide Management" consists of three parts: (a) processing of applications for pesticide registrations, (b) rules of quality control of pesticides, and (c) rules of safe application of pesticides.

According to the new regulations, the application for registration of any new pesticide must first be submitted to the Institute for the Controlling of Pesticides, an authoritative state organization subordinated to the Ministry of Agriculture. This Institute will review the submitted sections on bioactivity, phytotoxicity and residues on behalf of the Ministry of Agriculture. All the other data will be sent simultaneously to the Ministry of Chemical Engineering (MCE), the Ministry of Public Health (MPH) and Environmental Protection Agency (EPA) for joint review. The MCE will scrutinize the related production technology, analytical methods, waste management, etc. The MPH will probe into the toxicology problem in detail, and the EPA will examine the pesticide pollution on soil, water system, and air. Only after all the data are cleared for final approval from these Ministries can the formal registration be granted by the Ministry of Agriculture. (Step D-F)

In applying for registration of a candidate pesticide, there are still more steps to go through; the whole process is shown schematically below in Figure 1. (Step A-H)·

When an organization, be it industry, university or research institution, has a candidate pesticide, the first step after the minimum research is completed (Step A) is to file an application for a Preliminary Technical Appraisal Conference (Step B). At this conference, organized by the appropriate national or local authorities, an steering appraisal committee will be set up. The committee is composed of representatives invited from national or provincial organizations of Science and Technology, Chemical Engineering, Agriculture, Environmental Protection, agriculture research institutions, medical institutions, occupational hygiene research institutions, Academy of Science, universities and other concerned bodies. These representatives will examine carefully all the reports which must include:

(a) Background
(b) Comparison of different synthetic routes
(c) Analytical methods
(d) Bioassay results of greenhouse and test plots
(e) Acute and subchronic toxicity data (it was affirmed recently (19,20) that subchronic teratogenic and mutagenic data should be included. For organophosphorus compounds, delayed neurotoxicity testing is required. Residue analysis in food crops, forage, poultry, animals and aquatics biota is also needed)
(f) Experimental waste disposal

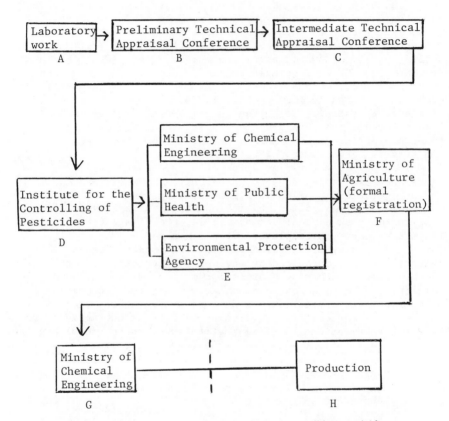

Figure 1.   Steps involved in the registration of a candidate pesticide

Only after all the reports are discussed, inquiries and
suggestions are made and the research work is evaluated as
acceptable and satisfactory will a Preliminary Appraisal
Certificate be drafted and signed by every representative and
the chief practitioners who are responsible for the reports. To
reach this stage usually take 1-3 years. With this formal step
completed, further financial support can be sought for additional
research work, which often takes 2-4 more years (field tests and
chronic toxicology tests require data for 2 successive years).
Then all reports have to submit into the authorities concerned
to request an Intermediate Technical Appraisal Conference. A
larger group of representatives from all the related fields will
be invited to examine and scrutinize all the details of the
reports. These reports should include the following:

(a) Pilot plant test runs
(b) Industrial analytical methods
(c) Bioassay in large demonstration and in productive field
     tests
(d) Subchronic and chronic toxicological data (subchronic-
     toxicity tests, 3 months; chronic-toxicity tests 2
     years. Chronic-toxicity tests include carcinogenicity
     and reproduction tests (3 generations), residue
     dynamics, and metabolism and degradation data
     (metabolism tests in animals will be required after
     1980)
(e) Formulation data
(f) Standardization of product, and impurities and residue
     allowances
(g) Waste management test runs on pilot plant
(h) Standard measures for occupational hygiene, detoxifi-
     cation and safety, the determination of ADI, (Acceptable
     Daily Intake), etc.
(i) Calculation of production costs

Many questions and requirements will be raised at this point.
Another Intermediate Appraisal Certificate has to be signed in a
similiar manner. Then all the documents, reports, and certifi-
cations are sent to the Ministries (Step D and E) for joint
review and approval. Only after the formal legistration is
granted (Step F) can the chemicals and equipments be disbursed
from the Ministry of Chemical Engineering (Step G). The design and
materialization of this project will take probably another 1-3
years before the new pesticide finally goes into production
(Step H). These rather complicated steps are set up to regulate
new pesticides properly and assure the utmost benefit to the
people. For example, after getting through Step A-C, the
required papers for the new insecticide Sha-Chung-Shuang (see
Table I) are now being submitted to the Institute for the
Controlling of Pesticides for further review (23).

The above procedure is applied primarily to the new pesticides or the old pesticides with new formulations. It will
gradually cover the re-registration of the older pesticides
already in use. Registration for imported pesticide with
reliable chronic toxicity data will be granted for field
application. However in the case of dichlozoline mentioned
above, Chinese workers confirmed its chronic-toxicity problems
with their own data and decided later to discontinue its use
in China.

As conditions require, the above sequence of processing can
be stopped at any step. Take Di-ku-Shuang (DKS) (see Table I)
for example. It is a new fungicide developed in the early
ꞌeventies by Sichuan Pesticide Research Institution. It has
an excellent systemic effect on bacterial leaf blight of rice,
which is ranked as one of the major crop diseases in China.
After much research work, its synthetic steps are well worked
out and its production cost is found as reasonable. Its acute
toxicity is rather low. (LD_{50} for mouse oral: 2250 mg/kg;
dermal: 150 mg/mg). It has a skin irritation as side effect.
Many formulations were tried, and a paste formulation was
developed to avoid the skin irritation during field application.
It passed the Preliminary Appraisal Conference. Then many
institutions cooperated to carry it through the following stages
of development, and intensive efforts were involved in the
succeeding two years to conduct pilot-plant runs, field tests
and chronic toxicity tests (including autopsy and histopathology
work on several non-rodent species). Results showed no carcinogenic effects, but unfortunately it was found during the reproduction tests to have a teratogenic effect on fetal mice (24).

Despite the great need for new
fungicides in China and the considerable investment and efforts
poured into it, the DKS project was resolutely interrupted at
Step C and its production was officially banned. Another new
insecticide Ming-Lin-Wei (an analogue of Herbicide no. 1, also
developed by Nankai University) had satisfactory effects on
paddy rice borer and was promising as a replacement for the
organochlorine insecticides. Its acute oral toxicity is LD_{50}:
500 mg/kg (rat), 100 mg/kg (mouse). Sub-chronic tests showed
a favorable low accumulation. But in teratogenic tests, there
is a phenomenom of incomplete calcification of mouse sternum
(25), so this new insecticide is withheld temporaily from
production until more data are obtained.

Some experts in China argued recently that overemphasis on
low toxicity might lead in the wrong direction, since low-
toxicity frequently refers to the acute-toxicity data, which
can not reflect the full picture in carcinogenic, teratogenic
and mutagenic characteristics. Some low-toxicity pesticides as

mentioned above (chlordimeform, dichlozoline, DKS, Ming-Lin-Wei) have chronic toxicity problems. Some highly toxic insecticides as aldicarb (Temik), carbofuran, monocrotophos (Azodrin) could have their toxicity reduced by improvement in formulation and still be used safely in the fields. This is the reason why acute toxicity should not be the prime criterion in pesticide regulation today.

Perspective

It seems clear now that modern toxicology has played a decisive role in the future development of the pesticide chemistry (26,27,28). Besides oncogenicity and teratogenicity, the importance of which has been fully recognized, the metabolic pathways and the mutagenic effects on reproduction are also becoming the main criteria in assaying the toxicological aspect of all pesticides. The situation is further complicated by the fact that when a pesticide gets into contact with the surrounding environment, under the influence of sunlight and various enzymes existing in different living organisms (insects, mammals, plants, micro-organisms), various pathways will produce many degradation products and metabolites. Thus scores of new compounds with unknown toxicity will be derived from just one simple structure. Take two well known types of fungicides for example. Dithio-carbamates metabolize in plants to form ethylene thiourea (ETU) which is known for its carcinogenicity and teratogenicity (fortunately, ETU is very sensitive to UV light and is degradable under sunlight). Another fungicide, PCBA (Pentachlorobenzyl alcohol), causes no chronic toxic problem itself, but it can be changed by microbial metabolism in the soil into 1,2,5,6-tetrachlorobenzoic acid, which is phytotoxic to following crops planted in the same field. Thus we notice that in such a diversified metabolic pathways in different living organisms, any one of these metabolites that has a serious toxicity problem will probably jeopardize the practical application of this pesticide, even though the pesticide itself causes no toxicity problem at all.

Since the pests possess the peculiar ability to propagate tremendously and adapt swiftly to the changing environment, we should believe that pesticide chemistry with its flexibility and versatility still can play an important role in combating pests. The point at argument is not to discard all the chemical pesticides, as some people have advocated, but to design and produce more active, more selective and more biodegradable kind of new chemical pesticides. Chemical pesticides have suppressed or eliminated many of the lethal contagious diseases, thus saving millions of human lives, and contributed much to the greatly increased output of farm produce. In human history, chemical pesticides should be honored for their achievements instead of being totally discredited. Although during the progress of

pesticide chemistry, there have regrettably been some serious
adverse effects on the environment which were not fully realized
until many new analytical and toxicological techniques were
developed in the recent years. In the face of the discouragement
caused by the ever-stricter regulations and the high risks involved in the discovery and development of any new chemical
pesticides, we should be aware of some of the promising fields
(pheromones, juvenile hormones, pyrethroids, amino-acid fungicides, phytoalexins, etc.) which have shown sound progress.
There is also much room for improvement in application and
formulation technique for chemical pesticides to reduce the
environmental pollution, since the major portion of an applied
pesticide often hits nontarget organisms rather than target
pests. Any improvement in methodology to simplify the chronic
toxicity tests that can cut down the immense costs and amount
of scientific manpower involved will certainly encourage the
development of pesticide chemistry.

Modern agriculture needs a new generation of pesticides to
keep in pace with the demands of the modern world. Today
pesticide chemistry is entangled in so many fields of science
that it can no longer be undertaken by small groups of specialists
in just a few lines of study. Only through persistent and
intensive basic research in collaboration with all the scientists
concerned can we rise to the challenge ahead of us.

Literature Cited

1. Oversea Press Release from Xinhua News Agency, April 8,
    1980.
2. Groen, H. J. and Kilpatrick, J. A. "China's Agricultural
    Production, Chinese Economy, Joint Economic Committee,
    U.S. Congress, Vol. 1. Policy Performance, A compendium
    of Papers"; U.S. Government Printing Office: Washington,
    D.C., Nov. 9, 1978; p. 635.
3. Kiangsu Institute of Sericulture, Peking Institute of
    Zoology, Shanghai Institute of Organic Chemistry, Acta
    Entomology Sinica, 1974, 17, 290-302.
4. He, Bing-Lin, Yang, Hua-Zheng, Acta Chimica Sinica, 1960,
    26(1), 1-6.
5. Yang, Shih-Hsien, Chen, Tian-Chi, Lee, Cheng-Ming, Li,
    Yu-Kwei, Wang, Chin-Sun, Yan, Mun-Gong, Dung, Shi-Yang,
    Scientia Sinica, 1960, IX (7), 897; C.A. 54 19467 (1960).
6. Yang, Shih-Hsien, Chen, Tian-Chi, Lee, Cheng-Ming, Li,
    Yu-Kwei, Tung, Hsi-Yang, Kao, Shou-Yi, Tung, Sung-Chi,
    Acta Chimica Acta, 1962, 28(3), 187; C.A. 59 3758 h (1963).
7. Li, Zheng-Ming, Pesticide Industry Technical Journal, 1978,
    2, 16.
8. Jiang, Shao-Ming, Huan Ching Ko Hsueh (Environmental Science)
    1978, 12, 22.
9. Res. Inst. of Elemento-Organic Chemistry, Huaxue Tongbao
    (Chemical Review), 1974, 1, 37.
10. Sichuan Pesticide Res. Inst., Pesticides, 1973, 4, 21.

11. Shanghai Pesticide Res. Inst., Pesticide Research Informa-
    tion, 1974, 4; 1974, 6.
12. Data taken from an unpublished report by Guizhou Chem. Eng.
    Res. Inst., "A new nereistoxin-type insecticide Sha-Chung-
    Shuang" May, 1979.
13. Yang, Shih-Hsien, Chen, Tian-Chi, Tang, Chu-Chi, Jin,
    Gui-Yu, Liu, Tian-Lin, Zhang, Jin-Pei, Pesticide Industry,
    1966, 1, 28.
14. Res. Inst. of Elemento-Organic Chemistry, Pesticide
    Industry, 1979, 3, 34.
15. Carson, R., "Silent Spring", Hamish Hamilton, London, 1963.
16. American Insect Control Delegation, "Insect Control of
    People's Republic of China" (CSCPRC Report no. 2); National
    Academy of Sciences: Washington, D.C., 1977.
17. American Plant Studies Delegation, "Plant Studies in the
    People's Republic of China" (CSCPRC Report); National
    Academy of Science: Washington, D.C., 1975.
18. Klassen, W., "Notes on trip to People's Republic of China
    of U.S. Biological Control Team" July 5-31, 1979, BARC,
    USDA, unpublished.
19. Proposed Regulations of Experimental Methods for Pesticide
    Toxicology, (Zhejiang conference, 1978, unpublished).

20. Proposed Regulations for Pesticide Toxicology and Residues
    (ibid.).
21. Proposed Regulations of Toxicity Experiments for Fishes
    (including Embryotoxicity Testing Methods), (drafted by
    National Aquatic Department of the P.R.O.C., 1979,
    unpublished).
22. Regulations of Pesticide Management (April 1980, unpublished).
23. Toxicological data for Sha-Chung-Shuang complied by
    Gui-Yang Medical College, Shanghai 1st Medical College,
    Gui-Zhou Provincial Sanitary Station, Jian-Su New Medical
    College (May, 1979, unpublished).
24. Sichuan Provincial Sanitary Station, "The Toxicology
    Research on Di-Ku-Shuang (N,N'-methylene bis(1,3,4-
    thiadiazole-2-amine)" (August 1979, unpublished).
25. Res. Inst. of Elemento-Organic Chemistry, Nankai Univ.,
    "Ming-Lin-Wei Intermediate Technical Appraisal Conference
    Data" (Conference held in Zhengjiang, Jiang-Su in May, 1979,
    unpublished).
26. Yang, Shih-Hsien, Liu, Lun-Zhu, Li, Zheng-Ming, Chen, Ru-Yu,
    "Advance of Pesticides" Vol. II; Chem. Eng. Publ. Hse.:
    Peking, 1979; p. 11, p. 83.
27. Environmental Protection Agency, Registration of Pesticides
    in the United States (Proposed Guidelines).   Federal
    Register Part II (July 10, 1978) pp. 29690-29741.
28. Environmental Protection Agency, Pesticide Programs
    (Proposed Guidelines for Registering Pesticides in the
    U.S.); Hazard Evaluation : Humans and Domestic Animals.
    Federal Register, Part II (Aug. 22, 1978), pp. 37336-37403.

RECEIVED March 10, 1981.

Pesticide Regulation in Canada

DAVID J. CLEGG

Toxicology Evaluation Division, Food Directorate, Health Protection Branch,
Health and Welfare Canada, Ottawa, Canada, K1A 0L2

The main authority for regulation of pesticides in Canada
resides in two Acts of Parliament, the Pest Control Products Act,
administered by Agriculture Canada, and the Food and Drugs Act,
administered by Health and Welfare Canada. This dichotomy of
authority necessitates close cooperation between the two departments. In addition, various other acts address such aspects as
water quality, fisheries, wildlife, etc., with respect to pesticides.

Agriculture Canada is responsible for registration of all
pesticides sold in Canada. To comply with this task, the
Department enlists the cooperation of other government departments, as advisors on specific aspects of pesticide activity.
Prior to registering any active ingredient under the Pest Control
Products Act, Agriculture Canada obtains expert advice from
such Departments as Environment Canada, and from Fisheries and
Oceans, with respect to the potential ecological impact of the
chemical; from Health and Welfare Canada, with respect to
potential adverse human health effects and acceptable food
residues and from any other Department with expertise for input
on specific non-agricultural aspects of the use of the chemical.
On the basis of their own expertise on efficacy, use pattern
potential, etc., combined with the advice and recommendations
from the other Departments (if these are all favourable)
Agriculture Canada will then initiate registration of the
product.

The Food and Drugs Act provides the authority to Health
and Welfare Canada to promulgate regulations indicating the
maximum residue limits of pesticides permissible on food at the
time the food first enters into commerce. The regulations list
the permitted levels, and also indicate that any food with a
residue level not listed in the Regulations, but exceeding 0.1
ppm of the pesticide is considered to be adulterated, and hence
is not permitted for human consumption. This, of course, does
not preclude the listing of maximum residue limits below 0.1 ppm.

This brief introduction, covering the legislative responsi-

0097–6156/81/0160–0537$05.00/0
© 1981 American Chemical Society

bilities of the various departments has served to indicate where
Health and Welfare Canada fits into the pattern of pesticide
regulation. To meet the responsibilities in this pattern, the
work in Health and Welfare can be subdivided into two major
areas - the residue area, and the toxicology. The former, I
intend to touch on only superficially. In this area, data from
supervised field trials on the various crops which may be exposed
to the pesticide are assessed and evaluated and maximum residue
limits, proposed by the petitioner are either modified or accepted. According to the data on plant metabolism, maximum residue
limits for parent compound + metabolites may be proposed or even
on rare occasions separate maximum residue limits are proposed
for the metabolites. Analytical methodology for detecting residues is examined for feasibility, and such factors as the
ability to detect total residues (i.e. parent + metabolites) or
separate entities (e.g. EBDC's and ETU) are determined. Once
the maximum residue likely to be found is determined (even if
this is at the limit of analytical detection) and characterized,
this information is passed to the toxicologists, to aid in the
provision of an assessment of the safety of potential food
residues.

The toxicology evaluation can also be subdivided into two
major areas - 1) the data pertaining to safety of food residues,
and 2) the data pertaining to occupational exposure (e.g. by
formulators, mixer applicator, etc.) and bystander exposure.

There are no listed requirements for submission of toxicity
data to support registration in Canada. However, the Acts place
the onus on petitioners to submit data to support the safety of
their pest control products. Thus considerable latitude exists
with respect to the toxicology data base which may be required
for any particular compound.

In the early 1950's, the toxicology data base utilized in
assessing the safety in use of a pesticide was frequently
extremely limited. In reviewing old files from that era, a
recommendation for registration of a pesticide, still in use
today, was noted. This recommendation was based on a single
acute oral study, and a three week feeding study in rat - the
parameters measured being limited to body weight gain, haemat-
ology on 2 rats/sex/dose level, and histopathology of 6 organs
on all survivors. Survival was 100%, and hence 36 rats (three
dose levels using 6 male and 6 female rats/dose level) were
examined. I hasten to add that, today, this compound is supported by a wide range of studies, including two rodent lifetime studies, multigeneration reproduction studies, teratology
studies, mutagenicity studies, etc. etc. Thus a very marked
change in toxicity data requested by the Health Protection
Branch has occurred during the last 30 years.

Because of the absence of any specific legal requirements
for submission of particular types of studies, it is almost
impossible to trace the historical development of this pro-

gression towards data requirements requested to support safety
in use of pesticides today. However, certain trends can be
delineated.

By early 1960, pesticides used on food had been divided
into two major categories - those which left a significant
residue on the food, as determined by specified analytical
techniques, and those which did not. The toxicological requirements for these categories differed considerably. It was argued
that, if there was no detectable exposure via the food, minimum
toxicity data should be required for the latter group of compounds. These data comprised at least acute oral studies in two
species (one of which should be non-rodent) and two 90 day feeding studies (again, one of which should be in a non-rodent species). This was supplemented in 1965, to include absorption,
distribution and excretion studies in animals. In addition, depending upon the method of application of the pesticide, dermal
and eye irritation studies, sensitization studies acute dermal
studies, and inhalation studies were requested.

The toxicology data were then evaluated, and a "no observable toxicological effect level" (NOEL) was determined. A
large safety factor (usually 1000 fold, based on 10 fold for
interspecies variation, 10 fold for intraspecies variation, and
10 fold because of the limited nature of the data) was applied
to the NOEL, and this was designated as the "negligible daily intake" (NDI) for man. The NDI was considered to be a toxicologically insignificant exposure level, via the food.

To determine the possible exposure level via the food, the
"theoretical daily intake" (TDI) was calculated. This figure
was derived by utilizing data on the rate of disappearance of
the food item in Canada, provided by Statistics Canada, and
determining from these data, the consumption of the crop per
capita (assuming disappearance rate and consumption to be comparable). It was assumed that the total crop could have
residues present at about the level of analytical sensitivity
of the method of detection. Thus based on the consumption
level and the maximum residue level, an estimate of exposure
from each crop can be calculated. The total exposure from all
crops is the TDI.

If a comparison between the TDI, and the NDI indicated the
NDI to exceed the TDI, no further data were required prior to
recommending registration. In the converse situation, additional
data could be requested - e.g. increased analytical sensitivity
of the method; additional toxicity studies, etc.

In the situation where finite residues were detected, it
was still possible that the NDI would exceed the TDI, and hence
human exposure would be considered to be toxicologically
negligible. However, in the majority of cases, even in the
early 1960's, when permissible residues were listed in the Food
and Drugs Act and Regulations, additional studies were required.
At that time, these usually comprised 2 year studies in both a

rodent and a non-rodent species, together with data on metabolism, identifying the major animal metabolites. In the early
1960's multigeneration studies, and shortly thereafter,
following the thalidomide tragedy, first reported in 1961,
teratogenicity studies were added to the battery of test requested. Further changes occurred in the 1970's. Thus, in 1975,
the requirement for 2-year dog studies was reduced to one year -
a provision which is still extant - and which differs from the
U.S. requirement for 6 month dog studies. More recently, the
acceptable multigeneration study, which used to comprise 3
generation, with 2 litters, per generation has also been reduced
with reference to the minimum requirement. Today, 2 generations
and a minimum of 4 litters is requested. This again, differs
from the U.S.A. where 2 generations with one litter per generation
is acceptable.

In addition to the "major" studies mentioned above, changes
have occurred in the "minor" studies requested for submission.
Thus, for the organophosate insecticides potentiation studies
were requested in the 60's and well into the 70's. However,
since the protocol used for most of these studies range from
farcical to barely acceptable, emphasis in this area has decreased. Conversely the improving knowledge in the area of
delayed neurotoxicity testing has increased our requests for
additional studies on those compounds which may have toxic
potential in this field.

As things stand today, policy is in process of change. So
far I have concentrated on toxicity requirements for pesticides
in relation to food residues. In 1978, a separate unit was
established in the Environmental Health Directorate, which
serves to advise Agriculture Canada on the adequacy of toxicity
data to support safety in use, with reference to occupational
and by-stander exposures. Obviously, in this area of concern,
the presence or absence of residues in food is only a relatively
minor portion of the potential exposure of the individual.
Further, different routes of exposure such as dermal, or
inhalation exposure are of much more importance than exposure
via the oral route. Thus, the old concept of minimal toxicity
data for pesticides which do not leave residues in food is being
abandoned.

The absence of specific guidelines, or legal requirements
for toxicity data submission on specified studies places the
Health Protection Branch in the position that it can require
any type of study prior to establishing maximum residue limits,
or recommending registration to Agriculture Canada. This
provides the advantage that each compound can be considered
individually, and the data requirements can be tailored to
fit the compound.

This approach can be exemplified by comparing the toxicity
data base required for, let us say, a soil sterilant and an
insecticide applied by aerial spraying. In the former case,

the test material may be applied as a soil drill, the formulation
used being a dustless granular material. Thus operator and by-
stander exposure is minimal. Further the probability of food
residues is minimal. In contrast, the insecticide may require
mixing in a suitable solvent, prior to spraying; flagmen, etc.,
will be exposed and residues may occur on the crop. Drift may
also occur, resulting in by-stander exposure. Obviously, the
requirement for toxicity data in the former case, to assess
safety in use, will be much less than that required in the latter
case, since human exposure would be negligible, if it occurred
at all.

The basic differences between the U.S. requirements, and
those in Canada apply mainly to the durations of studies. At
the present time, dog studies of 90 days duration are accepted
in Canada, for thosecompounds where human exposure via the
food, is negligible. However, where maximum residue limits are
requested, a one year dog study is requested, as opposed to the
U.S. requirement for a study of 6 months minimum duration. This
position was promulgated in the mid 1970's and was based on a
brief internal examination of previously submitted dog studies.
However, since this decision was made, the minimum number of dogs/
sex/dose level, utilized in toxicity studies, has increased from
4 to 6; the parameters investigated in dog studies have increased
in number and the accuracy and sensitivity of tests to determine
these parameters has increased. With these factors in mind, the
Health Protection Branch is considering analysing results of
more recent studies to determine the necessity for requiring the
1 year study, as opposed to the 6 month study.

A second type of study, where Canadian and U.S. requirements
differ, is the multigeneration reproduction study. In EPA, I
understand the requirement is for 2 generations with 1 litter/
generation, as a minimum. Following an analysis of some 70
reproduction studies, HPB has also reduced the number of
generations required to two. However, a total of 4 litters is
required, at least one litter being produced in both generations.
Thus, HPB requires more data than the U.S.A.

Changes in Canadian regulations during the last decade which
were initiated because of adverse toxicity data generated on the
active ingredient, contaminants in the technical material, or
metabolites of the active ingredient, have largely resulted
because of concerns relating to potential carcinogenicity. Maximum residue limits were rescinded for monuron, and for chlordimeform, because of carcinogenicity studies, which indicated
positive effects. In the case of chlordimeform considerable
additional data is known to have been generated, but since it
has been withdrawn from use in Canada, the only present use
being on cotton, these data have not been reviewed by HPB.

Several of the organochlorine pesticides have been
implicated as potential human carcinogens as a result of the
induction of hepatocellular carcinomas in mice. Unlike the

official position in the U.S.A. Canadian toxicologists accept
the hypothesis that there are instances where a "no effect"
level can exist for a carcinogen. In the case of the mouse
hepatocellular carcinoma, liver hypertrophy appears to be a
necessary precursor for nodular hyperplasia, and subsequent
hepatocellular carcinoma formation. Until the mid '70's, the
liver hypertrophy noted in mouse was considered to be a functional
hypertrophy, rather than a true toxicological effect, and hence,
although it was considered in the overall evaluation, it was not
attributed the concerns which we now realize are merited. This
change in position resulted in marked reductions in maximum
residue limits in foods, and in cancellation in use, in some
cases due to the potential hazards from formulator/applicator
exposures.

Use of lead arsenate as a pesticide was also cancelled on
most crops - the basis for the cancellation being two-fold -
the irreversibility of neurological damage resulting from lead
exposure of children and the potential for lead-induced renal
tumors noted at high levels in rodents resulted in a requirement
to reduce the total lead exposure from all sources, and secondly
the occupational exposure data indicates the potential for human
skin cancer resulting from arsenic exposure.

All the ethylene bis-dithiocarbamate (EBDC) maximum residue
limits in food were cancelled, and replaced on a restricted
number of crops, at lower level, because of concern regarding
the potential for exposure to the metabolite, ethylene thiourea
(ETU) which was generated particularly during cooking of crops
bearing EBDC's. ETU is a thyroid carcinogen, but again thyroid
hyperplasia appears to be a necessary prerequisite for tumor
formation. Since "no observable effect levels" can be assessed
for the thyroid hyperplasia the possibility of a "no effect"
level for the carcinogenicity of ETU exists. However, since our
knowledge of reversibility of thyroid hyperplasia is still the
subject of basic research, Canadian regulations under the Food
and Drugs Act do not permit the presence of any ETU in food
crops.

Although maximum residue limits of leptophos were considered
to be insufficient to be of toxicological importance the estimates of potential formulator/applicator exposure resulted in the
deregistration of this compound, because of the potential to
induce delayed neurotoxicity in humans.

Legislation has also been enacted with respect to levels of
contaminants permitted in technical active ingredients. An
obvious example is the limit of 0.1 ppm tetrachlorodibenzo-p-
dioxin, permitted as a maximum level in technical 2,4,5-T,
which incidentally is still registered for use by Agriculture
Canada.

A further example, in the more recent past, is the requirement for less than 1 ppm of n-nitrosodipropylamine in trifluralin.
As yet, the limits for nitrosamine content of other dinitro-

analine pesticides have not been set.   It is, however, under very
active review at this time.

Whilst no specific legal requirements exist with regard to
studies which must be submitted to support safety in use of
pesticides, there are unofficial minimum requests which must be
met.   Thus, a petition received some 8 years ago which contained
one page of "toxicity data" indicating that "over 13,000 animals
have been exposed to the test material without any adverse.
effects.   The menstrual cycle of rodents was unaffected" was
obviously rejected.   Not only was the absence of supporting data
unacceptable - the scientific integrity of any petitioner be-
lieving rats have a menstrual cycle must be in some doubt!!

The basic minimum data base, which might be considered as
acceptable would include acute oral and dermal LD_{50}'s, acute
inhalation LC_{50} eye, and skin irritation studies; feeding studies
of at least 90 days duration on adequate numbers of animals of
at least two species, one of which would be a non-rodent species,
absorption, distribution and excretion data in one of the species
used in the 90 day studies, mutagenicity screening studies,
possibly a multigeneration reproduction study, and any special
studies indicated by the chemical structure of the compound under
test.

Provided the test compound does not result in any residues
exceeding 0.1 ppm on food crops, applicator exposure is minimal,
and the limited toxicity data indicate that the negligible daily
intake (i.e. that dose below which intake, on a daily basis for
a lifetime is considered to be toxicologically insignificant -
usually a dose level of 1/1000th of the "no observable effect
level" on the most sensitive toxicological parameter examined, in
the most sensitive species) will not be exceeded it is possible
that registration would be recommended.   The **tendency is**, however, to require considerably more data than the minimum acceptable package described above.

Looking ahead to the future, I would suggest that three
phases are likely to be distinguishable.   Firstly, the data
base requested as minimum data will be expanded to include
rodent carcinogenicity studies on all pesticides.   In addition,
data on reproductive effects, and on teratogenicity will also be
included in minimum requirements.   The second and third phases
are less predictable but are distinct possibilities.   As the
evaluation and extrapolation of shorter term carcinogenicity
screening studies becomes more precise a reduction in the
requirements for lifetime carcinogenicity studies should, I
believe occur.   This would comprise phase 2.   However, concomitant with this reduction, development of reliable methodology
in the areas of behavioral toxicology, and immuno-toxicology is
also a reasonable postulate.   As our ability to extrapolate
such data increases then since the present state of knowledge
indicates that, at least in the case of behavioral toxicology,
sensitivity of tests appears to exceed the sensitivities of

present routine tests it is extremely likely that such tests will become requirements for inclusion in the basic data needed for safety evaluation.

Overall therefore, it is likely that in the future minimal requirements for toxicology data in Canada will increase in the near future, but hopefully will stabilize in quantity required, with a change in emphasis with regard to the types of tests requested.

RECEIVED March 9, 1981.

Regulatory Aspects: A Summary

PHILIP C. KEARNEY

U.S. Department of Agriculture, Building 050 BARC West,
Beltsville, MD 20705

The emerging importance of toxicology in pesticide registration policy has not only depended on major scientific achievements in the discipline of toxicology, but also in the growing
involvement of various segments of society who are concerned
about risk assessment and public health. These societal groups
have also had a major impact on the use of toxicolgical data to
force regulatory agencies to develop more sophisticated methods
of risk/benefit assessment. This session of the symposium
examines some of these public forces that comment on pesticide
regulations, and then examines regulations in various countries
of the world.

Industry is concerned how the intent of the authority
granted to the Environmental Protection Agency (EPA) by Congress
in the 1972 amendments to the Federal Insecticide, Fungicide and
Rodenticide Act (FIFRA) are being mandated by a growing number of
and changes in regulations, rules and guidelines. The whole
process of pesticide registration is perceived as becoming
infinitely more complex. A major problem in the reregistration
process is the large number of compounds involved (1400 individual pesticides and 33,000 formulated products encompassing some
15 classes of pesticides). Because of the enormity of the regulatory task, reglations have been proposed and adopted over a
period of six to eight years; the process is not yet complete. A
number of fundamental changes were made:

(1) The negligible residue concept was abandoned;

(2) Studies on non-target organisms (i.e. environmental
studies) were required;

(3) Pesticides were to be classified for general or
restricted use; and,

(4) An assessment of applicator or user hazard was
required.

As a result of the change in regulations, the use of a number of pesticides has been restricted, and, in some cases, registrations have been canceled. The registration process has become lengthy and much more complicated. Industry proposes some recommendations that could ultimately benefit private industry and the public. These recommendations include the use of certified summaries, clarification of the objectives of the EPA Good Laboratory Practices program, modification of the current proposed toxicology guidelines and acquisition of a stable, authoritative body of scientific personnel to review toxicology data.

The environmental groups, represented by the Environmental Defense Fund (EDF), are concerned about the accuracy of toxicology data being provided to the Environmental Protection Agency to do the risk/benefit analysis. The environmentalists want comprehensive testing and assessment of pesticides before they are widely disseminated into the environment. The reason for these concerns are based on some unacceptable laboratory practices that have been found in several drug and pesticide laboratories. They strongly support uniform requirements for testing because of past inaccuracies and encourage good laboratory practices. EDF favors a preventative approach to pesticide registration based on suggested evidence from animal studies rather than actual harm to human populations. They believe the societal costs are too great to risk less than near absolute safety in pesticide registration.

To a lesser extent consumer groups have had an impact on toxicology as it relates to pesticides. The Consumer Federation of America, the nation's largest consumer organization, and other consumer groups have not been highly active in pesticide regulatory affairs, but sometimes consumer and environmental interests work jointly to comment on Federal regulatory policies. They closely scrutinize proposed guidelines published in the Federal Register. Consumer groups work hard to assess and report the views of ordinary people, particularly their pocketbook interests, and gain credibility from this grass roots connection. Joint industry, professional, and consumer seminars and work sessions are being convened in related fields (e.g., Food Safety Council) because they agree that Government is seriously deficient in its handling of their interests. Consumer groups are interested in working with scientific societies, like ACS, who want to explore the possibility of frank and open meetings with consumer advocates who have the same long range interests as industry and science in safe, effective, and reasonably priced pesticides.

Before discussing regulation of pesticides, some discussion of the use of animal data in assessing human risk is necessary, since it forms the basis of many judgments from both societal and regulatory groups. There are three fundamental methods of estimating potential human risk as a result of exposure to toxic substances. These are (1) epidemiological studies, (2) animal

tests, and (3) short-term or _in vitro_ analyses such as studies of
DNA damage or mutagenesis. Of these the greatest confidence is
placed upon epidemiological studies in humans. Due to problems
in obtaining human epidemiological data, we are left then with
the next alternative -- the use of other mammalian species as
human surrogates. The rationale for using animals in toxicological testing are based on the fact that mammals are anatomically,
physiologically and biochemically similar, have similar health
and disease manifestations and causes, and respond similarly to
exogenous chemical, biological, and physical agents. The
validity of animal data in assessing human risk must consider
factors such as test animal species, route of administration,
dose levels, adequacy and uniformity of pathological examinations, disease prevalence in control animals, and false negatives
or false positive statistical analyses. The five most important
biological parameters which best detemine the potency or virulence of an animal carcinogen are number of species affected,
number of tissue sites affected, latency periods, dose-response
relationships, and nature or severity of lesions induced. All of
these factors, plus others, must be considered in extrapolating
animal to human risk situations.

In the United States, the Environmental Protection Agency
(EPA) is responsible for pesticide regulation, and more specifically for assessing the degree of risk posed by the numerous
pesticides available, and determining what level of risk, tempered by benefit, society ought to accept. Risk/benefit analysis
is the chief tool established in the basic pesticide law, the
Federal Insecticide, Fungicide and Rodenticide Act (FIFRA) for
reaching regulatory decisions. The Rebuttable Presumption
Against Registration (RPAR) process is the review mechanism by
which risks and benefits are measured. Animal studies form the
primary basis for assessing pesticide risk. To a limited extent,
human epidemiological data are used to judge risk, although prudent public administration cannot wait to act until an effect is
observed in a human population. EPA is concerned, however, that
extrapolation from animal data to human risk is, while
scientifically supported, still full of uncertainty. One of the
most troublesome areas is extrapolating cancer data in animals to
arrive at risk assessment in humans. Reliable short-term bioassays or appropriate batteries of short-term tests offer one of
the best and perhaps most expedient methods for conducting risk
assessments on active ingredients, and possibly inerts and
contaminants in the future. Improved human exposure models will
also play a major role in future pesticide policy decisions.

A typical example of risk/benefit analyses by EPA was the
recent RPAR against the herbicide pronamide. Pronamide was found
to produce liver tumors in male mice. After careful evaluation
of the exposure levels in humans, the number of excess tumors to
be expected from this exposure, the monetary loss if the registration of pronamide were canceled, and the availability of

substitute herbicides, EPA proposed that the registration of pronamide should be continued after some changes in use pattern and method of application.

Pesticide registration requirements in Europe today do not differ basically from those in the United States or elsewhere on an international basis. The key administrative processing of pesticide registration lies with the Ministries of Agriculture as compared to EPA in the U.S. Administration of the European requirements may be somewhat more flexible. For example, a checklist developed by various countries may be used to provide full or provisional registrations without carcinogenicity or chronic toxicology data, when all available data, including mutagenicity and residue tests, are favorable. Harmonization of registration standards is achieved by the Council of Europe through publication of Pesticides. From a toxicology standpoint, a few countries in Europe have issued detailed testing guidelines in the field of toxicology. It is anticipated that some countries may soon adopt similar, more stringent toxicology testing requirements. One trend that is apparent in Europe is a tendency to require more ecological testing. One difference existing between European countries deals with setting tolerances for pesticide residues in crops.

Canadian pesticide authority resides in two Acts of Parliament: the Pesticide Control Products Act administered by Agriculture Canada, and the Food and Drugs Act administered by Health and Welfare Canada. Agriculture Canada is responsible for registration of all pesticides sold in Canada. It obtains expert advice from a number of other governmental agencies. Health and Welfare Canada promulgates regulations on the maximum residue limits of pesticides permissible on food at the time it first enters commerce. The major toxicological evaluation of pesticides can be subdivided into two major areas: (1) data pertaining to safety of food residues, and (2) data pertaining to occupational exposure and bystander exposure. Theoretical daily intake (TDI) values and negligible daily intake (NDI) values are used to estimate exposure and set registration policy. Although no specific legal requirements exist for studies which must be submitted to support safety in use of pesticides, certain unofficial minimum requests must be met in Canada. These include acute oral and dermal LD_{50}'s; acute inhalation LD_{50}'s; eye and skin irritation studies; feeding studies of at least 90 days duration on adequate numbers of animals of at least two species, one of which would be a non-rodent species; absorption, distribution and excretion data in one of the species used in the 90 day studies; mutagenicity screening studies, and any special studies indicated by the chemical structure of the compound under test.

Pesticide regulations in the People's Republic of China began to take form in the last five years. A series of national conferences held in 1976 and 1978 produced two papers entitled "Proposed Regulations of Experimental Methods for Pesticide

Toxicology" and "Proposed Regulations for Pesticide Toxicology and Residues" that form the basis of the regulatory program. An application for a new pesticide progresses through a review system involving an Institute for Controlling Pesticides, followed by simultaneous review by the Ministry of Chemical Engineering, Ministry of Public Health, and the Environmental Protection Agency. Formal registration is granted by the Ministry of Agriculture. As an example, a new fungicide (DKS) is traced through its testing and review at various stages of development. This promising fungicide was officially banned when a large scale study suggested statistical evidence for health effects.

RECEIVED March 10, 1981.

WORKSHOPS

Discussion Groups and Workshops: A Report

MARGUERITE L. LENG

Health and Environmental Sciences, The Dow Chemical Company,
Midland, MI 48640

As a new feature in special conferences sponsored by the
Division of Pesticide Chemistry, eight workshops were organized
to discuss topics related to our main theme: The Pesticide
Chemist and Modern Toxicology. Capacity crowds attended several
sessions of particularly timely interest while smaller groups
discussed more specialized topics. Participation was often
lively and those who attended the workshops enthusiastically
endorsed this informal means for sharing points of view on
current issues.

Leaders and recorders were appointed on an ad hoc basis from
those who expressed interest in a specific topic. Notes from the
discussion groups were drafted into summaries prior to the end of
the conference, and were distributed for review and comment by
others who attended the sessions. We trust that the following
brief reports present the main conclusions drawn by the participants in the various workshops.

EPA Guidelines for Good Laboratory Practice in Hazard Evaluation

A group of 45 participants devoted 3 hours on Tuesday
evening to discuss EPA's Proposed Good Laboratory Practices
Guidelines for Toxicology Testing (Federal Register, April 18,
1980, pp. 26373-26385). These guidelines would be applicable to
studies conducted to meet requirements of EPA's Proposed Guidelines for Registration of Pesticides - Hazard Evaluation: Humans
and Domestic Animals (Federal Register, August 22, 1978,
pp. 37336-37403).

An outline of nine points developed during the discussion
was submitted by Dr. Gerald G. Still to EPA on August 5, 1980 as
comments on these proposed GLP guidelines on behalf of the Division of Pesticide Chemistry through the Division's Committee on
Chemistry and Public Affairs. The comments centered around the
following areas of concern to the group.

Multiple GLP Standards. Laboratories conducting toxicology
studies cannot comply with three or four different GLP standards

for testing products regulated under different laws. Consistent standards must be adopted by various government agencies to avoid the unnecessary costs of duplicating studies, maintaining separate facilities, and keeping different records for varying lengths of time to meet different requirements.

Many laboratories are already conducting studies in compliance with FDA's Good Laboratory Practice Regulations for Non-clinical Laboratory Studies (Federal Register, December 22, 1978, pp. 59986-60025). These standards were first proposed by FDA on November 19, 1976, and became final on June 20, 1979, after more than two years of widespread review, comment and consideration from various sectors of the scientific, regulatory, and industrial communities. The FDA regulations for testing drugs and food additives are reasonable and should be adopted by EPA without change for studies required for registration of pesticides under the Federal Insecticide, Fungicide, and Rodenticide Act (FIFRA). In fact, the FDA standards for testing may be law for studies in support of pesticide food tolerances which are regulated under the Food, Drug and Cosmetics Act.

To further compound the problem, test laboratories are also faced with EPA's Proposed Health Effects Test Standards for Toxic Substances Control Act Test Rules and Proposed Good Laboratory Practice Standards for Health Effects. These proposed test standards were published separately for studies on chronic health effects (Federal Register, May 9, 1979, pp. 27334-27375) and for acute, subchronic, mutagenic, teratogenic, reproductive, and other health effects (Federal Register, July 26, 1979, pp. 44054-44093). Testing guidelines have also been drafted by the Interagency Regulatory Liaison Group (IRLG) for acute toxicity studies and other studies (Federal Register, August 21, 1979, pp. 49015-49016).

Although EPA's guidelines for testing pesticides are still only proposed, they are in effect being implemented by Agency reviewers and by some industry toxicologists anxious to proceed with their research programs. [Editorial note: Subsequent to this conference, EPA published a data call-in program for all registered pesticides (Federal Register, October 7, 1980, pp. 66736-66740). This proposed rule listed a number of rigid rejection criteria to be used by registrants for judging the validity of toxicological studies conducted in accordance with standards of acceptance in effect at the time they were originally submitted to FDA (or EPA since 1971).]

International Implications. GLP guidelines have also been developed by the international Organization for Economic Cooperation and Development (OECD). The OECD Principles of GLP were provisionally accepted by the OECD Chemicals Group High Level Meeting in May 1980 (and were reviewed by the OECD Expert Group on GLP in Washington in September 1980). These principles closely parallel the FDA regulations for toxicology testing but also encompass other types of studies such as environmental testing.

Participants in the conference discussion included representatives from Europe who expressed concern about the acceptability to EPA of studies done in countries other than the United States of America, and vice versa, the acceptability of U.S. studies elsewhere.

Concern was also expressed over potential conflict between proposed EPA procedures for inspection of laboratories conducting toxicology studies in other countries and national laws regarding foreign inspection. Further clarification is needed on requirements for certification or accreditation of personnel in foreign laboratories, and on the qualifications needed for quality assurance officers.

[Editorial note: Mr. Douglas Costle, Administrator of the EPA has pledged that studies conducted according to OECD guidelines would be acceptable to EPA. An international GLP seminar is scheduled for May 14-15, 1981, in Rome. The preliminary program includes national and international aspects of GLP compliance issues, as well as function, responsibilities and training for quality assurance.]

Cost vs Benefit of Proposed GLP Guidelines. Compliance with EPA's proposed guidelines and excessive GLP requirements will not necessarily result in better data. The ultimate responsibility for the scientific quality of a study rests with the study director, technical staff, and management.

Compliance with meaningless, redundant, excessive GLP requirements is a waste of precious laboratory resources, and a waste of time for highly trained personnel. These excessive costs are particularly burdensome to laboratories where the workload exceeds the capabilities of the available facilities and/or scientific personnel, at a time when more and more studies will be required to meet escalating requirements for testing of chemicals.

The quality assurance aspects of EPA's proposed GLP requirements are estimated to add as much as 40% to the cost of studies. No evidence has been presented to show that the quality of data would be reduced by compliance with existing FDA GLP regulations in contrast to the more stringent and costly proposed EPA requirements. Additional burdensome regulations are not necessary, such as co-signing of original data entries, analysis of all lots of fortified diet, excessive retention time for archival records and samples of diet, etc.

Variations in Inspection Criteria. Uniform criteria are required for inspection of facilities and study records. Unfortunately, implementation of the regulations by various laboratories, and interpretation of the regulations by various inspectors can change with time. These practical problems already exist and will increase as testing facilities are subjected to inspection by officers from different government agencies, and possibly from foreign countries with different national inspection laws.

Metabolism/Pharmacokinetic Studies. The group also discussed the practicality of applying GLP principles to metabolism studies where methodology and operating practices would, of necessity, depend on the chemicals being tested. The group also questioned whether the proposed pharmacokinetic approach was appropriate for the evaluation of pesticides in contrast to human drugs, since such studies are not relevant for non-target organisms. The consensus of opinion was that the proposed exhaustive study in rats (Section 163.85-1) should be replaced by conventional pesticide studies on material balance, tissue residues, and metabolite identification.

Metabolite Significance - Analytical vs Toxicological

Interest was high among the 75 registrants who attended a 2-hour discussion on the toxicologic significance of major and minor metabolites of pesticides. In general, the mere presence of a detectable residue of a metabolite may have little relation to its toxicologic significance. Major metabolites may not make a significant contribution to the toxicity of a substance whereas minor metabolites can be biologically very significant.

Toxicity Testing of Metabolites. The group discussed the necessity of testing the toxicity of all metabolites of plant or animal origin, and of photoproducts and formulation impurities. Primary metabolites in plants are often basically the same as those in animals whereas photoproducts may be structurally quite different. In general, toxicity tests should not be needed for animal metabolites if testing of the parent compound does not reveal toxicological problems such as carcinogenicity or neurotoxicity. An exception would be if the pattern of metabolism in humans is found to be markedly different than that in the test organisms. On the other hand, unique plant metabolites and photoproducts whose chemical structures indicate potential toxicological concern should be synthesized and tested in parallel with the parent compound. However, their bioavailability to animals should also be considered in making this decision.

Analytical Significance of Metabolites. The classification of metabolites as major or minor is a convenient approach from an analytical standpoint but may have little relation to toxicological significance. Metabolites can also be classified as organosoluble, water soluble, nonextractable, and releasable by acid, base or enzyme action. Most studies are done on organosoluble metabolites because suitable analytic techniques are available. Development of analytic methods for water soluble and nonextractable "metabolites" may demonstrate that many are simply endogenous biochemical compounds into which the radioactive label has been incorporated. The limits for identifying minor metabolites should be consistent with the state of the art in analytical methodology. (See Waggoner, Biochemical Aspects, herein). In general, EPA will accept a residue method which measures a common moiety without quantification of individual metabolites.

Potential toxicity of metabolites may depend on activation reactions in resistant vs sensitive strains. Reactive metabolites generated in resistant plants are generally labile and would not likely be incorporated into animals in measurable levels. In animals, the resultant toxicity of a compound may depend on the proportion converted into active metabolites in sensitive vs resistant species of test animals. (See Gillette, Biochemical Aspects, herein).

Dosage Levels for Metabolism and Toxicology Studies. Administration of doses above those which saturate metabolic systems or the capacity-limited elimination rate in the test species can cause toxic effects which do not occur at lower dosage levels. This saturation can be the most sensitive indicator of overdosing and should be taken into consideration in choosing the Maximum Tolerated Dose (MTD) for toxicity studies. Toxicological effects generated in overdosed animals may simply be artifacts from which valid extrapolations to potential effects in humans cannot be made.

Under current requirements, chemicals with extremely low toxicity must be tested at unrealistically high doses. Instead, the group suggested conducting chronic toxicity studies at doses related to actual environmental exposure levels incorporating an adequate margin of safety (such as 100-fold). Consideration should also be given to actual biological concentrations of chemicals and to the possibility of biological magnification. (See Ramsey, Biochemical Aspects, herein).

Structure-Activity Relationships. Further research is needed on the relation between chemical structure and toxic responses. A computer program for pattern recognition might be helpful, but actual toxicity testing is still needed to confirm predictions based on structure.

Analytic Aspects of Pesticide Chemistry Research

Discussion by the 33 participants was facilitated by posting a list of topics related to pesticide analytical techniques. Among the topics discussed were advantages of various detectors, packing techniques for columns, specificity of analyses, interpretation of spectra, advantages of buying a complete unit such as GC-MS coupled to a data system, cleanup techniques, and methods for cleaving pesticide conjugates.

It is proposed to hold similar discussions on a regular basis at national meetings of the ACS, and possibly to organize a subdivision of the Pesticide Division for those interested specifically in analytical aspects of pesticide chemistry research.

High Performace Liquid Chromatography. The Radial Com-
pression Module (RCM) was found to be particularly useful for
analytical HPLC work and Gel Permeation Chromatography was discussed for cleanup of metabolites and bound residues. Two main
advantages for the electrochemical detector (Kissinger cell
Bioanalytical Systems, Lafayette, IN) were economical replacement
of the cell ($100) and good sensitivity (picogram levels for
chlorophenols in the oxidative mode). Good sensitivity and
selectivity were also reported for the Tracor photoconductivity
detector and the Technicon post-column derivatization-fluorescent
detection system. Improved sensitivity to 250 dpm was reported
for Radioactive Monitoring (RAM) detectors such as those
currently available from Bertholde and C.A.I.

Gas-Liquid Chromatography (GLC). Among the topics discussed
was a new technique of using packed columns interfaced with a
Packard capillary column for cleanup and analysis, or with
Fourier Transform-IR detection for metabolite identification to
0.5-1.0 micrograms. New packed columns such as Supelco 1240DA
permit direct gas chromatography of polar compunds with minimum
derivatization but have limitations of low load and low operating
temperatures (180-200°C).

Mass Spectrometry. HPLC-MS was thought to be particularly
useful in conjunction with selected ion monitoring (SIM).
Pyrolysis-GC-MS has been successful in the analysis of bound
residues. Negative Chemical Ionization (NCI) is more specific
than EC-GC for the detection of chlorinated aromatic compounds,
and permits specific detection of the 2,3,7,8- isomer of tetra-
chlorodibenzo-p-dioxin (TCDD). In GC-MS analysis of metabolites,
spectral interpretation is facilitated by use of ^{13}C labeled
compounds if the $^{13}C/^{12}C$ ratio is known. The group concurred
that adding a data system to an existing GC-MS system is not as
good as buying a complete GC-MS-DS unit.

General Techniques. Cleanup of analytical samples can be
done conveniently with Waters Sep-Paks or equivalents available
from Extrelut or Merck, and they can be regenerated. The major
problem in isolation, characterization, and analysis of pesticide
conjugates is ensuring 100% cleavage of many conjugates.

Re-entry Standards - Relevance to Farm Worker Safety

EPA's February 1980 draft guidelines for re-entry data were
discussed by 15 participants concerned about possible requirements for additional data on residues of pesticides under field
conditions. These draft guidelines are still being revised to
incorporate comments made by EPA's Scientific Advisory Panel and
others who attended a workshop conference in Tucson in February
1980.

The re-entry interval is defined as the time between last
application and the time a worker can enter a field and work for
more than one-half hour without protective clothing. Re-entry is

not to be confused with preharvest intervals which pertain to legal residues on food crops.

The new guidelines were developed by EPA as a result of problems concerning illness among farm workers in California, particularly among those exposed to parathion in treated citrus groves. California found that Federal standards for re-entry were inadequate and imposed more stringent requirements for pesticides which inhibit cholinesterase. Other states such as Florida found less need for re-entry standards.

EPA's proposed standards would apply to any pesticides that produce acutely toxic or chronic effects in all kinds of situations. In general, three sets of data would be needed to set federal standards for re-entry of farm workers into treated fields:

1. Determination of dislodgeable residues on foliage and fruit, i.e. those residues which can be removed by mechanical action (using specific equipment designed and developed by researchers in California).
2. Correlation between dislodgeable residues and external exposure (by as yet undefined means).
3. Relation between external exposure and toxic effects. (This is difficult to evaluate because, in some cases, no data are available on what constitutes an effect level by dermal exposure, nor on how to relate dermal absorption to dietary level.)

Of concern to the group was the proposal to require 90-day subchronic dermal toxicity studies in animals. Current requirements specify a 21-day rabbit dermal study which takes 6 months at a cost of about $40,000. The proposed 90-day study would cost about $100,000 and would provide no additional useful information. The question remains as to whether the study should be done with only the parent compound or with each formulated product due to potential effect of surfactants and solvents on the rate of absorption of the active ingredient(s) into the exposed subjects.

Participants also expressed concern over techniques for estimating dermal exposure such as the use of patches, and the interpretation of data generated from patch contamination. Patch exposure cannot be correlated with uptake by unprotected skin, nor with blood levels or urinary excretion (except in some cases such as the phenoxy herbicides). Among questions remaining are: how to extrapolate the ratio between acute dermal and acute oral doses in animals to man, whether re-entry data will be needed for every pesticide in every crop or crop grouping for every geographic location, and whether humans can legally be used for dermal absorption studies.

[Editorial note: Another workshop on methods for estimating exposure was held on October 29-31, 1980, in Hershey, PA. Revised proposed guidelines for re-entry requirements are to be published by EPA in the Federal Register, possibly in the spring of 1981.]

The RPAR Process - Is It Working?

About 30 participants met to discuss EPA's process of
Rebuttable Presumption Against Registration (RPAR) for evaluating
pesticides deemed to present an unreasonable risk to humans or
the environment. A lucid description of the mechanics of the
RPAR process, and the role of EPA's Scientific Advisory Panel
(SAP) in this process, was presented by Dr. Robert Neal, a member
of SAP. No EPA representative was present to field questions or
express EPA's viewpoint on this controversial topic.
Comments centered mainly on the advantages of having documentation of safety data and an improved or more complete definition of a wide variety of terms employed to evaluate safety,
risk, and benefits. Also discussed were items such as dose/
response data, occupational exposure of chemical manufacturer
employees, formulators, applicators, growers, fish and wildlife,
and the general public.
Problems related to the RPAR process were also discussed.
For the most part criticisms were related to the frustration of
industrial scientists who must accept the delays encountered in
the decision process. Usually product manufacturers must develop
and document reponses to an RPAR within a relatively short finite
period of time. On the other hand, there is no requirement or
time limit for an EPA decision to be made or published.
Some suggestions on ways to shorten the time required for
review of responses were discussed. Because of the large volume
and broad range of scientific disciplines covered, EPA may parcel
out various sections of responses to selected members of the
scientific community for objective review and opinion. However,
an individual reviewer may then see only a small part of the
picture and may not have an opportunity to review reports that
could have a direct relation to the reports reviewed. It was
recommended that reviewers should have a complete copy of
responses in order to develop a comprehensive opinion. Another
alternative would be to develop and maintain the expertise within
EPA to reach the decision point in a more timely manner.

EPA Guidelines for Subchronic Toxicity Testing

A group of 10 participants discussed EPA's proposed guidelines for subchronic toxicology studies (Sections 163.82-1 in
Proposed Guidelines for Registering Pesticides in the U.S.;
Hazard Evaluation: Humans and Domestic Animals, Federal
Register, August 22, 1978, pp. 37363-37366). Although final
rules for subchronic testing have not been issued, the group
anticipated considerable modification of the requirements.
In May 1979, EPA sponsored a workshop in Denver, Colorado,
to review their proposed guidelines for subchronic toxicity
testing. Attendance was limited to invited participating and
observer scientists from academia, public interest groups,

industry and government who are knowledgeable in subchronic testing procedures. To date, only draft reports of the workshop have been available for review.

The overall thrust of proposed revisions to the proposed guidelines for end-result subchronic studies is to decrease the number and complexity of required tests and to place more emphasis on the scientific judgment of the study director. The group enthusiastically supported the following recommended changes:

1. Reduction of the required histologic tissues sections from about 40 to 20.
2. Only one of paired organs need be examined.
3. Urine analyses would not be required.
4. The number of clinical chemistry tests would be reduced to 13.
5. Initiation of testing would be delayed until the animals were older.

The groups agreed that these revisions in the guidelines would significantly reduce the cost of subchronic studies without reducing the quality.

Mutagenicity Testing - Relevance to Carcinogenicity

A group of about 10 participants discussed the "battery" of acceptable tests to be used in the genotoxic evaluation of a new chemical/pesticide. Considerable concern was expressed about reliability and reproducibility of various test systems and reasons for these concerns were addressed. The test methods (assays) chosen should minimize variability. It was generally agreed that good predictors of carcinogenicity are currently available in the battery of mutagenicity assays.

Evaluation of new chemicals for genotoxicity should be done through a tier system, or battery or core of tests which evaluate all types of genetic damage including gene mutations, chromosomal damage and DNA damage. Specific assays and tests in a battery should include the Ames Test, Micronucleus Test, Sister Chromatid Exchange, Yeast Gene Conversion, Transformation, Mammalian Point Mutation Assays, DNA Damage with Eschericia coli, and Unscheduled DNA Synthesis.

Communicating Technical Information

A select group of seven assembled to discuss how to effectively exchange information among scientists and the general public. Individuals recounted some of their experiences and frustrations, admitting that communicating technical information to the public is difficult but important. Discussants sensed that the scientist has lost the confidence of the public and cannot be "trusted". To alleviate this problem, scientists should participate in community groups but must avoid an

adversary approach or "talking down" to the lay person. The
media was felt to be responsible for much misinformation by
distortion or selection of information so as to make it
"newsworthy".
  A recommendation was made that greater use should be made of
appropriate committees of the American Chemical Society for
interpreting scientific issues to the public where technological
problems are involved.

Acknowledgements

The conference committee wishes to thank the following participants who proposed the topics, led the discussions, and/or
provided notes on what was discussed at each workshop.

1. GLP Guidelines - J. Bart Miaullis, Stauffer Chemical
    Company; James Puhl, Mobay Chemical Corporation; Olav
    Messerschmidt, Velsicol Chemical Corporation; Enrico
    Knuesli, Ciba Geigy Ltd, Basel, Switzerland.
2. Metabolite Significance - Janice Chambers, Mississippi State
    University; Yousef Attalla, Velsicol Chemical Corporation;
    James R. Gillette, National Institute of Health Laboratory
    of Clinical Pharmacology.
3. Analytical Aspects - Barrie Webster, University of Manitoba,
    Canada; Hamdy Balba, Uniroyal Chemical; Allan Cessna,
    Agriculture Canada Research Station, Regina, Saskatchewan.
4. Reentry Standards - Marie Siewierski, Rutgers University;
    Gunter (Jack) Zweig, Environmental Protection Agency.
5. RPAR Process - Wendell (Bud) Phillips, Campbell Institute
    for Food Research; Richard Connizzaro, Thompson-Hayward
    Chemical Company; Robert Neal, Vanderbilt University.
6. Subchronic Toxicology Testing - Bobby Joe Payne, Toxicity
    Research Laboratories; Gordon S. Dean, Toxicity Research
    Laboratories.
7. Mutagenicity Testing - Jerry J. Carter, Carter Research
    Corporation; Robert W. Naismith, Pharmakon Laboratories.
8. Communicating Technical Information - Elvins Y. Spencer,
    Agriculture Canada Research Institute, at London, Ontario;
    Ben Luberoff, editor of CHEMTECH.

RECEIVED March 18, 1981.

INDEX

INDEX

Jacket design by Carol Conway.
Production by Candace A. Deren and Cynthia E. Hale.

The book was composed by Service Composition, Baltimore, MD,
printed and bound by The Maple Press Co., York, PA.